Bioenergetics of Wild Herbivores

Editors

Robert J. Hudson, Ph.D.
Professor
Department of Animal Science
University of Alberta
Edmonton, Canada

Robert G. White, Ph.D.
Professor
Institute of Arctic Biology
University of Alaska
Fairbanks, Alaska

CRC Press
Taylor & Francis Group
Boca Raton London New York

CRC Press is an imprint of the
Taylor & Francis Group, an **informa** business

FOREWORD

The concept of this book promises the realization of an overdue change in the approach to bioenergetics, to nutrition and husbandry, and thus to the management of wild herbivores: the final emancipation from rules and views based primarily on domesticated herbivores or on experimental animals held under unnatural conditions, necessarily impeding them behaviorally, physically, and psychically.

While wild herbivore populations are threatened in so many ways in most parts of the world, multidisciplinary scientific efforts have helped to show how wild species like red deer or oryx antelopes, in short, can change into astonishing subjects of human management yielding vast quantities of new data so much different from cattle or sheep.

It was, on the other hand, a long lasting mistake to look at wild herbivores in an extremely fractionated way. All too many ''models'' presented to the scientific public had to collapse because they were based on the narrow scope of monofactorial observations. Studies of bioenergetics today have to be based on the evolutionary background of a herbivore species and they have to consider the multiple influences and feedbacks from specific ecological systems to which a herbivore belongs.

Studies performed in this spirit have been teaching us striking and admirable strategies of herbivore survival and they have revealed the great importance of dynamic flexible systems of adaptation—all of which take their origins in the cell. They become permanent patterns finally as part of the genetic substance.

Our scientific attitude has changed as the scope of this book shows: multiple yet integrated sources of information, presented by competent authors with a critical and frequently self-critical attitude, especially to the methods of investigation. We all have to admit, from time to time, that we are far from the final truth. In herbivore bioenergetics, we are aware that all those species selecting plant material rich in cell content still keep more secrets than those species reveal which are able to digest cell wall, i.e., to dissolve plant fiber through their adopted microorganisms. This confession must increase our fascination of a striking co-evolutionary development of plants and herbivores over millions of years. Are we narrow-minded enough to ignore this aspect of evolution by accommodating all regions of the world for just a few ''productive'' species? This book is a combined scientific effort also to counter this short-sighted trend. An attitude of ''pure'', passive conservation needs little such information and will not protect the wonderful variety of herbivores and their specific but wide-ranging adaptations.

This book provides a great amount of detail but its expressive aim is to lead us to the whole animal, to a herd, to a population as integral parts of an ecological entity which in turn is the result of evolutionary forces. This we all must learn to promote and not to disturb. I trust this book will contribute.

Reinhold R. Hofmann
Professor of Comparative Anatomy of Domestic and Wild Animals
Fac. Veterinary Medicine and Animal Husbandry
University of Giessen
Giessen, West Germany

Giessen, August 1984

PREFACE

Bioenergetics is an emerging discipline which offers a more profound understanding of the ecology, behavior, and evolution of wild herbivores. Increasingly, bioenergetic principles have been applied in management since they provide insight into population dynamics and are relevant to manipulation of habitats and assessment of the impacts of resource development. Growing interest in the agricultural potential of wild herbivores has provided further impetus. In spite of this promise, there are few comprehensive syntheses of the concept and its application to wild herbivores. This volume attempts to fill this need.

The central concept of this discipline is that of the energy budget. Herbivores gain energy from the ingestion, digestion, and assimilation of forage. They use energy for physiological maintenance, thermoregulation, and activity. The surplus is allocated to growth and reproduction. In their daily lives, animals attempt to regulate these energetic transactions by physiological and behavioral means. Body size and functional anatomy determine the way they experience energetic tradeoffs and hence their behavior. On an evolutionary scale, energetic factors have contributed to shaping the present-day diversity of large mammals.

These general relationships are explored in the introductory chapter. This is followed (Chapter 2) with a description of how environments can be parametized in bioenergetic terms. The remainder of the volume is organized according to the major components of the energy budget. Chapters 3 to 5 deal with the supply of energy and nutrients including resource selection, forage intake, and digestive processes. Chapters 6 to 8 describe energy requirements for maintenance, activity, and thermoregulation. Chapters 9 and 10 deal with allocation of retained energy to growth and reproduction. Chapters 11 and 12 summarize ways to assess nutritional status and to predict the productive responses of wild herbivores in complex nutritional environments.

An attempt has been made to involve active contributors to the field as either authors or reviewers although this was constrained by inevitable problems of text space and busy schedules of potential contributors. The volume has been edited through exchange with other authors as well as external referees. Editorial comments on various chapters were provided by J. Bligh, R. Cameron, V. Geist, R. Hofmann, R. Kay, D. Klein, J. Kennelly, G. Mathison, A. Loudon, W. Mautz, W. Regelin, L. Renecker, K. Risenhoover, U. Seal, D. Shackelton, D. Spalinger, B. Stelfox, L. Verme, W. Watkins, and B. Young. To the authors and external referees, we extend sincere appreciation. Graphics for Chapters 1, 6, 9, and 12 were prepared by K. Davis, Mapping Concepts. Special recognition goes to Yasmin Hudson for her administrative assistance and enthusiastic support. Finally, we would like to thank Sandy Pearlman, Suzanne Lassandro, and other members of the CRC editorial and production staff for guiding the manuscript through to final publication.

R. J. Hudson
R. G. White

EDITORS

Dr. Robert J. Hudson, Ph.D., is Professsor of Wildlife Productivity and Management with the Department of Animal Science, University of Alberta, Edmonton. He received his B.Sc. (Agr) and Ph.D. degrees from the University of British Columbia. His current research both in Canada and East Africa is on the dynamics of native grazing systems and agricultural potential of indigenous herbivores. Dr. Hudson is Associate Editor (Endangered Species) for the interdisciplinary journal, *Biological Extinction*.

Dr. Robert G. White, Ph.D., is Professor of Zoophysiology and Nutrition at the Institute of Arctic Biology, University of Alaska, Fairbanks. He received a B.Ag.Sci. from The University of Melbourne, and M.Rural Sci. (Physiology) and Ph.D (Nutritional Biochemistry) degrees from The University of New England. His research interests are in the fields of nutrition and habitat relations of wild animals. He has published widely on the intermediary metabolism, energy, protein, mineral and water metabolism, feeding dynamics, aspects of ecology of these large animals and on the nutritional physiology of domesticated reindeer. Dr. White was a member of the US IBP Tundra Biome Program and pioneered the use of tame animals as wild animal analogs in remote northern research sites. His current research projects center on nutritional influences on reproduction and lactation in muskoxen and on the energy costs of migration in caribou.

CONTRIBUTORS

Graham W. Arnold, Ph.D.
Senior Principal Research Scientist
Division of Wildlife and Rangelands Research
CSIRO
Midland, Western Australia

Elizabeth A. Boomker, M.Sc.
Eugène Marais Chair of Wildlife Management
University of Pretoria
Pretoria, South Africa

Fred L. Bunnell, Ph.D.
Professor, Wildlife Ecology
Faculty of Forestry
University of British Columbia
Vancouver, British Columbia

Robert J. Christopherson, Ph.D.
Professor, Animal Physiology
Department of Animal Science
University of Alberta
Edmonton, Canada

Steven G. Fancy, Ph.D.
Institute of Arctic Biology
University of Alaska
Fairbanks, Alaska

Albert W. Franzmann, D.V.M., Ph.D.
Game Biologist
Alaska Department of Fish and Game
Soldotna, Alaska

Michael P. Gillingham, Ph.D.
Faculty of Forestry
University of British Columbia
Vancouver, British Columbia

N. Thompson Hobbs, Ph.D.
Wildlife Researcher
Colorado Division of Wildlife
Fort Collins, Colorado

Olav T. Oftedal, Ph.D.
Nutritionist
Department of Zoological Research
National Zoological Park
Smithsonian Institute
Washington, D.C.

Katherine L. Parker, Ph.D.
Research Associate
Faculty of Forestry
University of British Columbia
Vancouver, British Columbia

Mick A. Price, Ph.D.
Associate Dean
Faculty of Agricultural Forestry
University of Alberta
Edmonton, Canada

Charles T. Robbins, Ph.D.
Professor
Program in Wildlife Biology
Washington State University
Pullman, Washington

Charles C. Schwartz, Ph.D.
Game Biologist
Alaska Department of Fish and Game
Soldotna, Alaska

Wouter van Hoven, Ph.D.
Associate Professor
Eugène Marais Chair of Wildlife Management
University of Pretoria
Pretoria, South Africa

TABLE OF CONTENTS

Chapter 1

BODY SIZE, ENERGETICS, AND ADAPTIVE RADIATION

R. J. Hudson

TABLE OF CONTENTS

The ungulates (and subungulates = elephants) include approximately 203 surviving species belonging to 3 orders, 13 families, and 83 genera (Table 1). Their condylarth ancestors appeared in the Paleocene and formed at least five lines; *Cete, Meridungulata, Phenacodonta, Eparctocyon,* and *Tethytheria*.[2] The latter three gave rise to the modern ungulate orders *Perissodactyla, Artiodactyla,* and *Proboscidea,* respectively.

Perissodactyls (odd-toed ungulates) are a very ancient group with an origin dating 50 million years to the late Paleocene. They diversified and became the most numerous herbivore until the Miocene. The emergence of ruminant digestion by the advanced *Artiodactyla* presumably led to competition and ultimately the demise of this Order.[3]

Artiodactyls (even-toed ungulates) represent perhaps the most important mammalian radiation. The significant feature of their evolution since their appearance in the Eocene is the development of ruminant digestion. The earliest suborders, *Suina* (pigs, peccaries, and hippos) and *Tylopoda* (camels), show various degrees of anatomical elaboration of the digestive tract to accommodate microbial fermentation. The camelids developed a superior pregastric system of fermentation which parallels development of the true ruminant system. But the typical multicompartmented rumen traces its origin to the forest-dwelling tragulids. This digestive theme contributed to the progressive dominance of the *Ruminantia* in the Oligocene and Miocene.[4]

Proboscids (elephants) originated in the Eocene, perhaps from a semi-amphibious ancestor similar to (but apparently not) *Meriotherium,* and attained peak diversity in the Pliocene.[5] There is clear evidence of progressive adaptation to coarser diets in dentition and increasing body size. Pleistocene extinction dramatically reduced their ranks to two surviving species, the African and Asian elephants.

These three orders present remarkable diversity of size, form, and function. An important question relates to the forces forging this adaptive radiation. Another is whether evolution of ungulates can be viewed in terms of individually adaptive traits. Or, are there design constraints which cause such traits to be linked in adaptive suites?

The fitness of genes is tested through the interaction of their carrier, the phenotype, with environment. Ultimately, genetic success is determined by the complex relationship between reproductive rate and reproductive life. Proximally, fitness can be equated with the success of phenotypes in *avoiding predation, competing for mates,*[6-9] and *tranforming resources,*[10,11] We should not expect these factors to be equally important in all stages of evolution. Geist[6-9] argues that during dispersal into rich vacant niches, social factors should emerge as the main selective force while bioenergetic efficiency should be more important under conditions of greater density.

Collectively, the imperatives of security, social dominance, and efficient resource transformation confront wild herbivores with a number of potentially complex compromises. An important area of inquiry probes the adaptiveness of traits which determine the way animals experience these tradeoffs and hence conduct their daily lives. This question can be addressed in two ways.

The first is to examine the association of phenotypic traits with ecological circumstances.[12] The potential problem with this *a posteriori* method is one of confusing correlation with causation particularly when it is uncritically accepted that all traits are adaptive. Of course, neutral or even maladaptive traits may be present because evolutionary change may lag behind a changing environment. Certain traits may simply be genetically linked to others which have adaptive significance. The tendency to imagine adaptations where they do not exist has been termed the Panglossian Fallacy.[31]

The second approach is predictive rather than deductive. It is based on *a priori* consideration of the sorts of traits which would enhance fitness followed by the construction of testable hypotheses. There are several reasons for modest progress in this area. First, there is the uncertainty of how natural selection operates on quantitative traits whose phenotypic

Table 1
TAXONOMY OF EXTANT UNGULATES AND SUBUNGULATES

Taxonomic Designation	Approximate number Genera	Species
Order **Perissodactyla**		
Equidae	1	8
Tapiridae	1	4
Rhinocerotidae	4	5
Order **Artiodactyla**		
Suborder Suina		
Suidae	5	8
Tayassuidae	2	3
Hippopotamidae	2	2
Suborder Tylopoda		
Camelidae	3	4
Suborder Tragulina		
Tragulidae	2	4
Suborder Ruminantia		
Cervidae (incl. Moschidae)[a]	15	37
Giraffidae	2	2
Antilocapridae	1	1
Bovidae	43	123
Order **Proboscidea**		
Elephantidae	2	2
Totals	83	203

[a] Moschidae sometimes considered a distinctive family.

Adapted from Corbet, G. B. and Hill, J. E., *A World List of Mammalian Species*, Comstock Publ., Ithaca, 1980.

expression varies with environmental opportunities.[9] Second, the adequacy of widely applied optimizing criteria remains open to question (Chapter 3). Third, it is difficult to select an appropriate proxy for neo-Darwinian fitness given the divergent nature of selective forces.

This volume deals primarily with the success of large herbivores as resource transformers. Although a variety of nutrients are relevant to the efficiency of this transformation, energy is used as a convenient tangible currency. This chapter evaluates the linkages of bioenergetics, ecology, and social behavior, and probes their allometric basis in body size.

I. SIZE AND SHAPE

Compared with other mammals, ungulates and subungulates are relatively large.[2] Body weights of modern species (data available for 177) form a log-normal distribution. However, there are several discontinuities in untransformed distributions which allow identification of three or perhaps four groups (Figure 1). Most ungulates weigh less than 50 kg and are selective feeders. Medium ungulates range from 125 to 325 kg and include a variety of feeding styles. Large ungulates weighing up to 625 kg are mostly nonselective grazers. Megaherbivores are scattered upwards in weight distribution and include an assortment of browsers, grazers, and mixed feeders.

Sexual dimorphism in weight varies widely. In many forest-dwelling microherbivores, females are similar to or even slightly larger than males[14] but, in most medium and large species, males are larger because of the importance of size in social dominance and mating

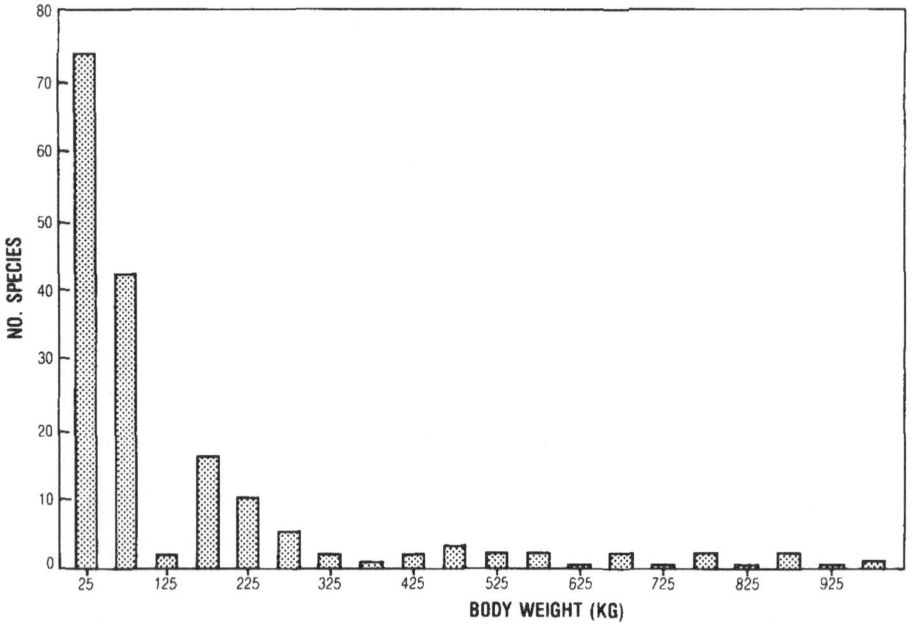

FIGURE 1. Frequency distribution of body weights in living ungulates. Parameters of the log-normal distribution are x = 216 kg, SD = 442, n = 177.

enhancement. In highly gregarious species with little social segregation, sexual dimorphism in size and coloration may again decline to confer the advantages of "hiding" in groups. Since males seem to be large for social reasons, females probably are close to the bioenergetic optimum size for each species and their weights rather than average weights appear most appropriate for allometric scaling.

A persistent theme in evolution has been to increase body size, a tendency which has been accompanied by consistent changes in body shape and adornment with horns, antlers, beards, or manes. Gradients of size and conformation form can be arranged with fossil as well as living forms at almost any taxonomic level although the clearest picture exists within phyletic lines.[6-9] These evolutionary changes are associated with the expansion of vacant habitats such as grasslands, deserts, floodplains, and tundra-steppes (Figure 2). A common feature is that new opportunities are afforded in habitats with a relative surfeit of potential forage. Typically, these habitats are open which encourages gregariousness to minimize the individual's risk of predation. Under conditions of forage abundance, herbivores are released from bioenergetic constraints so fuller expression of genetic potential for growth and development can be achieved. This allows natural selection to operate on traits which confer social advantage as mating becomes more competitive in a gregarious society. In addition to size, the premium placed on mating enhancement favors development among males of striking appendages for combat and display.

Conformational correlates of weight are related in the extreme to the requirements of support and locomotion. The two poles of the conformational gradient are recognized as Schlupfer and Laufer types.[15] The Schlupfer type is most primitive and is characterized by relatively heavy hind quarters and rounded back. The Laufer type is characteristic of more advanced ungulates living in open grassland. The forequarters are better developed and the spine is more rigid.

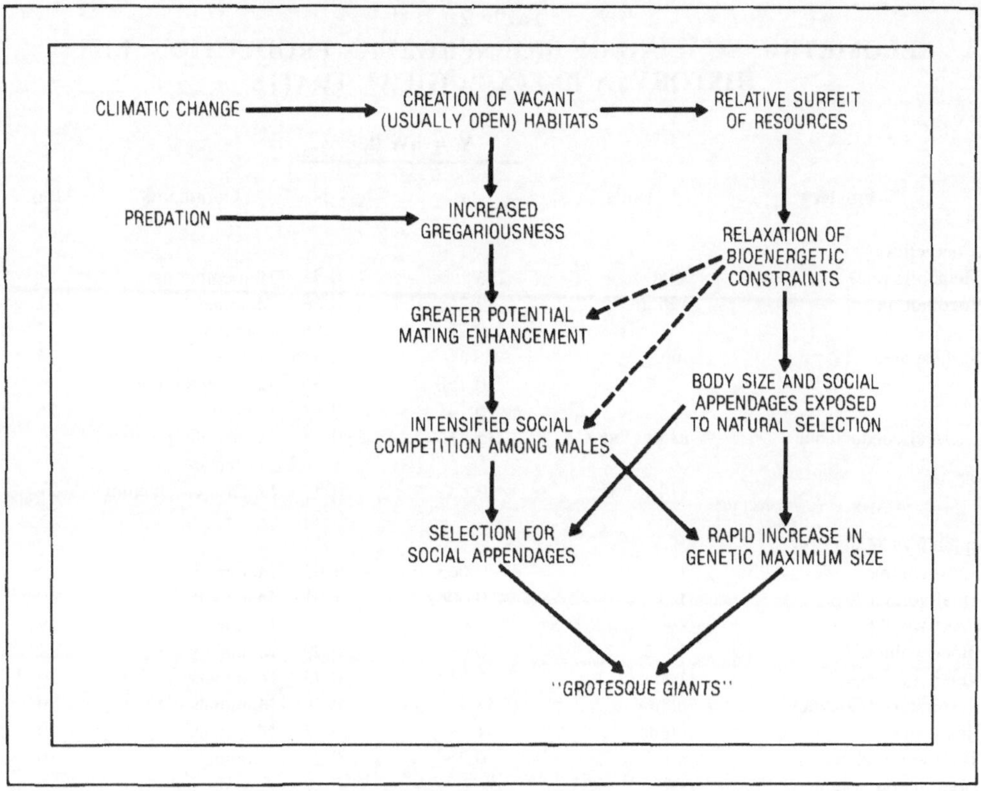

FIGURE 2. The Dispersal Theory.[8,9] Summary of factors operating on dispersal phenotypes which favor evolution of "grotesque giants".

Size-related gradients of form extend to the allometry of limb bones.[16] McMahon[17] offered a biomechanical explanation, noting that it is frequently impossible to design a large machine to work the same way as a small model while maintaining perfect isometry. He hypothesized that the consistent relationship between size and form might preserve elastic similarity while recognizing that genetic factors may preserve certain shapes within related taxonomic groups.

One of the most exciting proposals is that this geometric remodeling may dictate many metabolic phenomena (Chapter 6).[18,19] Allometric relationships fall into several quantitative clusters.[19] Physical support (e.g., skeletal weight) scales hyperallometrically. Control structures such as neural and endocrine tissues scale to less than unity, often approximately to surface area $\alpha W^{0.66}$). Capacities of organs which function volumetrically (lungs, heart, and digestive tract) scale approximately to $W^{1.0}$. Volume-rates (respiratory gas exchange, metabolic rate, cardiac output) scale to $W^{0.75}$. Finally, biological frequencies (heart rate, respiration rate) scale to $W^{-0.25}$. Many facets of the behavior, ecology, and life history of large herbivores also can be traced to body size as either ultimate cause or consequence.[20] The remainder of this chapter explores allometric scaling of selected traits (Table 2) and attempts to provide a functional interpretation.

II. ENERGY BUDGETS

Many of the effects of body size appear directly or indirectly related to bioenergetic phenomena. Energy budgets summarizing relevant transactions are comprised of three basic components: assimilation, expenditure, and retention. These components aggregate a number of parameters describing the flow of energy from forage to tissue (Figure 3).

Table 2
ALLOMETRIC SCALING OF BIOENERGETIC, PRODUCTION, LIFE HISTORY, AND ECOLOGICAL TRAITS

		$Y = aW\ (kg)^b$			
Parameters	**Units**	**a**	**b**	**Comments**	**Ref.**
Bioenergetics					
Metabolic rate	kJ/d	293	0.75	Homeotherms	21
Locomotion	kJ/km	12.26	0.66	Mammals	22
		10.77	0.684	Mammals	23
Surface area	m^2		0.625		24
		0.150	0.56	Large herbivores	25
		0.142	0.635	Cervids	25
Thermal conductance	$kJ/kg^{0.75}{\cdot}d{\cdot}°C$	35.81	−0.35	Temperate ungulates in winter	26
	$kJ/kg{\cdot}d{\cdot}°C$	10.83	−0.54	Small arctic mammals	27
			−0.50		28
Respiratory/Circulatory					
Lung volume	l	0.063	1.02	Mammals	29[a]
Tidal volume	l	0.0062	1.01	Mammals	29[a]
Heart weight	kg	0.0058	0.99	Mammals	29[a]
Blood volume	l	0.055	0.99	Mammals	29[a]
Ventilation rate	l/min	0.334	0.76	Mammals	29[a]
Respiratory frequency	b/min	53.5	−0.26	Mammals	29[a]
Heart rate	b/min	241	−0.25	Mammals	29[a]
Time to circulate total blood volume	min	0.35	0.21	Mammals	30
Digestive Function					
Fermentation contents	kg	0.116	1.03	Herbivores	31
		0.105	1.04	Ruminants	31
Fermentation rate	μmol gas/g·hr	708.78	−0.23	African ruminants	32
Feed intake	kg/d	0.058	0.80	African herbivores	33[b]
Rumination capacity	g cell wall/min	0.02	0.95—1.14	Ruminants	31
Parotid gland weight	g	2.98	0.75	Ruminants	34[b]
Growth and development					
Skeletal mass	g	61.45	1.09	Mammals	35
Antler growth	g/d	0.04	1.33	Cervids	27
Brain weight	kg	0.01	0.70	Mammals	29[a]
Postnatal growth	g/d	5.62	0.72	Mammals	36
		20.42	0.52	Ungulates	36
		10.33	0.71	Ungulates	37
Reproduction					
Birth weight	kg	0.214	0.79	Ungulates	37
		0.1698	0.83	Mammals	38
Gestation length	d	67.48	0.25	Mammals	39
		109.97	0.19	Ungulates	27
		120.31	0.16	Artiodactyls	40
Birth wt/gestation	kg/d	0.0035	0.60	Nonprimates	41
		0.0019	0.60	Ungulates	27
Total energy req. pregnancy	kJ	0.85	0.96	Ungulates	27
Peak milk yield	kg/d	0.0835	0.765	Nondairy mammals	42
		532	0.69	Mammals	42
	kJ/d	987.42	0.52	Ungulates	37
Mammary weight	kg	0.045	0.82	Mammals	42
Life History Phenomena					
Age at first reproduction	d	203.47	0.18	Mammals	43
		229.23	0.27	Artiodactyls	40
Birth rate	%	135.07	−0.33	Artiodactyls	40

Table 2 (continued)
ALLOMETRIC SCALING OF BIOENERGETIC, PRODUCTION, LIFE HISTORY, AND ECOLOGICAL TRAITS

Parameters	Units	a	b	Comments	Ref.
		$Y = aW\ (kg)^b$			
Longevity	d	2091.60	0.22	Artiodactyls	40
Life expectancy at birth	d	813.61	0.20	Artiodactyls	40
Popn. turnover time	yrs	1.02	0.29	Herbivores	44
Popn. doubling time	d	219.44	0.26	Mammals	45
Lifespan in captivity	d	4236	0.20	Mammals	46
Population cycles	yrs	6.53	0.25	Herbivores	44
Ecological Traits					
Speed	km/h	5.5	0.24	Mammals	47
Foraging radius	km	2.0	0.40	Ungulates	48
Home range	ha	2.30	1.02	Herbivores	49
		6.06	0.91	Temperate ungulates	49
		≈2.0	≈1.00	African ungulates	33[b]
Population density	no./ha	0.955	−0.75	Herbivores	50, 51
Population biomass	kg/ha	0.95	0.25	Herbivores	50, 51
Incisor breadth	cm		0.33	Ungulates	52
Foraging time	% 24 hr	24	0.08	Ungulates	33[b]
Foraging plus rumination time	% 24 hr	52	0.09	Ungulates	31[a]

[a] Cited.
[b] Data.

Retention (RE) is the net amount of energy available for productive processes such as growth and reproduction. Retention is related to the other energetic components by the deceptively simple equation:

$$RE = ME - H \tag{1}$$

But each component is complex and interdependent.

Metabolizable energy (ME) represents the amount of energy available to tissues for maintenance or production (Chapter 2). Only a proportion (about 60%) of the gross energy of forage is digested. Although the main determinant of this proportion is the physical and chemical composition of the diet, anatomical and physiological characters of the animal dictate the rate of passage and nature of the digestive milieu, hence the completeness of digestion. Further losses (about 15%) are through combustible gases (principally methane) and nonmetabolizable compounds in urine.

An equally complex component is heat production (H). Energy expenditures of resting animals in a thermoneutral environment can be factored crudely into the basal expenditures of tissues (usually measured as fasting metabolic rate) and that associated with digestion and metabolic conversion. The main determinant of fasting metabolism (F) is body weight but physiological state and environmental stimuli can be important. The heat increment (I) of feeding encompasses the work of eating, rumination, and digestion as well as the heat of anaerobic fermentation in the gut and heat arising from nutrient metabolism by body tissues.

The ability of animals to maintain positive energy balance in large measure determines their productivity and genetic contribution. Their capacity to do so in various environments is constrained ultimately by their anatomical design but they have a well-developed ability to regulate energy budgets by physiological and behavioral means.

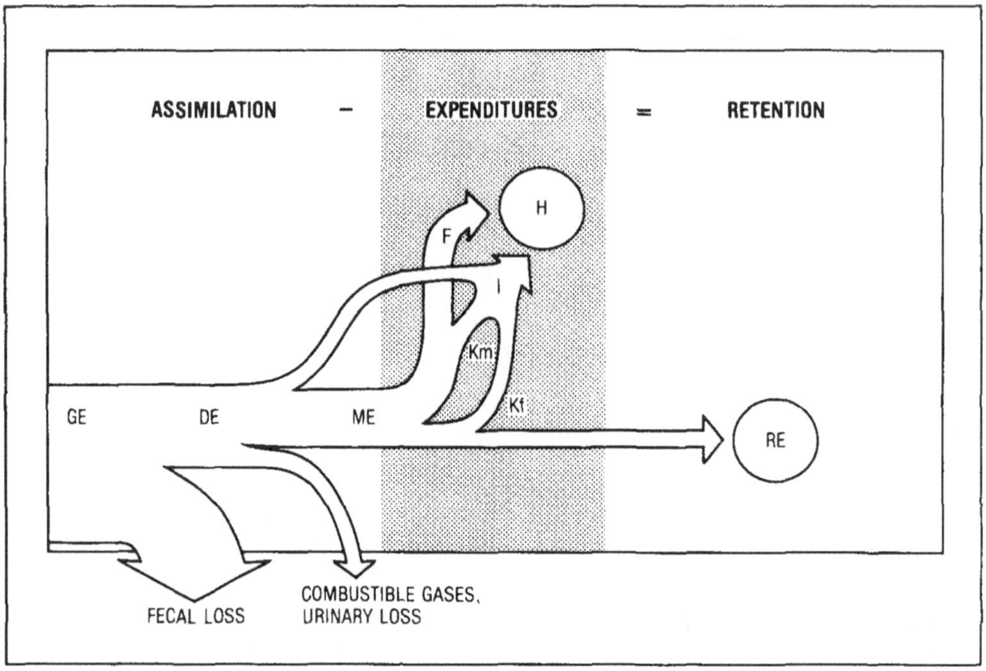

FIGURE 3. Energy budget during positive energy balance for a herbivore at rest in a thermoneutral environment.

A. Energy Costs

Tissue-related energy expenditures fall into three separable costs (maintenance, activity, and thermoregulation) which sum to give total requirements for ecological maintenance. Fasting metabolism is the base to which increments for activity and thermoregulation are added. The interspecies mean for mature mammals scales to $W^{0.75}$, a term which is often called metabolic mass (Chapter 6).

The net cost of horizontal movement in ungulates scales to $W^{0.68}$ similar to other eutherian quadrupeds (Chapter 7). On slopes, the net cost appears independent of weight and angle of ascent. However, the efficiency of recovering potential energy stored during ascent depends on both weight and incline. Large animals descending steep slopes appear to expend considerable energy to maintain posture.

Size is important in determining the energetic response of animals to the thermal environments. Heat loss of spherical objects is proportional to surface area ($\alpha W^{0.66}$) but because of size-related differences in shape and external insulation, heat loss scales empirically to a lower exponent in homeothermic animals (Chapter 8). Thermal inertia can be used to cope with heat as well cold, particularly in those species which have labile body temperatures.

Average daily energy costs of free-ranging animals represent the sum of basal metabolism, activity, and thermoregulation, all of which scale somewhat differently to body weight. The expectation that the exponent will be less than 0.75 has been confirmed for small mammals[53] and probably applies to ungulates.

B. Digestive Capacity

While expenditures scale to fractional exponents of body weight, fermentation contents of the digestive tract scale close to unity with only small differences between ruminants and other mammalian herbivores.[31, 54] There is a tendency within ruminants for selective feeders to have smaller digestive capacities than roughage grazers.[55] Digestive tract fill as a per-

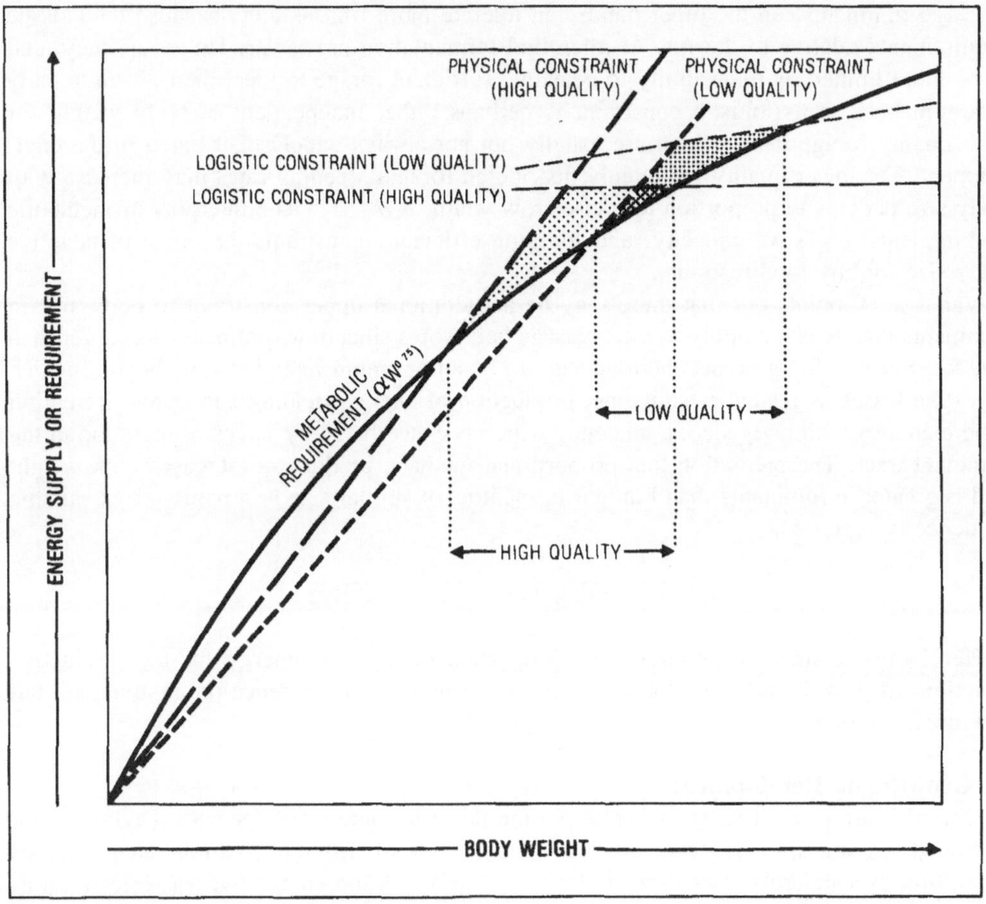

FIGURE 4. Constraints on adaptive body size of herbivores exploiting high and low quality forage resources.

centage of total capacity also tends to be greater in large nonselective grazers (Chapter 5).

Daily assimilation of energy depends on turnover rate as well as fermentation contents. Mean retention time, one measure of turnover, appears related to body weight in a linear manner.[27] For any forage type, high turnover rates tend to be associated with high fermentation rates.[31]

C. Bioenergetic Constraints

The most important consequences of allometric scaling of energy requirements and digestive capacity relate to the ability of animals to utilize forages of varying qualities (Figure 4). Small animals have low absolute requirements relative to larger ones but higher requirements per unit body weight. However, since digestive capacity is related approximately to $W^{1.0}$, they must select higher quality forage. More precisely, ruminants must be sufficiently selective to support fermentation rates which scale to $W^{-0.25}$ ($W^{0.75}/W^{1.0}$). Hoppe[32] confirmed this prediction in a study of African ruminants.

The yield of metabolizable energy is limited ultimately by the finite maximum rate of anaerobic fermentation which places a lower limit on the possible size of animals relying on this mode of digestion.[56] Herbivores reaching this theoretical limit have only partial dependence upon the end-products of microbial digestion (Chapter 5) and at least some species may have adaptively lower metabolic rates (Chapter 6).

Large ruminants, on the other hand, can tolerate more fibrous diets because their tissue requirements relative to the rate of microbial fermentation are lower. However, they tend to be more limited by their ability to consume sufficient forage to meet their absolute daily requirements.[56] This logistic constraint is perhaps rather independent of body weight for high quality forage items which are usually but not always widely distributed in the environment. For lower quality but evenly distributed forages, feeding rates may increase with body size perhaps in proportion to incisor row width ($\alpha W^{0.33}$).[52] The interplay of metabolic requirements, digestive capacity, and foraging efficiency constrains the range of adaptive body size for any food resource.

Van Soest[31] points out that there may be an additional upper constraint to body size in ruminants that does not apply to monogastric herbivores since few ruminants are larger than 800 kg. For any forage type, there appears to be an optimum retention time beyond which digestive losses as methane reduce net production of energy-yielding substrates. Retention time increases with body size, a tendency which is exacerbated by selective retention in the ruminant tract. The prediction that proportional methane production increases with weight and is greater in ruminants than hindgut fermentors of similar size is supported by available evidence (Chapter 5).

III. PRODUCTIVE FUNCTIONS

Herbivores retain a surprisingly low proportion of ingested energy for the productive functions of growth and reproduction. This is a penalty of dependence on an abundant but low quality food source.

A. Growth and Development

One measure of relative growth rate is time taken to mature (Chapter 9). Taylor[57] demonstrated striking similarities in the growth processes of different mammalian genotypes when time is standardized as "metabolic time" ($\alpha W^{0.27}$) and cumulative variables such as live weight or food consumed are scaled to mature weight ($\alpha W^{1.0}$).

Eutherian mammals exhibit growth rates which scale in a manner consistent with Taylor's hypothesis.[36] Similar scaling of growth and metabolic rate is not surprising since respiratory metabolism supplies ATP which is used in building macromolecules. However, the relationship breaks down under fine-grained analysis of closely related species. Compared with other mammals, ungulates express very high growth rates which seem to scale to a power less than that typical for other taxa. The adaptive significance of high growth rate in ungulates as a group and variation among individual species can be traced to the importance of (1) quickly passing through the vulnerable early postnatal period when predation pressures are high and (2) attaining sufficient size to cope with hardships imposed by seasonal environments.

Body composition scales weakly with body size. The distribution of muscle reflects biomechanical changes in the shape of the skeleton with increasing weight.[58] But from a bioenergetic standpoint, the most important compositional aspect is the amount of body fat, the principal energy depot. The propensity of animals to fatten (% carcass fat) varies widely among species. This is not surprising given the variety of purposes served by fat depots[59,60] and widely varying opportunities for energy storage afforded by seasonal pulses of forage production. Absolute amounts of fat, therefore, scale very crudely to $W^{1.0}$.

Along with their relatively low metabolic rate, this should enable large animals to wait out periods of acute starvation as long as these are preceded by ample opportunities to accumulate reserves.[19] This may be a partial explanation for Bergmann's Rule[8,52] since northern environments tend to be more strongly seasonal. However, it probably does not extend to the high arctic where periods of forage abundance are too short and seiges of winter too long to make this a viable strategy. The Svalbard reindeer has apparently adapted

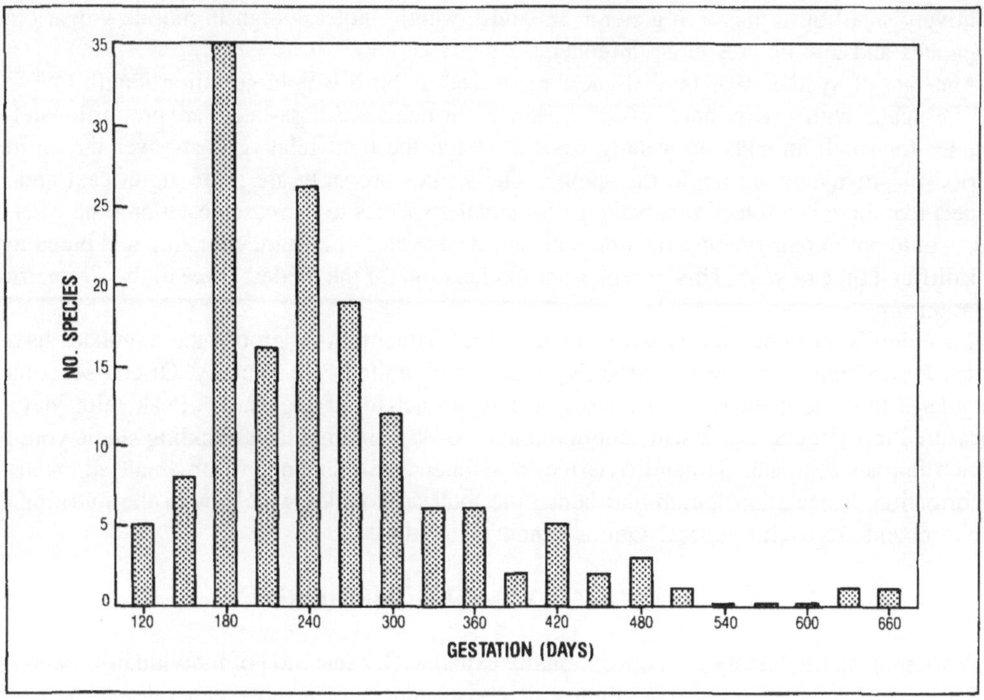

FIGURE 5. Frequency distribution of gestation lengths of ungulates.

by reduction in body size but deviates from general allometric rules in the high proportion of body fat.[61] This carries a penalty and may be possible only in predator-free environments.

B. Reproductive Processes

Gestation length is important because it potentially determines the number of reproductive cycles per year. In strongly seasonal environments where parturition must coincide with suitable foraging conditions, natural selection should favor gestation lengths which allow synchronization with the seasons. Gestation lengths of ungulates range widely from less than 120 days to more than 669 days (Chapter 10). The distribution is bimodal (Figure 5) with peaks at 150 to 180 and 210 to 240 which is consistent with the importance of staying in phase with the astronomical calendar. However, considering ruminants alone, short gestations which potentially permit two breeding cycles per year are more frequent in temperate species (mainly caprines) rather than African ruminants which could capitalize on the typical bimodal distribution of tropical rainfall.

Gestation length scales rather weakly to maternal size. Ungulates[37] and particularly artiodactyls[40] show lower slopes than other mammals. Since young are born at different stages of development, a better relationship might be expected with birth weight. But the relationship is poor both between and within orders. A much clearer relationship is obtained with neonatal brain weight, an index of the altricial/precocial status of the organism. Neural development is similar in most organisms and may be the rate limiting factor of fetal development and perhaps postnatal growth.[62]

Birth weights are relatively large in ungulates to counteract heavy predation on neonates. Since birth weights scale close to the metabolic exponent, Geist[63] expresses neonatal mass as a proportion of calculated basal metabolism. The mean is 0.84 g/kJ with highest values in the saiga antelope and smallest in the pigs. Rupricaprines and smaller deer also have relatively small birth mass. In general, this birth weight index is high in mobile gregarious ungulates and low in "resource defenders".

The rate of synthesis of fetal tissues, expressed as birth weight/gestation length ($W^{0.79}/W^{0.19}$), scales with an exponent of 0.60.[37] Hence, the demands of gestation are proportionately greater for small animals on a daily basis although the total relative costs over the entire period of pregnancy are about the same.[27] Daily rates probably are more significant and I expect that there is a selective advantage for smaller species to increase gestation time where this would not impair synchronization with seasonal cycles or require courtship and breeding at a difficult time of year. This provides one explanation for the modest slope of the allometric relationship.

Lactation is an important aspect of maternal investment. As a group, the ungulates have opted for prolonged care of few offspring rather than limited care of many. Of course, some ungulates have adopted relatively r-selected reproductive strategies.[63-65] Peak milk yields measured in energetic terms scale approximately to $W^{0.52}$ in ungulates suckling single young which implies a greater demand (relative to maintenance metabolism) on small animals.[37] Information on lactation length and hence the total cost varies widely with the nutritional environment, so useful generalizations cannot be made.

IV. LIFE HISTORY AND DEMOGRAPHY

Variations in life history are conventionally explained by selection of individually adaptive traits. However, such traits may simply be consequences of a more general selection for body size.

A. Vital Parameters

Lifetime reproductive success depends on age at first breeding, interbirth interval, litter size, and reproductive life. These parameters are aggregated in various ways and the literature is rather incomplete for ungulates. Western[40] has assembled data on the age at first reproduction and birth rate for African species. For artiodactyls, the age at first reproduction scales to $W^{0.27}$, identical to the slope for physical maturation. Birth rate (litter size/interbirth interval) scales to $W^{-0.35}$. Interbirth intervals of multiparous species tend to be longer than similarly sized uniparous species, thereby narrowing the contrast in birth rates.

Whereas birth rates decline in larger animals, lifespan increases.[19] The relationship for mammals[46] scales to $W^{0.20}$ with the intercept varying widely among orders. For African artiodactyls, the exponent is not significantly different.[40]

The balance of reproductive and mortality rates determines reproductive output and rate of population increase. The actual relationship between birth and death rates is described crudely by a hyperbolic function.[66] Hence, contrasts in maximum rates of increase are expected to be less strongly influenced by body weight. Empirical evidence suggests that the balance favors higher rates of increase in small species with population doubling times[45] scaling to $W^{0.26}$.

B. Demographic Strategies

Schedules of natality, mortality, and dispersal are considered collectively as life history strategies. In general, wild herbivores have long lifespans, high adult survival, low fecundity, and late reproductive maturity. Complex age and sex-specific schedules and population structures make determination of optimal life tables difficult.[67]

Attempts to understand the effects of natural selection on demographic parameters have been dominated by the principle of r- and K-selection. Animal populations are characterized by potential rates of increase (r) and equilibrium levels (K). Small animals have higher reproductive performance which is only partially offset by lower survival rates. Their population turnover and higher rate of increase enable them to rapidly respond to changing environmental conditions. Species exploiting more stable environments are termed ''K-strategists'' because of their ability to compete well in dense populations near carrying capacity.

It is common to view r- and K-selection as the two ends of a continuum.[66] However, Jones[68] noted that certain intermediate rates of increase are associated with considerable instability. In this domain, many generations are required to restore equilibrium following perturbation. But there is no information to determine how important this may be over the range of parameters exhibited by ungulates.

Another important question is concerned with optimal allocation of resources among the competitive demands of maintenance, growth, and reproduction.[67, 69] Fitness is equated with lifetime reproductive performance defined as net recruitment of offspring to the breeding population. This is not the same as high reproductive rates since raising young in any year may impair ability to reproduce subsequently. Females hedge the probability that young can be raised successfully in any one year against the probability of the future reproductive success. Reproductive risks should be taken in inverse proportion to residual reproductive value. Young females should be less likely to breed and abandon calves more readily if environmental conditions are poor. As females approach the end of their reproductive life, they should take increasingly greater risks to raise young. Therefore, Clutton-Brock[70] wondered why fecundity of aged females often declines. The most likely reason is that the ability of senile animals to acquire resources is impaired so even though a greater proportion of available nutrients is invested in offspring, the absolute amount must be lower.

Since the risk of reproduction is inversely related to body reserves, it is not surprising that fecundity is related more strongly to body weight at the time of conception than to age.[71] Weight thresholds for reproduction probably differ among species. Species with relatively high population turnover rates should take greater risks by breeding younger and at proportionally lighter weights. Predictions on age at first breeding are confirmed by empirical evidence but insufficent work has been done on weight thresholds.

A final aspect of life history evolution is density-dependence of vital schedules. Typically, herbivores (especially larger ones) do not express density dependence until they approach carrying capacity.[72] Part of the reason for this nonlinearity is related to the age structure of expanding and stable populations. However, natural selection (so-called *theta* selection) may favor this pattern.

V. BEHAVIORAL ECOLOGY

The interrelationship of body size, ecology, behavior, and bioenergetics is complex. Classifications of ecoethological traits have been offered[73,74] and several authors[15, 64] have drawn attention to key relationships.

A. Gregariousness

A dominant factor favoring gregariousness is the risk of predation.[75] In open habitats, this risk declines in proportion to group size but not necessarily equally for all members since predation tends to be concentrated on the periphery of herds. Although this is the dominant advantage of group membership, the chance that the predator may kill at all may be influenced by gregariousness. Large groups may be easier to discover but more difficult to stalk and to concentrate on individual prey.

Weighted aganist the individual's risk of predation are factors related to resource acquisition. In the security of a group, an individual may redirect activity from vigilance to foraging.[76,77] But increasing group size may eventually result in feeding interference[78] or local forage depletion which requires the group to be more mobile.[73]

Costs and advantages and hence optimal group sizes vary with predation pressure and forage supply.[79] Largest groups are expected in productive grasslands where other individuals rather than vegetation provide cover and where forage is available in a dense carpet which reduces the rate of forage depletion. As primary productivity declines and food items become widely dispersed, optimal group sizes decline.

Animals that have been injured or are otherwise in poor condition often retire from groups and move to dense cover. A similar response is observed in males whose body reserves have been depleted during the rut.[80] The explanation appears to lie both in foraging efficiency and predation risk. Weakened animals experience greater feeding interference because of their usually subordinate status. Since predators often test prey before selection, escaping detection can be a more effective antipredator strategy than hiding within groups.

Body size may influence the range of optimal group sizes. Smaller ungulates living in dense cover can use crypsis and may be disadvantaged by the presence of other individuals. Very large ungulates may be able to defend themselves and their young either individually or in small groups. The greatest benefits of gregariousness may accrue to intermediate-sized ungulates living in open habitats where communal vigilance and use of other individuals as cover offer particular benefits.

B. Male Mating Systems

The most striking feature of social organization is the male mating system. Relationships with body size are based primarily on the tendency of large species to occupy open habitats. Since the objective of males is to gain access to estrous females, their interrelationships are dictated proximally by the gregariousness, cohesiveness, and spatio-temporal distribution of female groups. Ultimately, these characteristics can be traced to environmental determinants.

One of the most important determinants is environmental seasonality. In temperate and arctic environments, breeding seasons must be sharply defined to ensure that young are dropped at a time of year that supports lactation of the dam and provides time for young to mature sufficiently before their first winter. Consequently, breeding is concentrated in a brief but intense rut which occurs a gestation-length before the optimal birth date. Seasonal harem formation is expected where females live in small cohesive groups (e.g., *Cervus*). Rank dominance is expected where females live in larger loose aggregations (e.g., *Bison*). A roving strategy is appropriate where females are more widely distributed (e.g., *Odocoileus*).

In tropical environments with less breeding synchrony, mating territories are common in spite of the relatively low mating enhancement they confer. Owen-Smith[81] considers this to be a low-cost, low-benefit strategy, pointing out that mating territories may be less energetically demanding. One reason why mating territories may work in the tropics is that the spatial distribution of estrous females "averages" with the longer breeding season and therefore corresponds more predictably with the quality of a territory.

Not surprisingly, ungulates exhibit considerable plasticity in response to environmental conditions. Uganda kob may either form leks or more widely spaced mating territories. *Cervus* forms harems where stags can control females but tends individual females in forests where this is more difficult. Social systems tend to be species-typical rather than species-specific.

But the adaptiveness of these patterns will remain idle speculation until the costs and benefits of various breeding systems are more fully documented.[81] Male mating systems bring the interplay of security, social, and bioenergetic needs into focus and reveal the incompleteness of current knowledge.

C. Home Range

Since home ranges expand and contract in relation to food supply, it is not unreasonable to expect them to scale to energy requirements and hence to metabolic weights.[82] However, empirical results for temperate herbivores suggest an exponent closer to unity.[49] The simplest explanation lies in its similarity with scaling of digestive capacity but other factors related to foraging efficiency and/or mobility may be relevant.

Pennycuick[48] argues that mobility may dictate the pattern of habitat heterogeneity that animals can exploit efficiently. Only highly mobile animals can take advantage of widely dispersed sources of water, cover, and forage. On the basis of allometric scaling of basal metabolism and energetics of locomotion, Pennycuick[48] calculated that the foraging radius (i.e., maximum distance animals travel between separated resources) should scale approximately to $W^{0.40}$. For circular home ranges, the area should scale to $W^{0.80}$ (i.e.,πr^2). This is still less than the empirical exponent for temperate herbivores, but the ability of large animals to use more coarsely grained environments may inflate this value. Another factor is that empirical scaling includes characteristics of the environment as well as herbivores.

D. Resource Selection

In spite of major theoretical developments, current concepts of prudent foraging have proven difficult to apply, especially to large herbivores which display a number of characteristics not predicted by existing theory (Chapter 3). Conventional optimization models are of several types. First are those dealing with the logistics of foraging.[83] But since herbivores often are constrained by digestive capacity, their decision process might better be framed in terms of balancing nutrients within the constraint of digestive tract fill.[84] Obviously there are circumstances where both factors are important and mixed models are appropriate.

In the simplest case, consider an animal faced with a choice of two resources which differ in quality and availability (Figure 6). If the animal selected only relatively abundant low quality forages, it may be limited to a certain level of intake by gut fill. On the other hand, if it selected rare high quality foods it likely would be limited by its ability to ingest an amount which would meet its daily requirement. If the animal chose a combination of forages, total consumption could increase both for physiological reasons (associative effects of feeds) and for logistic reasons (in searching for one type of plant, the other would be discovered fortuitously). Possible combinations map the consumption frontier.

The optimum strategy depends upon the ability of animals to extract nutrients from each forage type. This is represented by the value exchange line which has a slope reflecting relative digestible (or metabolizable) energy yield. The optimum is the point of tangency.

Optima for selective and nonselective feeders differ for two reasons. Firstly, the consumption frontier depends upon body size, feeding apparatus, and digestive capacity of each herbivore for the two forage classes. Secondly, the relative value of forages depends upon specific digestive adaptations.

More comprehensive application of these concepts have been offered by Belovsky[85] and Case.[56] However, in addition to problems with a currently limited data base, there is doubt that optimizing criteria are necessarily appropriate in complex stochastic environments.

E. Activity Budgets

Activity budgets of herbivores are conservative. They spend most of their time either feeding or resting and very little in other activities, particularly those which are energetically expensive. Animals forage longer when food is limited but such behavioral adaptation is constrained by the need for rumination, rest, and other essential activities.

Attempts have been made to scale feeding time to body size in expectation that greater absolute food requirements would be associated with longer feeding periods. It is true that the heaviest commitment to foraging is found in the elephant, the largest terrestrial herbivore.[86] But in interspecific comparisons, foraging time scales rather weakly to body weight.[33] Although forage requirements scale to metabolic weight, large animals often can feed more

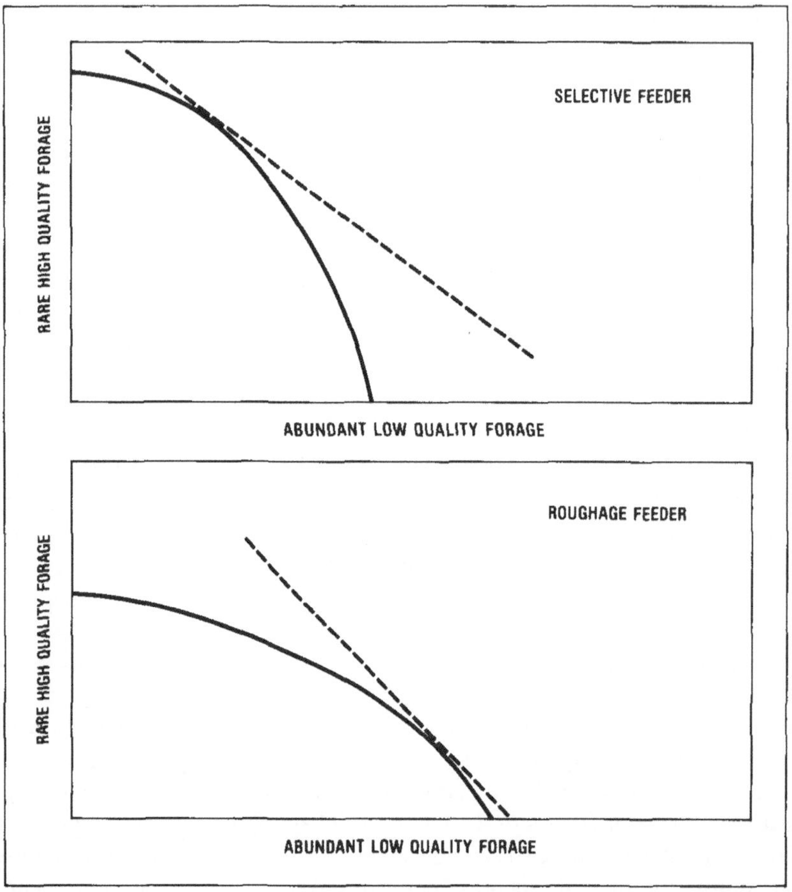

FIGURE 6. Optimal choice of selective and roughage feeders between rare high quality forage items and abundant low quality forage items. Optimal combinations are defined at the point of tangency of the value exchange line and the consumption-possibilities frontier.

rapidly because of their larger bite size. Clutton-Brock and Harvey[52] suggested that since the width of the incisor row scaled to $W^{0.33}$, relative feeding rates on uniform swards should scale similarly. On uniform monospecific pastures, feeding time should therefore scale at no greater than $W^{0.42}$ ($W^{0.75}/W^{0.33}$). But in the diversity of the natural world, small animals probably use their additional time to forage more selectively.

Many other factors contrive to give such a modest slope. These data pool monogastric/ruminants and browsers/grazers. Geist[64] predicted that monogastric herbivores would spend more time foraging than ruminants of comparable size which divide their nutritional activities between foraging and ruminating. Leuthold[74] generalized that browsers may require more time to ingest the same amount of food as similarly sized grazers. Other aspects of the activity budget are covered in Chapter 3.

VI. GRAZING SYSTEMS DYNAMICS

A. Biomass and Productivity

In response to latitudinal patterns in plant production, animal biomass and productivity change in a predictable manner. Polar deserts support very low densities of animals providing little sustained yield. With amelioration of temperatures southward, productivity indexes increase to rather impressive levels on the southern fringe of the boreal forest. Continuing southward, moisture rather than low temperatures constraints primary production and animal densities and productivity decline towards the global desert belt. The diversity and productivity of grazing systems again increase towards the equator with improved available moisture providing the most spectacular herbivore assemblages in the world. But as tree cover increases in the tropical rain forests, productivity of terrestrial herbivores again declines.

Coe et al.[87] provide clear empirical relationships of biomass and production of large mammalian herbivores in African savannas with primary production and its main determinant, mean annual precipitation. Correlations using single species were not high, presumably because of competitive interactions within animal communities. A number of factors would be expected to complicate these relationships. For example, periodic droughts may maintain lower populations than predicted on the basis of mean annual precipitation. Carrying capacity may be modified by soil and drainage patterns and a variety of factors relating to temporal and spatial variation in habitats and surface water. This positive correlation is restricted to herbaceous plant communities. In forested habitats, the relationship may be inverse as greater amounts of primary production are tied up as structural carbohydrate which is refractory to microbial fermentation or elevated beyond the reach of terrestrial species.[88]

But even in the tropics, marked contrasts are evident. Eisenberg[89] noted that ungulate biomasses for selected national parks in Southern Asia range from only 3 to 29 kg/ha. These values are considerably lower than the African grazing systems which may be 50 to 100 kg/ha. He feels that the discrepancy is probably related to the evolutionary histories of the two areas. The ungulate fauna of Asia is typified by browsers or mixed feeders in contrast to the diversity of grazers characteristic of East African savannas. Lower productivity also is found in the American tropics where ungulate biomass densities are 3 to 5 kg/ha. This appears due to the low diversity of ungulates resulting from widespread extinctions of endemic species.

Biomass is related in a complex way to the characteristics of members comprising the grazing community. One of the most fundamental linkages is between size, metabolic rate, and forage requirements. Because of allometric scaling of energy requirements, large animals place less demand on their environment kilogram for kilogram, than do smaller ones. Also, large animals are able to use a lower quality but usually more abundant forage resource. Hence, large animals tend to achieve higher biomass ($\alpha W^{0.25}$) but lower numerical densities ($\alpha W^{-0.75}$) than smaller ones.[50] The product of population density and metabolic demand per individual tends to be independent of body size. Calder[20, 51] considers this to represent a tradeoff between space and time and to provide an allometric basis for group sizes in herbivores.

While the ability to accumulate biomass is greater in large animals, the rate of production in relative terms is lower. Kleiber[21] predicted that these two tendencies would balance to give approximately the same net productivity irrespective of body size. However, evidence points to the existence of an optimum (in natural environments) at intermediate weights.[90] The elephant, because of its longevity and low maintenance metabolism, has a tremendous capacity to accumulate biomass but its rate of production is low. The meadow vole has tremendous ability to produce new tissue but a limited ability to accumulate biomass due to high metabolic costs and small fat reserves. The white-tailed deer is of intermediate weight where capacity for production and biomass accumulation achieve an optimal balance. Of course, productivity cannot be equated to fitness but it is relevant to management and consumptive use of wild populations.

B. Plant-Herbivore Interactions

Perhaps the most incisive view of grazing systems dynamics is offered by a model of trophic interactions developed by May[91] and elaborated by Noy-Meier[92] and Caughley.[93] The system is modeled by two linked equations describing a population of plants (V) and a population of herbivores (H):

$$dV/dt = a_1V(1 - V/K) - Hc_1(1 - e^{-d_1V}) \quad (2)$$

$$dH/dt = H[-a_2 - c_2(1 - e^{-d_2V})] \quad (3)$$

Where

K = equilibrium biomass of ungrazed vegetation
a_1 = maximum intrinsic rate of plant growth
a_2 = rate of decline of herbivores when $V = 0$
c_1 = rate of consumption of vegetation when not limiting
c_2 = demographic scope of herbivores
d_1 = grazing efficiency of herbivores
d_2 = demographic efficiency of herbivores

Depending upon the values of coefficients, the system may reach a constant equilibrium, enter a stable limit cycle, or crash to extinction. The amplitude of oscillations increases with the potential rate of increase and demographic efficiency of the herbivore. The period of the oscillation increases with demographic traits characteristic of larger animals. There is evidence that the period of grazing systems in which elephants are dominant herbivores may be as long as several hundred years.[93] Application of demographic information for elephants in East Africa suggests 70 years.[94] The period is about 20 years for the most medium-sized temperate ruminants, 10 years for snowshoe hares and 4 years for microtine rodents. These cycles are poorly documented; therefore, it is not possible to determine exactly how they scale to body size. However, Calder[44] feels that existing information is consistent with an allometric exponent of 0.25.

VII. SYNTHESIS

Much of the diversity within ungulates, and indeed all mammals, may be related simply to the problem of maintaining functional similarity with increasing size.[24] Whether this is to maintain dynamic[16] or elastic[17] similarily is of little consequence. This variation must be extracted before specific adaptations can be disclosed.

A. Size as an Adaptive Strategy

Many bioenergetic, productive, life history, and ecological traits of wild herbivores, like other organisms, scale allometrically. Large body size conveys many advantages including the ability to use a lower quality but more abundant food resource, greater mobility, reduced risk of predation, and access to important resources such as food and mates. But large size also carries penalties including inability to subsist on rare food items and reduced maturation and reproductive rates. Thus, larger is not necessarily better in all environmental settings.

This dominant influence of body size perhaps calls for a more epigenetic view of evolution.[95] Life history phenomena may be allometric consequences rather than individual objects of natural selection. However, considerable (presumably adaptive) variation remains when the effect of body weight is removed. Western[40] considers size to be a primary adaptation and argues that determination of whether animals are specifically adapted should be based upon the divergence of a trait from that expected on interspecific allometric grounds.

At the same time, body size may have a rather indirect effect on particular traits (Figure 7). For example, metabolic requirements may be more directly related to shape (specifically cross-sectional area) than to size per se. Since metabolism provides energy for productive functions, growth rates are most closely linked to metabolic rate. In turn, growth rates of suckling young are proximal determinants of the demands of lactation which may influence

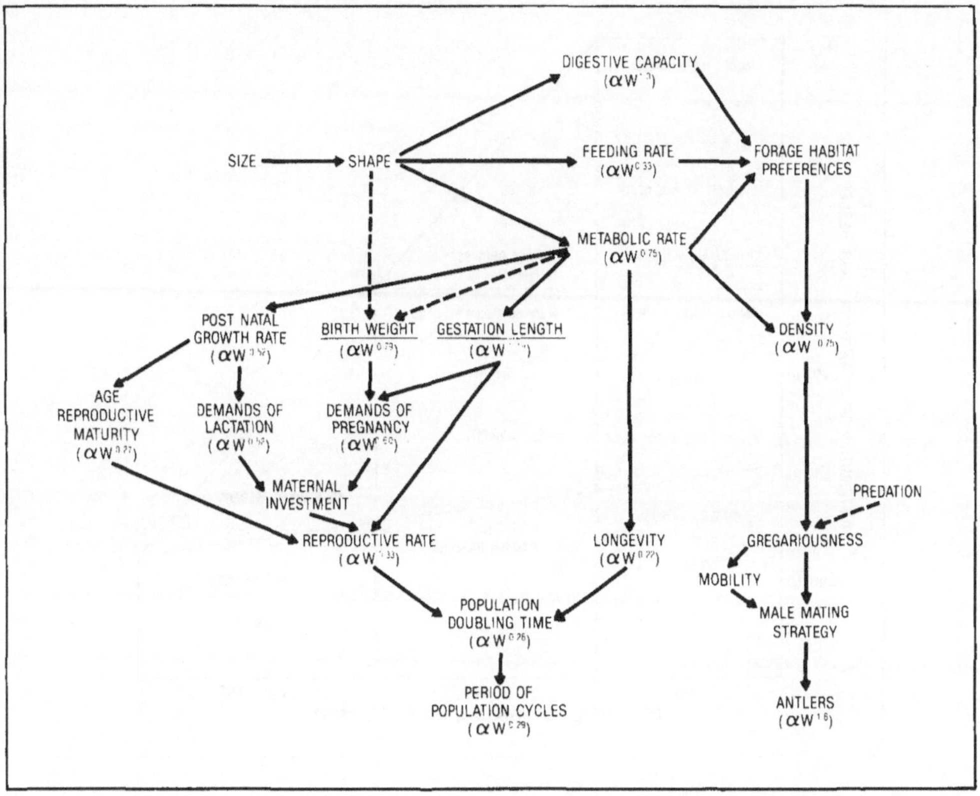

FIGURE 7. Possible linkages underlying allometric scaling of physical, metabolic, life history, and ecological traits of large herbivores.

the frequency of reproduction and other population phenomena. One illustration of spurious allometric scaling relates to antler size in the *Cervidae*.[96] Antler size scales hyperallometrically. However, this relationship turns out to be an indirect consequence of the tendency of larger-bodied species to mate in large groups.

Another problem is that allometric scaling is most pronounced among widely divergent taxa. Strong relationships (steep slopes) among subfamilies degrade at the generic level and may be nonexistent at the species level. For example, ponies and draft horses differ vastly in size but gestation lengths, reproductive rates, and lifespans differ very little. The extent to which allometric rules apply to speciation must remain an open question.

B. The Large Herbivore Niche

Since the natural world is discontinuous offering, it seems, relatively discrete niches, traits tend to fall out in adaptive suites which are evident in convergent evolution of different phyletic lines. Geist[63] notes that "rhinos" have evolved six times, "horses" four times, "pigs" three times and "deer" twice since the Eocene. Although there have been many evolutionary experiments, certain basic designs have worked better than others.

One of the most useful attempts to order these patterns is based on digestive anatomy which reflects the feeding niche of large herbivorous mammals.[55] Ruminants can be arranged in three major categories; namely, roughage grazers, selective feeders, and intermediate mixed feeders (Chapter 5). Specific anatomical and functional adaptations are required for efficient exploitation of plant tissues which are relatively uniform and usually of low digestibility (e.g., monocotyledonous plants) or more differentiated with at least a portion of rapidly fermentable material (e.g., dicotyledonous plants).

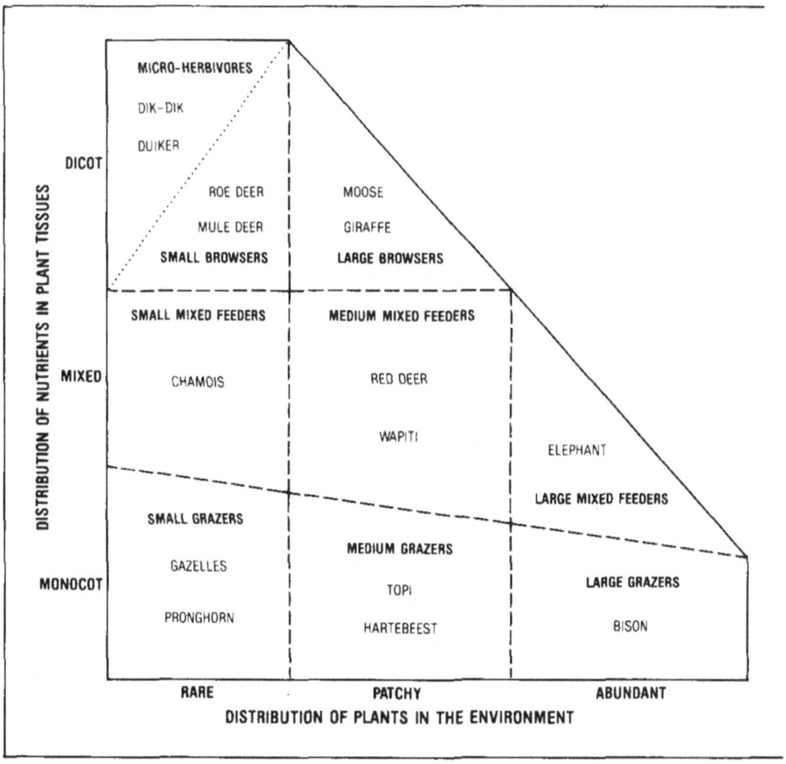

FIGURE 8. Feeding niche of ungulates defined by the distibution of nutrients in plant tissues and distribution of plants in the environment.

Attempts have been made to correlate these digestive traits with body size.[34] In general, frequency distributions of body size are lowest for selective feeders and highest for roughage grazers but there are so many exceptions that such a generalization is not particularly useful. Several small-bodied species such as the reedbuck and sheep are unexpectedly nonselective while the largest ruminant (giraffe) and the largest cervid (moose) are concentrate selectors.

The ability of ruminants to be very small and for small ruminants to use coarse forage may depend upon adaptive lowering of metabolic rate. The existence of large selective feeders requires a more complex explanation. Body size of selective feeders normally is constrained by the logistics of foraging so the association of resource use with body size is really an environmental correlation between the quality of food items and their abundance and distribution. Giraffe and moose simply use resources which are highly clumped. In these situations, feeding rate is not seriously limited by availability and large body size may facilitate movement between widely spaced but dense food patches.

To accommodate this important factor, a classification of the herbivore niche should include the distribution of food items as well as the structure of plant tissues (Figure 8). Forage distribution has temporal, spatial, and vertical components which are relevant to resource partitioning and adaptive radiation. However, much of the variation can be captured by considering food items to be aligned along the gradient: rare, patchy, and abundant. There is an environmental correlation between the distribution (abundance) and tissue structure (quality) of forage plants; dicotyledonous plants are usually rare but may occur in patches while monocotyledonous plants often form dense swards. The "periodic table" of herbivore niches is more triangular than square.

Within each forage class, herbivores can be aligned along a gradient of body size. In general, larger species are less selective and consume lower quality diets. Hence, animals preferring dicotyledonous plants cannot always be considered "selective feeders". Small ruminants within this group feed on fruits and buds. Larger members consume foliage and twigs. The diameter of twigs at the point of browsing is roughly related to body size. Among grazers, small species forage mainly on fresh grass shoots and rangeland forbs. Medium sized species (e.g., *Damaliscus*) select for grass leaves while larger species (e.g., *Bison*) forage without discriminating among plant parts. Mixed feeders gain flexibility by using all forage classes but may digest each less well than browsers or grazers. There is a tendency for large mixed feeders to select proportionately more graminoids.

Although this two-dimensional summary accounts for much of the interspecific variation, an additional gradient describes the permanence and grain of resources used by herbivores. Nomadic and migratory species probably warrant distinctive classifications.

C. Future Directions

As information accumulates from field studies on wild herbivores, more probing questions of adaptive radiation can be addressed. From comparative analyses, patterns largely based on *a posteriori* methods have been established. Only recently have predictions been constructed and tested experimentally. This chapter has reviewed evidence that many aspects of the biology of wild herbivores have a bioenergetic basis. Fortunately, the heavy representation of herbivores among agricultural animals provides a base for understanding bioenergetic principles. But, domestic stock represent only a small sample of a diverse assemblage of wild species. The following chapters attempt to compile what is known about the bioenergetics of wild ungulates and to lay the foundation for a more quantitative and predictive approach to understanding and managing this important faunal resource.

REFERENCES

1. **Corbet, G. B. and Hill, J. E.,** *A World List of Mammalian Species*, British Museum Natural History, Comstock Publ., Ithaca, N.Y., 1980.
2. **Eisenberg, J. F.,** *The Mammalian Radiations: An Analysis of Trends in Evolution*, University of Chicago Press, Chicago, 1981.
3. **Janis, C.,** The evolutionary strategy of the *Equidae* and the origins of rumen and cecal digestion, *Evolution*, 30, 757, 1976.
4. **Kingdon, J.,** *East African Mammals: An Atlas of Evolution in Africa*, Academic Press, London, 1979.
5. **Maglio, V. J.,** Origin and evolution of the *Elephantidae*, *Trans. Am. Phil. Soc.*, 63, 1, 1973.
6. **Geist, V.,** On the relation of the social evolution and dispersal in ungulates during the Pleistocene, with emphasis on the Old World Deer and the genus *Bison*, *Quat. Res. (N.Y.)*, 1, 283, 1971.
7. **Geist, V.,** A comparison of social adaptations in relation to ecology in gallinaceous bird and ungulate societies, *Annu. Rev. Ecol. Syst.*, 8, 193, 1977.
8. **Geist, V.,** *Life Strategies, Human Evolution, Environmental Design: Toward a Biological Theory of Health*, Springer-Verlag, New York, 1978.
9. **Geist, V.,** On the evolution of ice age mammals and its significance to an understanding of speciations, *ASB Bulletin*, 30, 109, 1983.
10. **Calow, P.,** Ecology, evolution and energetics: a study of metabolic adaptation, *Adv. Ecol. Res.*, 10, 1, 1977.
11. **Townsend, C. R. and Calow, P., Eds.,** *Physiological Ecology: An Evolutionary Approach to Resource Use*, Blackwell Scientific, Oxford, 1981.
12. **Clutton-Brock, T. H. and Harvey, P. H.,** Comparison and adaptation, *Proc. R. Soc. London*, 205, 547, 1979.
13. **Gould, S. J. and Lewontin, R.,** The spandrels of San Marco and the Panglossian paradigm: a critique of the adaptionist's programme, *Proc. R. Soc. London, Ser. B.*, 205, 581, 1979.
14. **Ralls, K.,** Mammals in which females are larger than males, *Q. Rev. Biol.*, 51, 245, 1976.

15. **Estes, R. D.,** Social organizations of the African *Bovidae*, in *The Behavior of Ungulates and its Relation to Management*, Geist, V. and Walther, F., Eds., IUCN Publ. NS 24, Morges, 1974, 166.
16. **Alexander, R. McN.,** Allometry of the limbs of antelopes (*Bovidae*), *J. Zool. London*, 183, 125, 1977.
17. **McMahon, T. A.,** Allometry and biomechanics: limb bones in adult ungulates, *Am. Nat.*, 109, 547, 1975.
18. **Calder, W. A., III,** Scaling of physiological processes in homeothermic animals, *Annu. Rev. Physiol.*, 43, 301, 1981.
19. **Lindstedt, S. L. and Calder, W. A., III,** Body size, physiological time, and longevity of homeothermic animals, *Q. Rev. Biol.*, 56, 1, 1981.
20. **Calder, W. A., III,** Ecological scaling: mammals and birds, *Annu. Rev. Ecol. Syst.*, 14, 213, 1983.
21. **Kleiber, M.,** *The Fire of Life*, Wiley-Interscience, New York, 1961.
22. **Cohen, Y., Robbins, C. T., and Davitt, B. B.,** Oxygen utilization by elk calves during horizontal and vertical locomotion compared to other species, *Comp. Biochem. Physiol.*, 61A, 43, 1978.
23. **Taylor, C. R., Heglund, N. C., and Maloiy, G. M. O.,** Energetics and mechanics of terrestrial locomotion. I. Metabolic energy consumption as a function of speed and body size in birds and mammals, *J. Exp. Biol.*, 97, 1, 1982.
24. **McMahon, T.,** Size and shape in biology, *Science*, 179, 1201, 1973.
25. **Moen, A. N.,** *The Biology and Management of Wild Ruminants*, Part 1, CornerBrook Press, Lansing, 1980, chap. 1.
26. **Parker, K. L. and Robbins, C. T.,** Thermoregulation in ungulates, in *Bioenergetics of Wild Herbivores*, Hudson, R. J. and White, R. G., Eds., CRC Press, Boca Raton, Fla., 1985, chap. 8.
27. **Robbins, C. T.,** *Wildlife Nutrition and Feeding*, Academic Press, New York, 1983.
28. **McNab, B. N.,** On the ecological significance of Bergmann's Rule, *Ecology*, 52, 845, 1971.
29. **Schmidt-Nielsen, K.,** Problems of scaling: locomotion and physiological correlates, in *Scale Effects in Animal Locomotion*, Pedley, T. J., Ed., Academic Press, London, 1977, chap. 1.
30. **Stahl, W. R.,** Scaling of respiratory variables in mammals, *J. Appl. Physiol.*, 22, 453, 1967.
31. **Van Soest, P. J.,** *Nutritional Ecology of the Ruminant*, O & B Books, Corvallis, Oregon, 1982.
32. **Hoppe, P. P.,** Rumen fermentation and body weight in African ruminants, *Proc. XIII Int. Congr. Game Biologists*, Wildlife Society and Wildlife Management Institute, Washington, D.C., 1971, 141.
33. **Owen-Smith, N.,** Factors influencing the composition of plant products by large herbivores, in *Ecology of Tropical Savannas*, Vol. 42, Huntley, B. J. and Walker, B. H., Eds., Ecological Studies, Springer Verlag, Berlin, 1982.
34. **Kay, R. N. B., Englehardt, W. V., and White, R. G.,** The digestive physiology of wild ruminants, in *Digestive Physiology and Metabolism in Ruminants*, Ruckebusch, Y. and Thivend, P., Eds., MTP Press, Lancaster, England, 1979, 743.
35. **Prange, H. D., Anderson, J. F., and Rhan, H.,** Scaling of skeletal mass to body mass in birds and mammals, *Am. Nat.*, 113, 103, 1979.
36. **Case, T. J.,** On the evolution and adaptive significance of postnatal growth rates in terrestrial vertebrates, *Q. Rev. Biol.*, 53, 243, 1978.
37. **Robbins, C. T. and Robbins, B. L.,** Fetal and neonatal growth patterns and maternal reproductive effort in ungulates and subungulates, *Am. Nat.*, 114, 101, 1979.
38. **Leitch, I., Hytten, F. E., and Billewicz, W. Z.,** The maternal and neonatal weights of some mammalia, *Proc. Zool. Soc. London*, 133, 11, 1959.
39. **Blaxter, K. L.,** Protein metabolism and requirements in pregnancy and lactation, in *Mammalian Protein Metabolism*, Vol. 2, Munro, H. N. and Allison, B. J., Eds., Academic Press, London, 1964, 173.
40. **Western, D.,** Size, life history and ecology in mammals, *Afr. J. Ecol.*, 17, 185, 1979.
41. **Payne, P. R. and Wheeler, E. F.,** Comparative nutrition in pregnancy and lactation, *Proc. Nutr. Soc.*, 27, 129, 1968.
42. **Hanwell, A. and Peaker, M.,** Physiological effects of lactation on the mother, *Symp. Zool. Soc. London*, 41, 297, 1977.
43. **Hafez, E. S. E., Asdell, S. A., and Blandau, R. J.,** Propagation: Mammals, in *Biology Data Book*, Vol. 1, Altman, P. L. and Dittmer, D. S., Eds., Fed. Am. Soc. Exp. Biol., Bethesda, 1972, 138.
44. **Calder, W. A., III,** An allometric approach to population cycles of mammals, *J. Theor. Biol.*, 100, 275, 1983.
45. **Frenchel, T.,** Intrinsic rate of natural increase: the relationship with body size, *Oecologia*, 14, 317, 1974.
46. **Sacher, G. A.,** Relationship of lifespan to brain weight and body weight, in *The Lifespans of Animals*, Wolstenholm, G. E. W. and O'Connor, M., Eds., Little, Brown, Boston, 1959, 115.
47. **Gunther, B. and Morgodo, E.,** Theory of biological similarity revisited, *J. Theor. Biol.*, 96, 543, 1982.
48. **Pennycuick, C. J.,** Energetic costs of locomotion and a concept of "foraging radius", in *Serengeti, Dynamics of an Ecosystem*, Sinclair, A. R. E. and Norton-Griffiths, M., Eds., University of Chicago Press, Chicago, 1979, 164.
49. **Harestad, A. S. and Bunnell, F. L.,** Home range and body weight — a reevaluation, *Ecology*, 60, 389, 1979.

50. **Damuth, J.,** Population density and body size in mammals, *Nature (London)*, 290, 699, 1981.
51. **Calder, W. A., III,** A tradeoff between space and time: dimensional constants in mammalian ecology, *J. Theor. Biol.*, 98, 393, 1982.
52. **Clutton-Brock, T. H. and Harvey, P. H.,** The functional significance of variation in body size among mammals, in *Advances in the Study of Mammalian Behavior*, Eisenberg, J. R. and Kleiman, D. G., Eds., Spec. Publ. 7, American Society of Mammalogists, Shippensburg, Pa., 1983, 632.
53. **Mace, G. M.,** The Evolutionary Ecology of Small Mammals, Ph.D. thesis, University of Sussex, Brighton, 1979.
54. **Parra, R.,** Comparison of foregut and hindgut fermentation in herbivores, in *The Ecology of Arboreal Folivores*, Mongomery, G. G., Ed., Smithsonian Inst. Press, Washington, D.C., 1978, 205.
55. **Hofmann, R. R.,** *The Ruminant Stomach: Structure and Feeding Habits of East African Game Ruminants*, East African Monographs in Biology, Vol. 2, East African Literature Bureau, Nairobi.
56. **Case, T. J.,** Optimal body size and an animals's diet, *Acta Biotheor.*, 28, 54, 1979.
57. **Taylor, St. C. S.,** A relation between mature weight and time taken to mature in mammals, *Anim. Prod.*, 7, 203, 1965.
58. **Berg, R. T. and Butterfield, R. M.,** *New Concepts of Cattle Growth*, Sydney University Press, Sydney, 1976.
59. **Pond, C.,** Morphological aspects and ecological significance of fat distribution in wild vertebrates, *Annu. Rev. Ecol. Syst.*, 9, 519, 1978.
60. **Young, R. A.,** Fat, energy and mammalian survival, *Am. Zool.*, 16, 699, 1976.
61. **Riemers, E., Ringberg, T., and Sorumgaard, R.,** Body composition of Svalbard reindeer, *Can. J. Zool.*, 60, 1812, 1982.
62. **Sacher, G. A. and Staffeldt, E. F.,** Relation of gestation time to brain weight for placental mammals: implications for the theory of vertebrate growth, *Am. Nat.*, 108, 593, 1974.
63. **Geist, V.,** On the reproductive strategies in ungulates and some problems of adaptation, in *Evolution Today: Proc. Int. Congr. System. Evol. Biol. 2*, Scudder, G. G. E. and Reveal, J. L., Eds., Carnegie-Mellon University, Pittsburgh, Pa., 1981, 111.
64. **Geist, V.,** On the relationship of social evolution and ecology in ungulates, *Am. Zool.*, 14, 205, 1974.
65. **Geist, V.,** On the evolution of reproduction potential in moose, *Nat. Can.*, 101, 527, 1974.
66. **Pianka, E. R.,** *Evolutionary Ecology*, 3rd ed., Harper & Row, New York, 1983, chap. 5.
67. **Goodman, D.,** Life history analysis of large mammals, in *Dynamics of Large Mammal Populations*, Fowler, C. W. and Smith, T. D., Eds., Wiley-Interscience, New York, 1981, 415.
68. **Jones, J. M.,** The r-K selection continuum, *Am. Nat.*, 109, 320, 1975.
69. **Stearns, S. C.,** Life history tactics: a review of ideas, *Q. Rev. Biol.*, 51, 4, 1976.
70. **Clutton-Brock, T. H.,** Reproduction effort and terminal investment in iteroparous animals, *Am. Nat.*, 123, 212, 1984.
71. **Hamilton, W. J. and Blaxter, K. L.,** Reproduction in farmed red deer. I. Hind and stag fertility, *J. Agric. Sci.* (Cambridge), 95, 261, 1980.
72. **Fowler, C. W.,** Density-dependence as related to life-history strategy, *Ecology*, 62, 602, 1981.
73. **Jarman, P. J.,** The social organisation of antelope in relation to their ecology, *Behaviour*, 48, 215, 1974.
74. **Leuthold, W.,** *African Ungulates*, Springer-Verlag, Berlin, 1977.
75. **Bertram, B. C.,** Living in groups: predators and prey, in *Behavioural Ecology: An Evolutionary Approach*, Krebs, J. R. and Davis, N. B., Eds., Blackwell Scientific, Oxford, 1978, 64.
76. **Underwood, R.,** Vigilance behavior in grazing African antelopes, *Behaviour*, 79, 8, 1982.
77. **Berger, J., Daneke, D., Johnson, J., and Berwick, S. H.,** Pronghorn *(Antilocapra americana)* foraging economy and predator avoidance in a desert ecosystem: implications for the conservation of large mammalian herbivores, *Biol. Conserv.*, 25, 193, 1983.
78. **Pulliam, H. R.,** The principle of optimal behaviour and the theory of communities, *Perspect. Ethol.*, 3, 311, 1976.
79. **Gosling, L. M. and Petrie, M.,** The economics of social organization, in *Physiological Ecology: An Evolutionary Approach to Resource Use*, Townsend, C. R. and Calow, P., Eds., Blackwell Scientific, Oxford, 1981, 315.
80. **Geist, V.,** Adaptive behavioral strategies, in *Elk of North America, Ecology and Management*, Stackpole Books, Harrisburg, 1982, 219.
81. **Owen-Smith, N.,** On territoriality in ungulates and an evolutionary model, *Q. Rev. Biol.*, 52, 1, 1977.
82. **McNabb, B. K.,** Bioenergetic and the determination of home range size, *Am. Nat.*, 97, 133, 1963.
83. **Pyke, G. H., Pulliam, H. R., and Charnov, E. L.,** Optimal foraging: a selective review of theory and tests, *Q. Rev. Biol.*, 52, 137, 1977.
84. **Westoby, M.,** An analysis of diet selection by large generalist herbivores, *Am. Nat.*, 108, 290, 1974.
85. **Belovsky, G. E.,** Diet optimization in a generalist herbivore: the moose, *Theor. Pop. Biol.*, 14, 105, 1978.
86. **Wyatt, J. R. and Eltringham, S. K.,** The daily activity of the elephant in the Rwenzori National Park, Uganda, *East Afr. Wildl. J.*, 12, 273, 1974.

87. **Coe, M. J., Cumming, D. H., and Phillipson, J.,** Biomass and production of large African herbivores in relation to rainfall and primary production, *Oecologia (Berlin),* 22, 341, 1976.
88. **Eisenberg, J. F. and McKay, G. M.,** Comparison of ungulate adaptations in the New World and Old World tropical forests with special reference to Ceylon and the rainforests of Central America, in *The Behavior of Ungulates and its Relation to Management,* Geist, V. and Walther, F., Eds., IUCN Publ. NS 24, Morges, 1974, 585.
89. **Eisenberg, J. F.,** The density and biomass of tropical mammals, in *Conservation Biology,* Soule, M. and Wilcox, B. A., Eds., Sinauer, Sunderland, Mass., 1980.
90. **Laws, R. M., Parker, I. S. C., and Johnstone, R. C. B.,** *Elephants and their Habitats: The Ecology of Elephants in North Bunyoro, Uganda,* Clarendon Press, Oxford, 1975, chap. 10.
91. **May, R. M.,** *Stability and Complexity in Model Ecosystems,* Princeton University Press, Princeton, 1973.
92. **Noy-Meier, I.,** Stability of grazing systems: an application of predator-prey graphs, *J. Ecol.,* 63, 459, 1975.
93. **Caughley, G.,** The elephant problem — an alternative hypothesis, *East Afr. Wildl. J.,* 14, 265, 1976.
94. **Hudson, R. J.,** Modelling the future of the African elephant *(Loxodonta africana), Biol. Extinction,* 1, 1, 1982.
95. **Western, D. and Ssemakula, J.,** Life history patterns in birds and mammals and their evolutionary interpretation, *Oecologia (Berlin),* 54, 281, 1982.
96. **Clutton-Brock, T. H., Albon, S. D., and Harvey, D. H.,** Antlers, body size and breeding group size in the *Cervidae, Nature (London),* 285, 565, 1980.

Chapter 2

FORAGE AND RANGE EVALUATION

C. C. Schwartz and N. T. Hobbs

TABLE OF CONTENTS

The fundamental question of range evaluation is how well (or how poorly) habitats provide for the needs of animals who live there. In answering this question, bioenergetics provides a currency for quantifying resource supply in units commensurate with animal demand, thereby allowing us to move stepwise from rudimentary estimates of food habitats, to more resolute predictions of animal nutritional status and habitat carrying capacity (Figure 1).

Evaluating food resources for herbivores from a bioenergetic standpoint is complex because we must not only describe how much food is present in a habitat, but must also assess its relative value. Early work on food limitation of animal populations[1,2] focused on predators. Their prey tended to be rare but energy and nutrients were invariably concentrated within those prey items. Thus, secondary consumers were limited by food scarcity and were largely unaffected by variation in the quality of foods they consumed. Following this line of reasoning, Hairston et al.[3] observed the abundance of herbage relative to herbivores eating it and postulated that primary consumers could not be food limited in a "green world". Since then, it has become clear that herbivore populations can be strongly limited by their food supply, but not necessarily by the quantity of herbage per se. Regardless of its abundance, herbage contains assimilatable energy and nutrients variably diluted by nonnutritious fiber. As a result, herbivores are limited by relative rather than absolute shortage of food.[4] Moreover, it has become clear recently that just as predators risk injury in capturing their prey, herbivores must sort among plants armed to the stomates with defense compounds.

Thus, describing food for herbivores requires answers not only to questions of "How much?", but also "How valuable?" and even "How safe?" Here we review knowledge on evaluation of food supplies for wild ruminants. We focus on describing methods for forage and range evaluation and synthesizing information obtained by those methods. We do not attempt a comprehensive synthesis of such a voluminous topic but instead refer the reader to other reviews (Table 1) for specifics on topics we cover more generally.

I. FORAGE — WHAT DO THEY EAT?

Herbivores choose their food from a large array of possible choices; in most ecosystems they consume only a small fraction of the "green world" available to them. Thus, all herbage is not food. Early work on evaluating food supplies for herbivores emphasized animal food habits; that is, distinguishing between herbage (plants which could be eaten) and forage (plants which were eaten). Although this approach has been criticized for focusing myopically on taxonomic units not meaningful to consumers,[17] it nevertheless allowed development of a fabric of theory on herbivore feeding strategies in relation to many aspects of their anatomy, physiology, and ecology (see Chapter 3). It also provided the foundation for quantifying food supplies in estimates of carrying capacity (Figure 1).

How have food habits been described? Perhaps the oldest technique is sampling ingesta from the digestive tract of the animal. Stomach sampling is useful because it allows direct inference to the population of interest by sampling its members. Problems with the technique include the necessity to sacrifice animals to obtain samples (but see Reference 18) as well as potential errors resulting from differential digestion of ingesta,[19-23] inability to quantify composition of samples,[24-30] and poor precision.[19,31]

Use of fecal analysis has solved some of these problems, but offers dilemmas of its own. It is appealing because it permits almost unlimited sampling; it does not affect the animals sampled; it is relatively cheap, and contract laboratories are available for conducting the analysis.[5,32,33] However, there remain serious questions about the method's accuracy; fecal analysis has sometimes failed to reliably estimate composition of known diets fed to herbivores.[23,30,34,35] Reasons for this inaccuracy are believed to include differential digestion of plant epidermis,[36-41] destruction of some plant species during sample preparation,[42] differences in identification ability among observers,[5,35,43] and problems with the reliability of

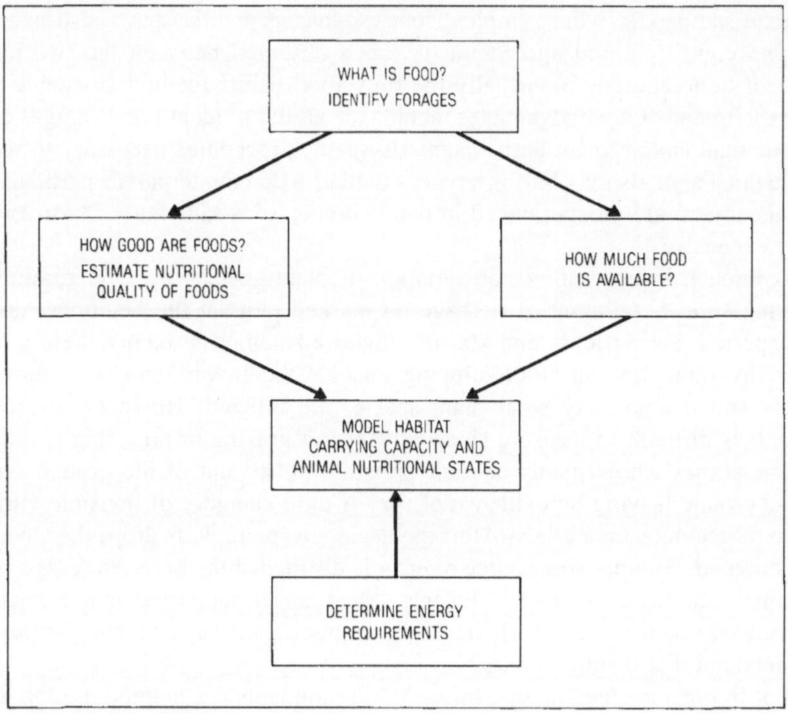

FIGURE 1. Importance of evaluation of forage quantity and quality in predictions of habitat value on a nutritional basis.

Table 1
IN-DEPTH REVIEWS ON GENERAL TOPICS COVERED IN THIS CHAPTER

Topic	Ref.
Food habits: methods	5
Food habits: findings	6
Forage quality: chemical methods	7
Forage quality: in vitro methods	8—9
Forage quality: in vivo methods	10
Forage quality: plant defenses	11
Forage quantity	12—15
Carrying capacity	16

epidermal density as an estimate of dry weight composition.[30,44,45] Although development of correction equations based on in vivo or in vitro digestions can improve accuracy of the fecal analysis procedure,[42,46,47] such corrections may entail their own biases when there are associative effects among digested forages.[35]

Direct observation of wild animals has been widely used to determine botanical composition of diets of herbivores. Two methods are prevalent, the feeding site technique[48] and observation of animals from a distance.[49] Unlike rumen and fecal analysis, these techniques allow determination of where foods are chosen. When herbivores use a variety of habitats, this advantage can be critical. The methods require little equipment. However, they are highly biased[50] and at best offer qualitative descriptions of diets.

Close observation of tame animals offers an alternative to watching wild ones from a distance. Dryweight composition of diets has usually been based on bite counts and hand-plucked forage samples,[51-56] but samples from esophageal fistulas have also been used to estimate bite weights.[57] Good agreement has been observed between the two methods.[57] Observation of tame animals is virtually the only food habits method amenable to use in controlled experiments. Other advantages include the ability to identify all forages consumed and to differentiate among plant parts eaten. However, procedures necessary to rear, train, and maintain tame animals may bias inferences to their wild counterparts[53] particularly when experimental subjects are inexperienced in plant communities sampled.[58,59] Moverover, the technique is expensive.[53]

Another approach in estimating consumption of plant species by wild ruminants is to estimate or measure the amount of herbage on marked plots at the beginning and end of each feeding period. Springfield[60] and Martin[61] reviewed methods used in determining animal consumption by plot clipping. Plot clipping enables the investigator to determine both quantitatively and qualitatively what plant species are utilized. However, use of rapidly growing plants is difficult to measure since evidence of grazing or browsing is obscured by regrowth. Sometimes whole plants or their parts are pulled out of the ground rather than cut off above ground leaving no evidence of use. A third category of invisible utilization is consumption of deciduous materials. After the deciduous plant parts drop, the amount taken cannot be estimated. An opposing source of error is distinguishing between forage consumed by study animals and losses to other influences. Most ranges supported an unknown number of other herbivores, and it may be difficult to distinguish between grazing or browsing of wild ruminants and that of other animals.

Cafeteria or free-choice feeding of confined wild ruminants is a method used to determine preferences for certain species of plants. The animals select their forage from a number of equally accessible species made available in approximately equal amounts. Forage is fed in bunks, in lots, or in pure-stand pastures. Smith and Hubbard[62] and Smith[63] used cafeteria feeding to determine plant preferences of mule deer (*Odocoileus hemionus*). Advantages of this method included: (1) plant species can be ranked according to animal preferences, (2) effect of unequal plant densities and their effort on preferences are controlled, and (3) it is possible to determine preferences for plant species not commonly eaten in the field. This method is criticized because the rate of intake may be disturbed by the presence of the observer and clipping of plants or growing species in pure-stands may alter animal preferences. Another problem associated with cafeteria feeding is the number of man-hours involved in collecting large quantities of forage.

Evaluation of ruminant food preferences associated with forage consumption is often an integral component of food habits studies. A preferred food has been defined as one which occurs proportionately more frequently in the diet of an animal than in the environment.[64] This definition has given rise to a plethora of indexes of dietary preference which attempt to weight the proportions of items in the diet by some measure of their availability.[65-69] In their rudimentary form, these indexes can usually be reduced to a simple fraction where the percentage of an item in the diet is divided by the biomass, density, or frequency of the item in the habitat.

For several reasons, such arithmetic manipulations contribute little to understanding diet choices. It is reasonable to assume that any herbivore will select some foods in an exploratory way.[70] That is, it will constantly test the food potentials of its environment and act on the basis of past tests.[71] If during that testing the animal consumes an exceedingly rare species, then that species would be calculated as being "highly preferred" *simply because it was rare*.

Moreover, common species in the environment tend to be equally available to the grazing animal despite differences in their density or biomass.[72] This is the case because those species are so abundant that searching time is virtually zero for all of them. However, despite their equal abundance in items of search time, species of lower biomass or density are calculated as "preferred."

There are also acute problems with comparison of preference indexes measured in one area where there are differences in mosaics of forage. Preference for one species is affected by availability of other species.[67,72] Thus, a plant that is highly preferred when surrounded by unpalatable neighbors might be rejected when its associated plants were eaten more often. Finally, the preference index suffers from severe sampling problems.[73,74]

Although the preference index has shortcomings, the phenomena it attempts to measure, "selectivity", is doubtlessly real. The process of food selection by herbivores is treated in Chapter 3. Here, we will proceed with discussion of evaluation of the quality of those foods.

II. FOOD QUALITY — HOW GOOD IS WHAT THEY EAT?

For herbivores, the value of a plant is related to its digestible protein and digestible energy content. To understand the nutritional complexity within a plant community, it is important to review the structural dynamics of a plant starting at the cellular level. Plant cells are surrounded by a relatively inflexible wall, which is largely composed of the carbohydrate cellulose along with hemicellulose and lignin. The structural components of the cell wall, referred to as cell-wall constituents,[75-77] form an envelope that contains the cytoplasm (cell contents). The cell contents contain proteins and soluble carbohydrates, both being important nutrients. The cell wall provides a support structure for the plant. In young growing tissues, which have to expand and do not support weight, the cell wall is thin and flexible. This condition is characteristic of the leaves and fruits of grasses and other flowering plants. In older tissues, particularly those bearing weight, the cell wall becomes thickened and hardened and cytoplasm volume decreases. This is the situation in the culms of tall grass and stems of woody plants. For a feeding herbivore, the division of feed into cell contents and cell walls is important because it affects the availability of protein and energy. Mammalian herbivores do not manufacture the digestive enzyme cellulase which is necessary to break down the cell wall. Ruminants are able to digest cellulose because of their symbiotic relationship with rumen microorganisms which do produce cellulolytic enzymes.[78] Cellulose digestion is a slow process particularly if the cell wall is lignified.[79] The cell wall, therefore, constitutes a physical obstacle to the extraction of the cell contents.

Once the cell wall is digested, it is available as a source of energy. Additional energy and protein is obtained from the cell contents. Therefore, herbivores must extract both protein and energy from a forage supply containing various amounts of obstructive carbohydrates (cell wall).

Forage analysis procedures have focused on describing the protein content, digestibility, and fiber characteristics of plants. Many forage analysis systems have been developed for foods eaten by domestic herbivores, primarily grasses and cultured legumes. In many instances, the chemical and physical characteristics of those forages differ enormously from foods consumed by wild herbivores. Consequently, application of some techniques developed for domestic animals can lead to erroneous inferences when applied to forages consumed by wild herbivores.

Forage evaluation procedures fall into two major groups, those which measure characteristics of forage within the animal (in vivo procedures) and those conducted in artificial, laboratory settings (in vitro procedures). Laboratory analysis can be further divided into systems of chemical evaluation and in vitro simulation of digestive processes.

Evaluating the relationships between energy contained in plants and its assimilation within herbivores requires understanding of how energy is partitioned within the animal. Energy, the capacity to do work, is commonly measured as the heat released on combustion, expressed in calories or joules (1 cal = 4.184 J). Gross energy (or combustion energy) in plant samples is routinely about 18.8 kJ/g dry matter,[80, 81] but may be higher when plants contain high levels of fats or oils. However, variation in gross energy content is of little significance to the animal.

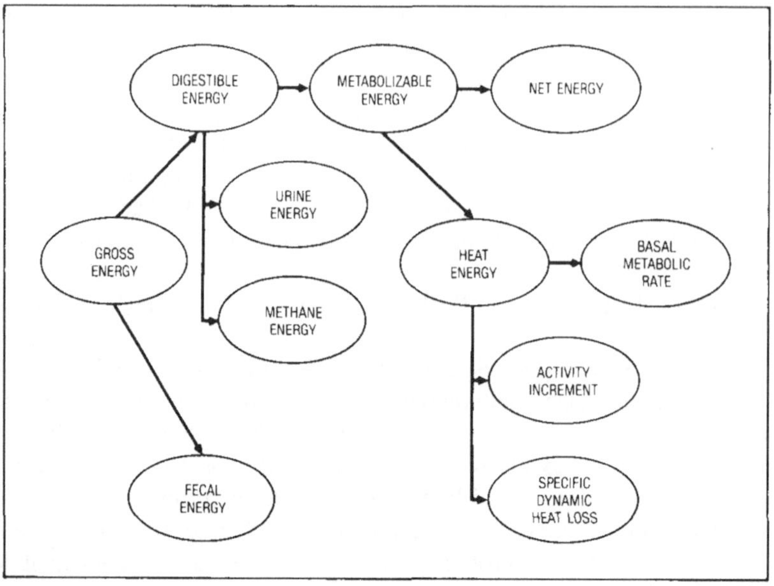

FIGURE 2. Illustrations of energy partitioning in ruminants.[82]

Instead, variation in assimilation of energy determines the energy value of forages to herbivores. The traditional scheme of energy partitioning within the animal divides gross energy ingested into digestible, metabolizable, and net energy fractions (Figure 2). The digested proportion of gross energy in foods of wild herbivores is highly variable, ranging from 25 to 85%. The portion of that digested energy which is metabolized is more consistent, averaging 75 to 90%.[83] Net energy as a fraction of energy metabolized from forage depends on the productive state of the animal and net energy for maintenance is usually about 50 to 75% of metabolizable energy.[83]

Energy partitioning within the animal is evaluated with in vivo balance trials. Methods for balance trials have been described in detail elsewhere;[10,84,85] however, in brief, they entail feeding known amounts of forages of known composition and collecting all products for the animal, urine, feces and gases. Because the principle energy loss as gases is methane which is difficult to measure directly, it is usually estimated indirectly. For foods of wild herbivores, energy in methane as a fraction of energy metabolized averages less than 1% for nonruminants[86,87] and is usually 2 to 10% for ruminants.[88-91] Urine energy losses are approximately equal to methane energy loss but may be higher for animals on certain diets such as browse.

The fraction of forage energy or materials which is digested can be expressed in two ways:

$$\text{Apparent digestion coefficient} = 100 \times \frac{\text{intake amount} - \text{feces amount}}{\text{intake amount}}$$

True digestibility, in contrast, is estimated as

$$\text{True digestion coefficient} = \frac{\text{intake amount} - (\text{feces amount} - \text{endogenous amount})}{\text{intake amount}}$$

The important difference between these two coefficients is that true digestibility removes the influence of materials of animal origin added to the feces during digestion. Discrepancies

Table 2
TRUE DIGESTIBILITIES OF PROTEIN IN FORAGE DIETS OF WILD HERBIVORES

Species	Diet	True digestibility (%)	Ref.
Bighorn sheep	Grass hay	100	96
Moose	Browse	74	97
Elk	Grass and alfalfa hay	98	93
White-tailed and mule deer	Grass and alfalfa hay, browse	95	92
Caribou and reindeer	Lichens and concentrate ration	79	98—99

Table 3
RELATIONSHIPS BETWEEN IN VIVO (Y) AND IN VITRO (X) DRY MATTER DIGESTIBILITY FOR WILD HERBIVORES

Species	Equation	R^2	Ref.
White-tailed deer	$Y = 6.3 + 0.99X$	0.98	105
	$Y = -6.2 + 0.95X$	0.98	106
White-tailed and mule deer	$Y = -0.71 + 0.99X$	0.94	92
Mule deer	$Y = 1.3 + 0.90X$	0.80	81
	$Y = -23.5 + 1.3X$	0.71	108
Elk	$Y = 8.49 + 0.82X$	0.82	109
Moose	$Y = -24.7 + 2.7X$	0.83	97

between true and apparent digestibilities can be substantial depending on the magnitude of excretion of endogenous products contained in digestive enzymes, mucosal sloughings, and bacterial bodies. That amount, in turn, depends on physical and chemical characteristics of the forage.[79, 92-95] Although true energy digestibility is highly variable and depends on animal species and apparent digestibility of energy,[83] true protein digestibilities are much more constant (Table 2).

The primary advantage of in vivo balance trials for evaluating forage is the strong and direct inference they provide to energy assimilation within the animal.[5,100] However, digestion trials are time consuming and frequently require collecting large quantities of test forage. It can be extremely difficult to obtain the voluminous amounts of even common forages necessary to feed large herbivores like moose or elk; such collections can be prohibitive for rare forages eaten by selective feeders. Browse samples collected for digestion trials with mule deer and elk cost $5/kg to collect and process.[101] In addition, confinement necessary to collect animal products may influence digestive processes,[102] and forages clipped for feeding may not be representative of those eaten by free-ranging animals.[6,100,103,104]

In vitro digestion systems were developed to reduce the cost and time necessary for conducting complete balance trials. Predictions from in vitro digestions often agree closely with in vivo values (Table 3). Some bias may occur, particularly for forages of extremely high or low quality. In addition, because in vivo dry matter digestibilities of forage of wild herbivores are closely related to energy digested (Table 4), in vitro digestion of dry matter offers a useful index to energy digestion within the animal.[110]

Two in vitro procedures are widely used, the methods of Tilley and Terry[111] and Goering and Van Soest.[77] These procedures offer different estimates; the former being analogous to apparent digestibilities, the latter to true digestibilities. Divergence in predictions of the two techniques is particularly great for browse.[112] This is probably because of accumulation of

Table 4
RELATIONSHIP BETWEEN APPARENT DIGESTIBLE DRY MATTER (X) AND ENERGY (Y) FOR FORAGES OF WILD HERBIVORES

Species	Diet	Browse	R^2	Ref.
Elk	Grass hay	Y = −0.61 + 0.98X	0.98	95
Mule deer	Grass hay, browse, concentrate ration	Y = −1.48 + 1.04X	0.99	81
Mule deer and white-tailed deer	Grass hay, browse, concentrate ration	Y = −0.71 + 0.99X	0.94	92
Moose	Browse	Y = −4.35 + 1.1X	0.99	97
Bighorn sheep	Winter range grass hay	Y = 1.54 + 0.98X	0.99	96
Winter + summer range grass hay		Y = −2.69 + 1.04X	0.98	96
Snowshoe hare	Browse and concentrate ration	Y = −9.23 + 1.16X	0.97	86, 105

bacterial bodies elaborated during fermentation of browse cell solubles.[113] Frequently, dry-matter digestions of browses predicted by Tilley and Terry[111] are much less than their cell soluble fraction[51,81,114-117] which should be essentially 100% digestible.[92] Consequently, the Van Soest procedure[77] may offer more accurate predictions of energy availability within browse.[112, 118]

An important controversy regarding in vitro techiques has centered on whether values derived from digestions with inoculum from domestic species differ from those based on inoculum from wild species. Some workers reported significant donor effects[92,119-122] while others found no influence.[106,123,124] However, these findings are not as equivocal as they appear. All reports of significant donor effects lacked animal replications within species; those which found no effect were reliably replicated. In vitro estimates of digestibility based on inocula from individuals of the *same* species can be highly variable.[105,125] It appears that previous reports of species-effects could easily be a result of individual variation in inoculum source rather than true interspecific variation.

Effect or diet of inoculum donor on in vitro determinations is not so easily resolved. Diet has influenced digestibility estimates in some studies[92,125,127-130] but had no effect in others.[124,131,132] Although knowledge of rumen microbial ecology (Chapter 5) suggests several plausible mechanisms accounting for diet effects, it is possible that error in the technique is greater than the magnitude of those effects.[124] Highly soluble forages like browses, legumes, and immature grasses are relatively insensitive to variation in inoculum;[126,127] thus, diet effects may depend on the type of forage evaluated. Regardless of these arguments, there exists a distinct tradeoff between the possible gain in accuracy derived by acquiring rumen inocula from wild animals consuming the type of forage to be evaluated, and the loss of experimental control such acquistion often entails.[126]

Extent of digestion, measured in vitro or in vivo, is commonly interpreted as an index of energy value in forages. However, rates of forage digestion and passage may be more important to herbivores, particularly small ruminants, because those processes control rates of energy absorption and energy intake on most roughage diets.[81,133-135] Despite its significance to the animal, information on rate of digestion and rate of passage of wild herbivore forages is limited.

Systems for chemical partitioning of plant constituents into nutritionally meaningful fractions were developed as alternatives to digestion methods for forage evaluation. The primary advantage of chemical analysis is increased precision and reduced cost.[113] However, chemical procedures offer estimates which are nutritionally relevant only if they are related functionally or statistically to specific processes of voluntary intake and digestion.[113]

The Weende system of proximate analysis[136] has been largely abandoned because it partitioned plant tissue into components heterogeneous in composition and variable in digestibility.[113] Detergent analysis of forage[75-77,137-138] remedied many of the problems of the proximate system. Using these procedures, plant tissue can be divided into uniform fractions predictably related to digestibility.[113] The plant is first partitioned into neutral detergent fiber (NDF) or cell wall (containing cellulose, hemicellulose, cutin, and lignin) and neutral detergent solubles (containing sugars, organic acids, starch, lipids, crude protein, nucleic acids, and most inorganic material). Herbivores without microbial symbionts can digest only the soluble fraction; microbial enzymes are required to digest cell wall. Extent of cell wall digestion depends on its composition, particularly its lignification.[135,139]

Robbins[83] reviewed important changes in detergent methods which have been introduced as the methods evolved:

1. Sodium sulfite, originally added to the neutral detergent solution, was omitted because it removed some lignin for the cell wall.
2. NDF and acid detergent fiber (ADF) were determined on the same rather than separate samples to improve estimates of hemicellulose obtained by difference.
3. Neutral detergent was found to remove pectins; consequently NDF is not strictly equivalent with cell wall.
4. Important digestion inhibitors were found in cell solubles, thereby complicating the interpretation of the nutritional value of that plant fraction.

Dry matter digestion of grasses and legumes can be predicted with reasonable accuracy with regression equations based on detergent analysis.[77,138] However, digestion of browses and forbs consumed by wild herbivores do not fit these relationships well.[112]

A promising development in evaluating quality of forage for wild herbivores is use of infrared reflectance spectroscopy to predict plant chemical and physical characteristics.[141] The method works by developing multiple-regression relationships between reflected wavelengths and nutritional quality measures for a set of standard forages. Reflectance from test forages can then be used to predict their composition, digestibility, and level of intake (Table 5). Accuracy of those predictions depends on standard forages incorporating the same range of compositional variation as forages tested.[113] Precision of the infrared technique is better than would be normally expected with wet-chemistry procedures.[146]

Forages consumed by wild herbivores differ widely in their crude protein content, digestibility, and physical composition (Table 6). Shrubs and forbs tend to contain less cell wall than grasses, but have a larger lignin fraction within that cell wall. Consequently, they are frequently digested less completely but more rapidly than grasses because rate of forage digestion is inversely related to cell-wall content while extent of digestion is inversely related to lignification.[135]

These differences are related to plant growth characteristics. The primary meristem of grasses/sedges is intercalary; browses/forbs have apical meristems. As a result, a relatively smaller amount of browses/forbs is actively growing at any time compared with monocots. Because of the difference between growth from intercalary and apical meristems, and the proportion of the whole grass or browse plant actively growing, the average grass plant is likely to be more homogeneous in food value than is a browse plant, especially a woody one, which is composed of an assembly of parts of heterogeneous value.

Forage evaluation must consider antiquality factors as well as those components which enhance forage nutritional benefits. Plant secondary compounds have been examined for some species in relatively few studies. However, these demonstrate that antiquality factors can have profound effects on diet selection. Bryant and Kuropat[168] synthesized data which strongly suggested that in subarctic ecosystems animals as diverse as ptarmigan, hares, moose, and beaver do not select forage on the basis of quality factors, but instead select their diets to avoid antiquality factors in browse. Evidence from other herbivore systems suggested this phenomena of diet selection by avoidance is not an isolated event.[169-175]

Table 5
FINDINGS OF STUDIES IN INFRARED REFLECTANCE AS PREDICTORS OF FORAGE QUALITY

Measurement	Forage type	N	R^2 for prediction	Wavelength selected (n*M*) 1st	2nd	3rd	4th	Ref.
Crude protein	Grasses and legumes	87	0.98	2.164	2.084	2.254	1.610	140
	Diet samples[a]	101	0.92	2.158	2.138			5
		21	0.98	2.135	2.043	2.239		141
DMD	Grasses and legumes	9	0.92	1.942	2.030	1.178		142
IVDMD	Grasses and legumes	76	0.90	2.260	1.902	1.602	1.504	140
	Diet samples[a]	21	0.84	1.610	2.386	2.322	1.542	5
	Grasses and legumes	9	0.80	1.818	1.894			142
NDF	Grass and legumes	87	0.97	2.294	2.072	1.902	2.204	140
		9	0.30	1.278	1.694	2.146	2.282	142
ADF	Grass and legumes	87	0.92	1.666	1.492	1.854	1.558	140
	Diet samples[a]	21	0.90	1.687	2.259	1.811		141
	Grass and legumes	9						
Organic matter intake	Grass, forbs, shrubs	21	0.72	1.410	2.214			140
Dry matter intake	Grass and legumes	76	0.64	1.976	1.690	1.898		141
		9	0.84	1.818	1.898			142

[a] Esophageal fistula samples from cattle.

Much of the food herbivores consume, as well as the herbage they avoid, contains secondary compounds (Table 7). These compounds fall into two broad groups: compounds which inhibit digestion and compounds which are directly and acutely toxic. Digestion inhibitors, ubiquitous in herbivore foods, include phenolics[198] and terpenes.[199] Phenolics complex with proteins in the digestive tract forming molecules too large for absorption. Phenolics also interfere with starch and cellulose digestion by interfering with enzyme action.[200,201] The effectiveness of monoterpenes as digestion inhibitors is controversial. Long believed to interfere with digestion because of toxicity to rumen microbes,[202-204] monoterpenes have recently been shown to exert minor effects on digestion in vitro.[131,160] Unfortunately, these studies focused on soluble forages which, because of their low cell-wall content, needed little microbial action for digestion to occur. It does appear, however, that herbivores possess mechanisms for reducing inhibition of digestion caused by monoterpenoids.[205,206]

Acutely toxic compounds include nonprotein amino acids, alkaloids, sesquiterpenes, cyanogenic glycosides, and saponins. While these may deter herbivores, there are many examples of animals consuming forages which contain toxic compounds (Table 7). Some of these cases, however, are drawn from studies which have shown that consumers are selective and avoid those individual plants which contain high concentrations of the deterrent.[185,192,196]

III. FORAGE QUANTITY — HOW MUCH IS AVAILABLE?

Although the quality of individual food items has profound influence on the condition of individual herbivores, the quantity of forage ultimately determines how many animals can exist within a habitat.[207] The simplest method for estimating forage biomass is clipping and weighing sample plots.[12,208,209] Despite their simplicity, such procedures are often prohibitively time consuming at desired levels of precision.[210-213] Efforts to improve precision and reduce costs have emphasized various forms of double sampling where regression relationships are developed between dryweight of clipped samples and more easily measured

Table 6
FORAGE QUALITY VALUES FOR FOODS OF WILD HERBIVORES

(% dry matter)

Forage class	Season[a]	Location	Digestibility	Crude protein	Cell wall	Lignin	Ref.
Browse stems	Winter	Wyoming	30[c,g]				119
		World-wide		6.8			144
		Colorado	28[c,i]	5.1	59	24.0	145
		New York		6.3			146
		Michigan	49[b,e]	6.2	26	7.9	147
		Oregon			24	16.8	148
		New Hampshire	39[b,e]	7.2			151
		Colorado	33[c,g]	6.1			51
		Texas	28[b,f]	5.6			150
			40[b,f]		62		135
				7.8			151
		Arkansas		7.1	47		152
		Alaska	37[k]	8.1			153
		Texas	75[b,f]	36	9.0		135
			40[b,f]	8.7	65	14.6	150
	Summer	Virginia	40[b,f]	4.9			154
		World-wide		11.0			144
		Texas	40[b,f]		61		135
			27[b,f]	5.3			150
	Fall	South Dakota	26[c,e]				155
		Texas	29[b,f]				156
			29[b,f]	5.5	69	18.4	156
			40		62		135
Browse leaves	Winter	Colorado	38[b,f]	5.5	39	12.5	51
		Arizona	34[c,e]	11.4			157
		Texas	52[b,f]				135
		Arkansas[m]		7.3	47		152
		Texas[m]	14[c,e]	5.3	49	23.3	114
	Spring	Texas	79[b,f]		26	9.0	135
		Arizona	47[c,e]	14.0			157
		Texas		12.4			151
		Utah		17.0			158
	Summer	Texas	67[b,f]		35		135
		Arizona	29[c,e]	13.3			150
		Virginia	36[c,d]	13.3			154
		Utah		13.2			158
	Fall	Texas	89[b,f]				156
		Arizona	36[c,e]	15.4			157
		Utah		11.8			158
		Texas	68[b,f]		32	10.0	135
Browse stems and leaves	Winter	Texas	25[c,e]	6.1	57	18.6	114
			54[c,e]	13.9			159
		Utah	56[c,e]			160	
		Colorado	50[c,i]	9.9	37	11.9	145
		Utah, Colorado, Wyoming, Nevada		12.4			161
		Texas	56[b,f]	11.7	39	12.3	150
	Spring	Texas	33[c,e]	16.2	32	9.9	114
		Alberta, Canada		14.6			162
		Texas	57[c,e]	17.4			159
			67[b,f]	16.4	36	10.1	150

Table 6 (continued)
FORAGE QUALITY VALUES FOR FOODS OF WILD HERBIVORES

(% Dry matter)

Forage class	Season[a]	Location	Digestibility	Crude protein	Cell wall	Lignin	Ref.
	Summer	Texas	31[c,e]	9.7	34	12.0	114
		Alberta, Canada		11.7			162
		Texas	57[c,e]	15.0			159
			58[b,f]	10.5	34	11.8	150
	Fall	Texas	28[c,e]	8.0	35	13.6	114
		Alberta, Canada		8.1			162
		Texas	53[c,e]	15.7			159
		South Dakota	46[c,e]				155
		Texas	60[b,f]	9.8	36	11.2	150
		Alaska	19[c,j]				108
			58[b,f]				108
Conifer stems and leaves	Winter	Oregon			16	15.9	148
		Texas	25[c,e]	7.5	41	14.9	114
		New Hampshire	48[b,e]	8.2			149
		Michigan	60[b,e]	3.8	20	9.8	147
			41[b,e]	3.4			163
		Texas	44[b,f]	8.7	45	17.1	150
	Spring	Texas	27[c,e]	9.2	45	13.4	114
		Arkansas		8.0	33		152
		Texas	40[b,f]	8.3	53	17.6	150
	Summer	Texas	29[c,e]	6.8	45	16.4	114
			34[b,f]	9.8	49	18.4	150
	Fall	Texas	26[c,e]	7.1	47	17.8	114
		Michigan	45[c,e]	6.9	43	19.2	115
		Texas	41[b,f]	7.7	47	18.0	150
Fruits	Winter	Arizona	54[c,e]	7.0			157
		Texas	67[b,f,n]	15.9	53	9.1	150
	Spring	Arizona	52[c,e]	6.3			157
		Texas	57[b,f,n]	13.4	45	14.7	150
	Summer	Arizona	61[c,e]	7.2			157
		Texas	58[b,f,n]	9.8	41	16.8	150
	Fall	Arkansas, Texas		5.7	40		164
		Texas	68[b,f]				156
			27[c,e]	5.3	28	14.9	114
		Arizona	53[c,e]	6.6			157
		Texas	76[b,f,n]	29.5	39	7.4	150
Forbs	Winter	Texas	27[b,f]	6.3	75	14.6	150
		Colorado	27[c,g]	6.2	52	16.0	51
		Wyoming	46[c,g]				119
		Texas	49[c,e]	14.0	31		114
			54[c,e]	13.9			159
			27[b,f]				135
	Spring	Texas	68[b,f]	13.3	44	8.8	150
		Arizona	56[c,e]	7.8			157
		Texas	37[c,e]	14.8	34	6.5	114
		Alberta, Canada		20.9			162
		Texas	57[c,e]	17.4			159
		Arkansas		15.4	26		152
		Texas	52[b,f]				135

Table 6 (continued)
FORAGE QUALITY VALUES FOR FOODS OF WILD HERBIVORES

(% Dry matter)

Forage class	Season[a]	Location	Digestibility	Crude protein	Cell wall	Lignin	Ref.
	Summer	Texas	52[h,f]	8.6	54	10.9	150
		Arizona	56[c,e]	6.5			157
		Texas	46[c,e]	11.4	37	8.1	114
		Missouri		16.6			169
		Alberta, Canada		13.3			162
		Colorado	65[c,e]	13.9	26	4.8	116
		Texas	57[c,e]	15.0			159
			52[h,f]				135
	Fall	Texas	42[h,f]	7.1	62	11.8	150
			37[c,e]	8.7	42	10.7	114
		Alberta, Canada		9.3			162
		Texas	53[c,e]	15.7			159
		South Dakota	31[c,e]				155
		Texas	45[h,f]				156
			42[h,f]	12.0			135
Grasses and sedges	Winter	Texas	57[h,f]		65		135
		New Mexico		3.3	74	6.6	166
		Alberta, Canada		6.9			162
		Texas	34[c,e]	14.8	57	5.2	114
		Wyoming	44[c,g]				119
		Colorado	45[c,g]	4.2	71	4.8	51
		Texas	20[h,t]	5.2	85	10.3	150
	Spring	Texas	59[h,f]		70	4.0	135
		Alaska		5.4			153
		Arkansas		13.2	67		152
		New Mexico[o]		10.6	72	4.7	166
		Texas		5.3			167
		Alberta, Canada		18.8			162
		Texas	30[c,e]	13.3	67	3.8	114
			49[h,f]	11.1	71	5.7	150
	Summer	Texas	53[h,f]		70		135
		New Mexico[o]		8.2	72	5.1	166
		Colorado	67[c,i]	12.8			164
		Virginia	48[c,e]	7.6			154
		Alberta, Canada		14.5			162
		Texas	31[b,e]	9.4	67	4.9	114
			45[h,f]	7.2	77	6.3	150
	Fall	Texas	59		61	8.0	135
		Alaska	44[c,j]				128
			43[h,j]				128
		Texas	34[h,f]				156
		South Dakota	29[c,e]				155
		New Mexico[o]		5.6	7.1	5.6	166
		Alberta, Canada		10.5			162
		Texas	26[c,e]	9.8	69	7.1	114
		Texas	34[h,f]	6.6	85	10.3	150
Fungi and lichen	Spring	Alaska		19.8			153
	Summer	Virginia		7.8			154
	Fall	Alaska	34[c,i]				128
			51[h,j]				128

Table 6 (continued)
FORAGE QUALITY VALUES FOR FOODS OF WILD HERBIVORES

(% Dry matter)

Forage class	Season[a]	Location	Digestibility	Crude protein	Cell wall	Lignin	Ref.
Cactus	Winter	Texas	73[c,e]	6.9			159
	Spring	Texas	70[c,e]	8.4			159
	Summer	Texas	70[c,e]	6.7			159
	Fall	Texas	67[c,e]	7.5			159

[a] Inferred.
[b] In vivo by comparison or nylon bag method.
[c] In vitro.
[d] Chile, U.S.A., Israel, Australia, South Africa, Argentina, Venezuela.
[e] With deer.
[f] With goats.
[g] With elk.
[h] With sheep.
[i] With cattle.
[j] With reindeer.
[k] Technique and animal used are not stated.
[l] Animal used not stated.
[m] Dead leaves.
[n] With mushrooms also.
[o] Season inferred from vegetation class.

attributes of the plant. Such attributes include visual estimates of weight,[210,212-218] measures of plant diameter,[219,222] counts of twigs,[223-227] as well as electronic measures of resistance-capacitance,[227-235] attenuation of beta-emissions,[236,237] and spectro-reflectance.[238-240] Double-sampling procedures allow rapid estimation of the absolute abundance of food for herbivores, but are not particularly useful in describing biomass of individual plant parts or other herbage categories commonly selected by wild herbivores.

Estimates of the absolute abundance of forages for wild herbivores are summarized in Table 8. Absolute abundance, however, is often reduced by spatial and temporal variation in its availability; the food an animal can actually find and eat may be a small fraction of the absolute amount of food.[265] Availability is difficult to quantify. Depending upon the herbivore species and its environment, any number of biotic and abiotic factors can enhance or restrict forage availability. For animals in northern climates, snow depth, density, and hardness can greatly alter the quantity of forage available. On mule deer range in northern Colorado, proportional areas of summer, early winter, and late winter range were 125:23:1.[266] Moen and Evans[267] estimated that 30 cm of snow made 97% of potential food unavailable to white-tailed deer in hardwood forests in New York. Likewise, studies of caribou,[268-273] moose,[274-276] mountain goats,[277] and black-tailed deer[278-280] all emphasize the importance of reduction in forage availability during winter.

IV. NUTRITIONAL CARRYING CAPACITY

Extensive information on the quality and quantity of forages allows reliable comparisons of the value of those plants in providing for animal nutrional needs. However, habitats offer complex assemblages of forages and this complexity complicates evaluation of habitats for herbivores. Land resource managers have tried for years to measure the capacity of native ranges to provide for the nutritional needs of herbivores. The term carrying capacity or

Table 7
COMMON PLANT DEFENSES IN FOODS OF WILD HERBIVORES

Forage species	Consumer	Plant defenses	Mode of action	Ref.
Carex aquatalis	Brown lemming	Ca oxalates	Acute toxicity; decreased growth	176
	Tundra vole			177
Halogeton glomeratus	Deer mouse		Acute toxicity	178
Atriplex confertifolia	Kangaroo rat			179
Opuntia fulgida, O. engelmani, Acacia sp.	Pack rat			179
Traganum nudatum, Suaeda mollis	Sand rat			180
Artemesia tridentata	Pygmy rabbit	Terpenoids	Antimicrobial, reduce digestibility	181
	Sage grouse			182
	Mule deer			183
	Antelope			131, 18
Pinus ponderosa	Abert's squirrel	Alpha-pinene and other terpenoids	Teratogenic and acutely toxic	185
	Pocket gophers			186
Pseudotsuga menziesii	Black-tailed deer	Essential oils	Antimicrobial, reduce digestibility	148
Distichlis stricta and other grasses	Montane vole and other voles	Cinnamic acid	Inhibits reproduction	187
"Forbs", *Erodium*	California quail	Isolfavones	Estrogenic: inhibit repro. in dry years	188
Goaphandra corieaeae	South Indian leaf monkey	Alkaloids	Acute toxicity	171
Drypetes oblongifolia				
Phalaris arundinacea	Meadow vole			189
Leucaena glauca	Brown rat	Mimosine (nonprotein aa)	Acute toxicity	190
Astragalus	Pocket gopher	Seleno-amino acid	Acute toxicity	191
Pteridium aquilinum	Deer, sheep	Cyanogenic; tannins	Acute toxicity, reduce digestibility	192
Herium	Gazelle, hyrax	Cardiac glycoside	Acute toxicity	193
Amelanchier alnifolia	Mule deer	Cyanogenic	Acute toxicity	194
Laportea alatipes	Mountain gorilla	Stinging hairs	Extreme irritant	195
Galium sinense	Mountain gorilla	Hooks on plant	Physical abrasive	195
Alnus crispa	Snowshoe hare	Pinosylvin methyl ether	Deters feedings,?	196
Acacia sp.	Vervet monkey	Condensed tannis	Reduce digestibility	170
Hagenia abyssinica	Mountain gorilla			195, 197
Pygeum africanum	Mountain gorilla			195, 197
Alnus rubra	Black-tailed deer			148
Acer sp.	Elk			94
Epilobium angustifolia	Elk			94

grazing capacity is used to describe the capability a range has to support animals. Stoddart and Smith[281] defined grazing capacity as the maximum number of animals that can graze each year on a given area of range, for a specific number of days, without inducing a downward trend in forage production, forage quality, or soil condition. The first attempts to measure grazing capacity used the range reconnaissance method. Initially, this was a subjective judgment of the composition and density of vegetation obtained by riding over the management unit on horseback and estimating condition and carrying capacity.[283] Since that time, many methods have been proposed to measure carrying capacity. These include the "forage area" concept,[283] point observation method,[284] plant utilization methods,[285] and indicator plant species.[286] Variations of these approaches have been used in an attempt to quantify grazing capacity for the past 20 years.

Table 8
BIOMASS PRODUCTION OF HABITAT TYPE

Habitat Productivity

Habitat type	Location	Forage type	Biomass (kg/ha)	Ref.
Mixed grass prairie	Montana	Wheatgrass	1500	240
		Blue gramma	700	
		Threadleaf sedge	580	
Agropyron grassland	Oregon	Herbage	1002	242
Annual grassland	California	Forage	2040	243
Mountain grassland	Montana	Graminoids	474	244
Southwest exposure		Forbs	337	
Northwest exposure		Graminoids	357	
		Forbs	1038	
Ungrazed grassland	India	Grass	1149	245
		Legume	524	
		Other	354	
Grazed grassland	India	Grass	1059	
		Legume	433	
		Other	140	
Grazed grassland	Serengeti,	Grass	4572	246
Ungrazed grassland	Africa	Grass	5019	
Grassland	Argentina	Grasses and forbs	5320	247
Mixed grassland	Wyoming	Herbage	769	248
Mountain meadow	Idaho	Grasses	806	249
		Forbs	898	
Juniper-mixed grass	West Texas	Perennial grass	200	250
Grass: forb: shrub	Utah	Grass	375	251
		Forbs	501	
		Shrubs	542	
Mesquite shrub	New Mexico	Perennial grass	278	252
Arctic tundra	Alaska	Carpet shrubs	558	253
		Low shrubs	1908	
		High shrubs	1896	
		Graminoids	1614	
		Lichen	2631	
		Forbs	800	
Ponderosa pine-grassland	Montana	Cool-season grasses	1060	254
		Cool-season forbs	195	
		Other	45	
Ponderosa pine	Idaho	Annual forbs	76	255
		Perennial forbs	310	
		Annual grasses	42	
		Perennial grasses	40	
Oak-pine	New York	Tree leaves	3510	256
		Total tree stratum	7958	
		Shrub stratum leaves	310	
		Total shrub stratum	607	
		Herbs	22	
Longleaf pine	Louisiana	Herbage	2934	257
Slash pine	Louisiana	Herbage	2729	
Pine-hardwood	Louisiana	Herbage	644	258
		Browse	220	
Grand-fir/myrtle boxwood	Idaho	Shrub	128	259
		Forbs	196	
		Graminoids	46	

Table 8 (continued)
BIOMASS PRODUCTION OF HABITAT TYPE

Habitat Productivity

Habitat type	Location	Forage type	Biomass (kg/ha)	Ref.
Pinyon-juniper	Arizona	Grasses	100	260
		Forbs	46	
		Half-shrubs and shrubs	70	
		Perennial succulents	378	
Mixed conifer (8 years post-clearcut)	Arizona	Deer forage:		261
		Herbage	169	
		Browse	18	
		Elk forage:		
		Herbage	231	
		Browse	13	
		Total graminoids	205	
		Total forbs	599	
		Woody plants	38	
Mixed conifer (16 years post-clearcut)	Arizona	Deer forage:		
		Herbage	416	
		Browse	3	
		Elk forage:		
		Herbage	440	
		Browse	9	
		Total graminoids	597	
		Total forbs	689	
		Woody plants	11	
Douglas fir	British Columbia	Pinegrass	524	262
Ponderosa pine	Arizona	Herbage	209	263
Moderately burned ponderosa pine				
0 years post-burn		Herbage	157	
		Forage	154	
2 years		Herbage	672	
		Forage	651	
8 years		Herbage	682	
		Forage	653	
Severely burned				
0 years post-burn		Herbage	201	
		Forage	62	
2 years		Herbage	668	
		Forage	424	
8 years		Herbage	341	
		Forage	107	
Ponderosa pine 0—19% crown cover (CC)	South Dakota and Wyoming	Grasses and sedges	1005	264
		Forbs	227	
		Shrubs	289	
Ponderosa pine 20—39% CC		Grasses and sedges	364	
		Forbs	103	
		Shrubs	226	
Ponderosa pine 40—59% CC		Grasses and sedges	147	
		Forbs	35	
		Shrubs	85	

However, during this period, research to determine animal requirements and range supplies, and to develop methods to measure them, has led to new approaches in determining grazing capacity. Moen,[287] Robbins,[288] and Wallmo et al.[266] advocated estimating range carrying capacity on a nutritional basis. Nutritional carrying capacity is defined as the ratio of range nutrient supply divided by the nutrient demands of individual animals.[16] In this way, it acts to integrate information on forage quantity and quality into one estimate of habitat value. The approach is useful, because by expressing range supply in units of energy or nutrients, individual bioenergetic requirements can be considered in carrying capacity estimates. Estimates of nutritional carrying capacity based on forage of energy, nitrogen, and dry-matter have been derived for habitats of elk,[289] white-tailed deer,[290] mule deer,[266] and African ungulates.[291]

The primary limitation of existing models of nutritional carrying capacity is their failure to relate supportable animal density to individual animal nutritional status.[266,289] Simultaneous prediction of the number of animals a habitat can support and the condition of those animals depends on quantifying the functional relationship between the quality of forage and its biomass. Such relationships will allow estimation of the amount of forage at specific levels of nutritional quality, and permit reliable evaluation of range nutrient supply in relation to animal nutrient requirements. Future models of habitat carrying capacity must incorporate such estimates of range supply in order to effectively use the present highly qualitative estimates of animal nutritional requirements in evaluating ranges for herbivores.

REFERENCES

1. **Elton, C.,** *Animal Ecology*, Macmillan, New York, 1927.
2. **Lack, D.,** *The Natural Regulation of Animals*, Clarendon Press, Oxford, 1954.
3. **Hairston, N. G., Smith, D. W., and Slobodkin, L. B.,** Community structure, population control, and competition, *Am. Nat.*, 94, 421, 1960.
4. **White, T. C. R.,** The importance of a relative shortage of food in animal ecology, *Oecologia (Berlin)*, 33, 71, 1979.
5. **Holechek, J. L., Vavra, M., and Pieper, R. D.,** Methods for determining the nutritive quality of range ruminant diets: a review, *J. Anim. Sci.*, 54, 363, 1982.
6. **Van Dyne, G. M., Brocklington, N. R., Szocs, Z., Duek, J., and Ribic, C. A.,** Large herbivore subsystems, in *Grassland, Systems Analysis and Man*, Breymeyer, A. I. and Van Dyne, G. M., Eds., Cambridge University Press, London, 1980, 298.
7. **Osbourn, D. F.,** Principles governing the use of chemical methods for assessing the nutritive value of forages: a review, *Anim. Feed Sci. Technol.*, 3, 265, 1978.
8. **Osbourn, D. F. and Terry, R. A.,** In vitro techniques for the evaluation of ruminant feeds, *Proc. Nutr. Soc.*, 36, 219, 1977.
9. **Holechek, J. L., Vavra, M., and Pieper, R. D.,** Botanical composition determination of range herbivore diets: a review, *J. Range Manage.*, 35, 309, 1982.
10. **Schneider, B. H. and Flatt, W. P.,** *The Evaluation of Feeds through Digestibility Experiments*, University of Georgia Press, Athens, 1975.
11. **Rosenthal, G. A. and Janzen, D. H.,** *Herbivores: Their Interaction with Secondary Metabolites*, Academic Press, New York, 1979.
12. **Mannetje, L. T.,** Measuring quantity of grassland vegetation, in *Measurement of Grassland Vegetation and Animal Production*, Mannetje, L. T., Ed., Commonwealth Agricultural Bureaux, Farnham Royal, England, 1978.
13. **Tucker, C. J.,** A critical review of remote sensing and other methods for non-destructive estimation of standing crop biomass, *Grass Forage Sci.*, 35, 177, 1980.
14. **Rutherford, M. C.,** Plant based techniques for determining available browse and browse utilization: a review, *Bot. Rev.*, 45, 205, 1979.
15. **Mitchell, J. E.,** Analysis of forage production for assessments and appraisals, U.S. Department of Agriculture, Forest Service Gen. Tech. Rep. RM-98, Washington, D.C., 1983.
16. **Mautz, W. W.,** Nutrition and carrying capacity, in *Big Game of North America*, Schmidt, J. L. and Gilbert, D. L., Eds., Stackpole Books, Harrisburg, 1978, 321.
17. **Janzen, D. H.,** New horizons in the biology of plant defenses, in *Herbivores: Their Interaction with Secondary Plant Metabolites*, Rosenthal, G. A. and Janze, D. H., Eds., Academic Press, New York, 1979, 331.
18. **Wilson, D. E., Hirst, S. M., and Ellis, R. P.,** Determination of feeding preferences in wild ruminants from trocar samples, *J. Wildl. Manage.*, 41, 70, 1977.

19. **Davison, V. E.,** A field method of analyzing game bird foods, *J. Wildl. Manage.*, 4, 105, 1940.
20. **Norris, J. J.,** Botanical analysis of stomach contents as a method of determining forage consumption of range sheep, *Ecology*, 24, 224, 1943.
21. **Gaare, E., Sørensen, A., and White, R. G.,** Are rumen samples representative of the diet?, *Oikos*, 29, 90, 1977.
22. **Bergerud, A. T. and Russell, L.,** Evaluation of rumen food analysis for Newfoundland caribou, *J. Wildl. Manage.*, 28, 809, 1964.
23. **McInnes, M.,** A Comparison of Four Methods used in Determining the Diets of Large Herbivores, M.S. thesis, Oregon State University, Corvallis, 1977, 127.
24. **Hartley, P. H. T.,** The assessment of the food of birds, *Ibis*, 90, 361, 1948.
25. **Courtright, A. M.,** An Evaluation of Rumen Analysis as a Means of Determining the Food Habits of Caribou, M.S. thesis, University of Alaska, Fairbanks, 1959.
26. **Dirschl, H. J.,** Sieve mesh size related to analysis of antelope rumen contents, *J. Wildl. Manage.*, 26, 327, 1962.
27. **Heady, H. F. and Van Dyne, G. M.,** Prediction of weight composition from point sample on clipped herbage, *J. Range Manage.*, 18, 144, 1965.
28. **Scotter, G. W.,** Sieve mesh size as related to volumetric and gravimetric analysis of caribou rumen contents, *Can. Field Nat.*, 80, 238, 1966.
29. **Medin, D. E.,** Stomach contents analysis: collections from wild herbivores and birds, in Range and Wildlife Habitat Evaluation: A Research Symposium, U.S. Department of Agriculture, Forest Service Misc. Publ. 1147, Washington, D.C., 1970, 133.
30. **Westoby, M., Rost, G. R., and Weis, J. A.,** Problems with estimating herbivore diets by microscopically identifying plant fragments from stomachs, *J. Mammal.*, 57, 167, 1976.
31. **Anderson, A. E., Synder, W. A., and Brown, G. W.,** Stomach content analysis related to condition in mule deer, Guadalupe Mountains, New Mexico, *J. Wildl. Manage.*, 29, 352, 1965.
32. **Ward, A. L.,** Stomach content and fecal analysis, in Range and Wildlife Habitat Evaluation: A Research Symposium, U.S. Department of Agriculture, Forest Service Misc. Publ. 1147, Washington, D.C., 1970, 146.
33. **Scotcher, J. S. B.,** A review of fecal analysis techniques for determining the diet of wild grazing herbivores, *Proc. Grassl. Soc. Afr.*, 19, 131, 1979.
34. **Takatsuki, S.,** Precision of fecal analysis: a feeding experiment with penned Sika deer, *J. Mammal. Soc. Jpn.*, 7, 167, 1978.
35. **Gill, R. B., Carpenter, L. H., Baker, D. L., and Schoonveld, G. G.,** Fecal analysis to estimate mule deer diets, *J. Wildl. Manage.*, 47, 902, 1983.
36. **Crocker, B. H.,** A method for estimating the botanical composition of the diet of sheep, *N. Z. J. Agric. Res.*, 2, 72, 1959.
37. **Stewart, D. R. M.,** Analysis of plant epidermis in feces: a technique for studying food preferences of herbivores, *J. Appl. Ecol.*, 4, 83, 1967.
38. **Vavra, M., Rice, R. W., and Hansen, R. M.,** A comparison of esophageal fistula and fecal material to determine steer diets, *J. Range Manage.*, 31, 11, 1978.
39. **Sanders, K. D., Dahl, B. E., and Scott, G.,** Bite-count vs. fecal analysis for range animal diets, *J. Range Manage.*, 33, 146, 1980.
40. **Hansen, R. M., Peden, D. G., and Rice, R. W.,** Discerned fragments in feces indicates dietary overlap, *J. Range Manage.*, 26, 103, 1973.
41. **Voth, E. H. and Black, H. C.,** A histological technique for determining feeding habits of small herbivores, *J. Wildl. Manage.*, 37, 223, 1973.
42. **Varva, M. and Holechek, J. L.,** Factors influencing microhistological analyses of herbivore diets, *J. Range Manage.*, 33, 371, 1980.
43. **Dearden, B. L.,** Analysis of the Discernibility of Plant Species during Digestion, Ph.D. thesis, Colorado State University, Fort Collins, 1973.
44. **Houstad, K. M. and Donart, G. B.,** The microhistological technique: testing two central assumptions in southcentral New Mexico, *J. Range Manage.*, 31, 469, 1978.
45. **Marshal, J. W. and Squires, V. R.,** Accuracy of quantitative methods used for the botanical analysis of oesophageal fistula samples, *Trop. Grassl.*, 13, 140, 1979.
46. **Dearden, B. L., Pegau, R. E., and Hansen, R. M.,** Precision of microhistological estimates of ruminant food habitats, *J. Wildl. Manage.*, 39, 402, 1975.
47. **Pulliam, D. E. and Nelson, J. R.,** Determination of digestibility coefficients for quantification of fecal analysis with elk, in *North American Elk: Ecology, Behavior and Management*, Boyce, M. S. and Hayden-Wing, L. D., Eds., University of Wyoming, Laramie, 1979, 240.
48. **Lovaas, A. L.,** Mule deer food habits and range use, Little Belt Mountains, Montana, *J. Wildl. Manage.*, 22, 275, 1958.

49. **Buechner, H. K.,** Life history ecology and range use of pronghorn antelope in Trans-Pecos, Texas, *Am. Midl. Nat.*, 43, 257, 1950.
50. **Wallmo, O. C., Gill, R. O., Carpenter, L. H., and Reichart, D. W.,** Accuracy of field estimates of deer food habits, *J. Wildl. Manage.*, 37, 566, 1973.
51. **Hobbs, N. T., Baker, D. L., Ellis, J. E., and Swift, D. M.,** Composition and quality of elk winter diets in Colorado, *J. Wildl. Manage.*, 45, 156, 1981.
52. **Collins, W. B., Urness, P. J., and Austin, D. D.,** Elk diets and activities on different lodgepole pine habitat segments, *J. Wildl. Manage.*, 42, 799, 1978.
53. **Wallmo, O. C. and Neff, D. J.,** Direct observation of tamed deer to measure their consumption of natural forage, in Range and Wildlife Habitat Evaluation: A Research Symposium, U.S. Department of Agriculture, Forest Service Misc. Publ. 1147, 1970, 105.
54. **Daily, T. V., Hobbs, N. T., and Woodard, T. N.,** Experimental comparisons of diet selection by mountain goats and mountain sheep in Colorado, *J. Wildl. Manage.*, 48, 799, 1984.
55. **Schwartz, C. C. and Nagy, J. G.,** Pronghorn diets relative to forage availability in Northeastern Colorado, *J. Wildl. Manage.*, 40, 469, 1976.
56. **Schwartz, C. C., Nagy, J. G., and Rice, R. W.,** Pronghorn dietary quality relative to forage availability and other ruminants in Colorado, *J. Wildl. Manage.*, 41, 161, 1977.
57. **Wickstrom, M. L., Robbins, C. T., Hanley, D. E., and Parish, S. M.,** Food intake and foraging energetics of elk and mule deer, *J. Wildl. Manage.*, 48, 1285, 1984.
58. **Bartman, R. M., Alldredge, A. W., and Neil, P. H.,** Evaluation of winter food choices by tame mule deer, *J. Wildl. Manage.*, 46, 807, 1982.
59. **Bartman, R. M. and Carpenter, L. H.,** Effects of foraging experience on food selectivity of tame mule deer, *J. Wildl. Manage.*, 46, 813, 1982.
60. **Springfield, H. W.,** The grazed-plant method for judging the utilization of crested wheatgrass, *J. For.*, 59, 666, 1961.
61. **Martin, S. C.,** Relating vegetation measurements to forage composition by animals, in Range and Wildlife Habitat Evaluation: A Research Symposium, U.S. Department of Agriculture, Forest Service Misc. Publ. 1147, Washington, D.C., 1970, 93.
62. **Smith, A. D. and Hubbard, R. L.,** Preference ratings for winter deer forages from northern Utah ranges based on browsing time and forage consumption, *J. Range Manage.*, 7, 262, 1954.
63. **Smith, A. D.,** Feeding deer on browse species during winter, *J. Range Manage.*, 3, 130, 1958.
64. **Petrides, G. A.,** Principal food versus preferred foods and their relation to stocking rates and range condition, *Biol. Conserv.*, 7, 161, 1975.
65. **Chamrad, A. O. and Box, T. W.,** Food habits of white-tailed deer in South Texas, *J. Range Manage.*, 21, 158, 1968.
66. **Jacobs, J.,** Quantitative measurement of food selection: a modification of forage ratio and Ivlev's electivity index, *Oecologia (Berlin)*, 14, 413, 1974.
67. **Heady, H. F.,** Palatability and preference, *J. Range Manage.*, 17, 76, 1964.
68. **Cock, J. M. W.,** Assignment of preference, *J. Anim. Ecol.*, 47, 805, 1978.
69. **Chesson, J.,** Measuring preference in selective predation, *Ecology*, 59, 211, 1978.
70. **Pyke, G. H., Pulliam, H. R., and Charnov, E. L.,** Optimal foraging: a selective reveiw of theory and tests, *Q. Rev. Biol.*, 52, 137, 1978.
71. **Freeland, W. J. and Janzen, D. H.,** Strategies in herbivory by mammals: the role of secondary plant compound, *Am. Nat.*, 108, 269, 1974.
72. **Westoby, M.,** An analysis of diet selection by large generalist herbivores, *Am. Nat.*, 108, 290, 1974.
73. **Hobbs, N. T. and Bowden, D. C.,** Confidence intervals on food preference indices, *J. Wildl. Manage.*, 46, 507, 1982.
74. **Loehle, C. and Rittenhouse, L. R.,** An analyses of forage preference indices, *J. Range Manage.*, 35, 316, 1982.
75. **Van Soest, P. J.,** Use of detergents in the analysis of fibrous feeds. I. Preparation of fiber residues of low nitrogen content, *J. Assoc. Agric. Chem.*, 46, 825, 1963.
76. **Van Soest, P. J.,** Symposium on factors influencing the voluntary intake of herbage by ruminants: voluntary intake in relation to chemical composition and digestibility, *J. Anim. Sci.*, 24, 834, 1965.
77. **Goering, H. K. and Van Soest, P. J.,** Forage Fiber Analysis, U.S. Department of Agriculture Handbook No. 379, U.S. Department of Agriculture, Washington, D.C., 1970.
78. **Hungate, R. E.,** *The Rumen and its Microbes,* Academic Press, New York, 1966, 533.
79. **Van Soest, P. J.,** Structural and chemical characteristics which limit the nutritive value of forage, in *Forage Economics-Quality,* Harrison, C. M., Ed., American Society of Agronomy Spec. Publ. 13, Madison, Wisc., 1968.
80. **Garrett, W. N. and Johnson, D. E.,** Nutritional energetics of ruminants, *J. Anim. Sci.*, 57, 478, 1983.
81. **Milchunas, D. G., Dyer, M. I., Wallmo, O. C., and Johnson, D. E.,** *In vivo/in vitro* relationship of Colorado mule deer forages, *Colo. Div. Wildl. Spec. Rep.*, 43, 1978.

82. **Crampton, E. W. and Harris, L. W.,** *Applied Animal Nutrition,* 2nd ed., W. H. Freeman, San Francisco, 1969.
83. **Robbins, C. T.,** *Wildlife Feeding and Nutrition,* Academic Press, New York, 1983.
84. **Harris, L. E.,** *Nutritional Research Techniques for Domestic and Wild Animals,* Department of Animal Science, Utah State University, Logan, 1970.
85. **Church, D. C.,** *Digestive Physiology and Nutrition of Ruminants, Vol. 2, Nutrition,* D. C. Church, Publ., Oregon State University, Corvallis, 1981, 801.
86. **Holter, J. B., Tyler, G., and Walski, T.,** Nutrition of the snowshoe hare *(Lepus americanus), Can. J. Zool.,* 52, 1553, 1974.
87. **Kempton, T. J., Murray, R. M., and Leng, R. A.,** Methane production and digestibility measurements in the grey kangaroo and sheep, *Aust. J. Biol. Sci.,* 29, 209, 1976.
88. **Rogerson, A.,** Energy utilization by the eland and wildebeest, *Symp. Zool. Soc. London,* 21, 153, 1968.
89. **Thompson, C. B., Holter, J. B., Haynes, H. H., Silver, H., and Urban, W. E., Jr.,** Nutrition of white-tailed deer. I. Energy requirements of fawns, *J. Wildl. Manage.,* 37, 301, 1973.
90. **Mautz, W. W., Silver, H., and Hayes, H. H.,** Estimating methane, urine, and heat increment for deer consuming browse, *J. Wildl. Manage.,* 39, 80, 1975.
91. **Simpson, A. M., Webster, A. J. F., Smith, J. S., and Simpson, C. A.,** The efficiency of utilization of dietary energy for growth in sheep *(Ovis ovis)* and red deer *(Cervus elaphus), Comp. Biochem. Physiol.,* 59, 95, 1978.
92. **Robbins, C. T., Van Soest, P. J., Mautz, W. W., and Moen, A. N.,** Feed analyses and digestion with reference to white-tailed deer, *J. Wildl. Manage.,* 39, 67, 1975.
93. **Mould, E. D. and Robbins, C. T.,** Evaluation of detergent analysis in estimating nutritional value of browse, *J. Wildl. Manage.,* 45, 937, 1981.
94. **Mould, E. D. and Robbins, C. T.,** Nitrogen metabolism in elk, *J. Wildl. Manage.,* 45, 323, 1981.
95. **Mould, E. D. and Robbins, C. T.,** Digestive capabilities in elk compared to white-tailed deer, *J. Wildl. Manage.,* 46, 22, 1982.
96. **Hebert, D. M.,** Altitudinal Migration as a Factor in the Nutrition of Bighorn Sheep, Ph.D. thesis, University of British Columbia, Vancouver, 1973.
97. **Hjeljord, O., Sundstol, F., and Haagenrud, H.,** The nutritional value of browse to moose, *J. Wildl. Manage.,* 46, 333, 1982.
98. **McEwan, E. H. and Whitehead, P. E.,** Seasonal changes in the energy and nitrogen intake in reindeer and caribou, *Can. J. Zool.,* 48, 905, 1970.
99. **Jacobsen, E. and Skjenneberg, S.,** Some results from feeding experiments with reindeer, *Proc. Int. Reindeer and Caribou Symp.,* 1, 95, 1972.
100. **Streeter, C. L.,** A review of techniques used to estimate the *in vivo* digestibility of grazed forages, *J. Anim. Sci.,* 29, 237, 1969.
101. **Baker, D. L.,** unpublished data, 1984.
102. **Mautz, W. W.,** Confinement effects on dry matter digestibility coefficients displayed by deer, *J. Wildl. Manage.,* 35, 366, 1971.
103. **Theurer, C. B., Lesperance, A. L., and Wallace, J. D.,** Botanical Composition of the Diet of Livestock Grazing Native Range, University of Arizona Agric. Exp. Stn. Tech. Bull. 233, Tucson, Ariz., 1976.
104. **Lesperance, A. L., Clanton, D. C., Nelson, A. B., and Theurer, C. B.,** Factors Affecting the Apparent Chemical Composition of Fistula Samples, University of Nevada Agric. Exp. Stn. Bull. T18, Reno, Nev., 1974.
105. **Mautz, W. W., Walski, T. W., and Urban, W. E., Jr.,** Digestibility of fresh frozen versus pelleted browse by snowshoe hares, *J. Wildl. Manage.,* 40, 301, 1976.
106. **Palmer, W. L., Cowan, R. L., and Ammann, A. P.,** Effect of innocular source on *in vitro* digestion of deer foods, *J. Wildl. Manage.,* 40, 301, 1976.
107. **Palmer, W. L. and Cowan, R. L.,** Estimating digestibility of deer foods by an *in vitro* technique, *J. Wildl. Manage.,* 44, 469, 1980.
108. **Urness, P. J., Smith, A. D., and Watkins, R. K.,** Comparison of *in vivo* and *in vitro* dry matter digestibility of mule deer forages, *J. Range Manage.,* 30, 119, 1977.
109. **Mould, E. D.,** Aspects of elk *(Cervus canadensis nelsoni)* nutrition and associated analytical procedures, Ph.D. thesis, Washington State University, Pullman, 1980.
110. **Rittenhouse, L. R., Streeter, C. L., and Clanton, D. C.,** Estimating digestive energy from digestible dry matter in diets of grazing cattle, *J. Range Manage.,* 24, 73, 1971.
111. **Tilley, J. M. A. and Terry, R. A.,** A two stage technique for in vitro digestion of forage crops, *J. Br. Grassl. Soc.,* 18, 401, 1963.
112. **Wilson, A. D., Weir, W. C., and Torrell, D. T.,** Comparison of methods of estimating the digestibility of range forage and browse, *J. Anim. Sci.,* 32, 1046, 1971.
113. **Van Soest, P. J.,** *Nutritional Ecology of the Ruminant,* O & B Books, Corvallis, Oregon, 1982.

114. **Short, H. L.,** Forage digestibility and diet of deer on southern upland range, *J. Wildl. Manage.*, 35, 698, 1971.
115. **Ullrey, D. E., Youatt, W. G., Johnson, H. E., Fay, L. D., Purser, D. B., Schoepke, B. L., and Magee, W. T.,** Limitations of winter aspen browse for the white-tailed deer, *J. Wildl. Manage.*, 35, 732, 1971.
116. **Baker, D. L. and Hobbs, N. T.,** Composition and quality of elk summer diets in Colorado, *J. Wildl. Manage.*, 46, 694, 1982.
117. **Hobbs, N. T., Baker, D. L., and Gill, R. B.,** Comparative nutritional ecology of montane ungulates during winter, *J. Wildl. Manage.*, 47, 1, 1983.
118. **Sidahmed, A. E., Morris, J. G., Koong, L. J., and Radosevich, S. R.,** Contribution of mixtures of three chaparal shrubs to the protein and energy requirements of spanish goats, *J. Anim. Sci.*, 53, 591, 1981.
119. **Ward, A. L.,** *In vitro* digestibility of elk winter forage in southern Wyoming, *J. Wildl. Manage.*, 35, 681, 1971.
120. **Oldemeyer, J. L., Franzmann, A. W., Brundage, A. L., Anderson, P. D., and Flynn, A.,** Browse quality and the Kenai moose population, *J. Wildl. Manage.*, 41, 533, 1977.
121. **Bryant, F. C., Kothman, M. M., and Merrill, L. B.,** Nutritive content of sheep, goat, and white-tailed deer diets on excellent condition rangeland in Texas, *J. Range Manage.*, 33, 410, 1980.
122. **Blankenship, L. H., Varner, L. W., and Lynch, G. W.,** *In vitro* digestibility of south Texas range plants using innoculum from four ruminant species, *J. Range Manage.*, 35, 664, 1982.
123. **Carpenter, J. R. and Nino-Duponte, R. Y.,** Tropical browse plants: chemical composition and in-vitro digestibility using rumen fluid from goats, sheep, and cattle fed high grain or high roughage rations, *Proc. West. Sec. Am. Soc. Anim. Sci.*, 32, 188, 1981.
124. **Welch, B. L., Pederson, J. C., and Clary, W. P.,** Ability of different rumen inocula to digest range forages, *J. Wildl. Manage.*, 47, 873, 1983.
125. **Nelson, B. D., Ellzey, H. D., Montgomery, C., and Morgan, E. B.,** Factors affecting the variability of an in-vitro rumen fermentation technique for estimating forage quality, *J. Dairy Sci.*, 55, 358, 1972.
126. **Milchumas, D. G. and Baker, D. L.,** In vitro digestion — sources of within and between trial variability, *J. Range Manage.*, 35, 199, 1982.
127. **Calder, F. W.,** Effect of barley supplement to the ration of donor animals used in *in vitro* digestibility determinations, *J. Anim. Sci.*, 50, 265, 1970.
128. **Person, S. J., Pegau, R. E., White, R. G., and Luick, J. R.,** *In vitro* and nylon bag digestibilities of reindeer and caribou forages, *J. Wildl. Manage.*, 44, 613, 1980.
129. **Thomas, C. D. and Kroeger, P.,** In vitro digestibilities of plants in rumen fluids of Peary caribou, *Arctic*, 33, 757, 1980.
130. **Trudell, J., White, R. G., Jacobsen, E., Staaland, H., Kildemo, E., and Gaare, E.,** Comparison of some factors affecting the *in vitro* digestibility estimates of reindeer forages, in *Proc. 2nd Int. Reindeer and Caribou Symp.*, Reimers, E., Gaare, E., and Skjenneberg, S., Eds., Directoratet for Vilt og Ferskvannsfisk, Trondheim, Norway, 1980, 799, 262.
131. **Pederson, J. C. and Welch, B. L.,** Effects of monoterpenoid exposure on ability of rumen inocula to digest a set of forages, *J. Range Manage.*, 35, 500, 1982.
132. **Knipfel, J. E. and Troelsen, J. E.,** Interaction between inoculum donor diet and substrate in *in vitro* digestion studies, *Can. J. Anim. Sci.*, 46, 91, 1966.
133. **Short, H. L.,** Rumen fermentations and energy relationships in white-tailed deer, *J. Wildl. Manage.*, 27, 184, 1963.
134. **Ammann, A. P., Cowan, R. L., Mothershead, C. L., and Baumgardt, B. R.,** Dry matter and energy intake in relation to digestibility in white-tailed deer, *J. Wildl. Manage.*, 37, 195, 1973.
135. **Short, H. L., Blair, R. M., and Segelquist, C. A.,** Fiber composition and forage digestibility by small ruminants, *J. Wildl. Manage.*, 38, 197, 1974.
136. **Fonnesbeck, P. U.,** Estimating nutritive value from chemical analysis, *Proc. Int. Symp. Feed Composition in Animal Nutr. Req. and Computerization of Diets*, Vol. 1, Utah Agric. Exp. Stn., Logan, Utah, 1977, 219.
137. **Van Soest, P. J.,** Use of detergents in the analysis of fibrous feeds. II. A rapid method for the determination of fiber and lignin, *J. Off. Agric. Chem.*, 46, 829, 1963.
138. **Van Soest, P. J.,** Development of a comprehensive system of feed analysis and its application to forages, *J. Anim. Sci.*, 26, 117, 1967.
139. **Perzanowski, K.,** The effect of winter food composition on roe-deer energy budget, *Acta Theriol.*, 23, 451, 1978.
140. **Norris, K. H., Barnes, R. F., Moore, J. E., and Shenk, J. S.,** Predicting forage quality by infrared reflectance spectroscopy, *J. Anim. Sci.*, 43, 889, 1976.
141. **Ward, R. G., Smith, G. S., Wallace, J. D., Urquhart, N. S., and Shenk, J. S.,** Estimates of intake and quality of grazed range forage by near infrared reflectance spectroscopy, *J. Anim. Sci.*, 54, 339, 1982.

142. **Brooks, J., Anderson, M., and Urness, P. J.,** Infrared reflectance analysis of forage quality for elk, *J. Wildl. Manage.*, 48, 254, 1984.
143. **Shenk, J. S., Westerhaus, M. O., and Hoover, M. R.,** Analysis of forages by infrared reflectance, *J. Dairy Sci.*, 62, 807, 1979.
144. **Davis, A. M.,** The oxalate, tannin, crude fiber, and crude protein composition of young plants of some *Atriplex* species, *J. Range Manage.*, 34, 339, 1981.
145. **Kufeld, R. C., Stevens, M., and Bowden, D. C.,** Winter variation in nutrient and fiber content and in vitro digestibility of gambel oak (*Quercus gambelli*) and big sagebrush (*Artemisia tridentata*) from diversified sites in Colorado, *J. Range Manage.*, 34, 149, 1981.
146. **Bailey, J. A.,** Sampling deer browse for crude protein, *J. Wildl. Manage.*, 31, 437, 1967.
147. **Ullrey, D. E., Youatt, W. G., Johnson, H. E., Cowan, A. B., Covert, R. L., and Magee, W. T.,** Digestibility and estimated metabolizability of aspen browse for white-tailed deer, *J. Wildl. Manage.*, 36, 885, 1972.
148. **Radwan, M. A. and Crouch, G. L.,** Plant characteristics related to feeding preference by black-tailed deer, *J. Wildl. Manage.*, 38, 32, 1974.
149. **Mautz, W. W., Silver, H., Holter, J. B., Haynes, H. H., and Urban, W. E., Jr.,** Digestibility and related nutritional data for seven northern deer browse species, *J. Wildl. Manage.*, 40, 630, 1976.
150. **Blair, R. M., Short, H. L., and Epps, E. A., Jr.,** Seasonal nutrient yield and digestibility of deer forage from a young pine plantation, *J. Wildl. Manage.*, 41, 667, 1977.
151. **Blair, R. M. and Epps, E. A., Jr.,** Distribution of protein and phosphorus in spring growth of rusty blackhaw, *J. Wildl. Manage.*, 31, 188, 1967.
152. **Segelquist, C. A., Short, H. L., Ward, F. D., and Leonard, R. G.,** Quality of some deer forages in the Arkansas Ozarks, *J. Wildl. Manage.*, 36, 174, 1972.
153. **LeResche, R. E. and Davis, J. L.,** Importance of nonbrowse foods to moose on the Kenai Peninsula, Alaska, *J. Wildl. Manage.*, 37, 279, 1973.
154. **Conroy, M. J., Oderwald, R. G., and Sharik, T. L.,** Forage production and nutrient concentrations in thinned loblolly pine plantations, *J. Wildl. Manage.*, 46, 719, 1982.
155. **Uresk, D. W., Dietz, D. R., and Messner, H. E.,** Constituents of *in vitro* solution contribute differently to dry matter digestibility of deer food species, *J. Range Manage.*, 28, 419, 1975.
156. **Short, H. L.,** Nutrition of southern deer in different seasons, *J. Wildl. Manage.*, 39, 321, 1975.
157. **Short, H. L.,** Food habits of mule deer in a semidesert grass shrub habitat, *J. Range Manage.*, 30, 206, 1977.
158. **Tew, R. K.,** Seasonal variation in the nutrient content of aspen foliage, *J. Wildl. Manage.*, 34, 475, 1970.
159. **Everett, J. H. and Gonzales, C. L.,** Seasonal nutrient content in food plants of white-tailed deer on south Texas plains, *J. Range Manage.*, 34, 506, 1981.
160. **Welch, B. L. and Pederson, J. C.,** *In vitro* digestibility among accessions of big sagebrush by wild mule deer and its relation to monoterpenoid content, *J. Range Manage.*, 34, 497, 1981.
161. **Welch, B. L. and McArthur, E. D.,** Variation in winter levels of crude protein among *Artemesia tridentata* subspecies grown in a uniform garden, *J. Range Manage.*, 32, 467, 1979.
162. **Johnston, A., Bezeau, L. M., and Smoliak, S.,** Chemical composition and *in vitro* digestibility of alpine tundra plants, *J. Wildl. Manage.*, 32, 773, 1968.
163. **Ullrey, D. E., Youatt, W. G., Johnson, H. E., Fay, L. D., and Brent, B. E.,** Digestibility of cedar and jack pine browse for white-tailed deer, *J. Wildl. Manage.*, 31, 448, 1967.
164. **Short, H. L.,** Composition and squirrel use of acorns of black and white oak groups, *J. Wildl. Manage.*, 40, 479, 1976.
165. **Torgenson, O. and Pfander, W. H.,** Cellulose digestibility and chemical composition of Missouri deer foods, *J. Wildl. Manage.*, 35, 221, 1971.
166. **Pieper, R. D., Kelsey, R. J., and Nelson, A. B.,** Nutrient fertilized and unfertilized blue gramma, *J. Range Manage.*, 27, 470, 1974.
167. **McAtee, J. W., Scifres, C. J., and Drawe, D. L.,** Digestible energy and protein content of Gulf cordgrass following burning or shredding, *J. Range Manage.*, 32, 376, 1979.
168. **Bryant, J. P. and Kuropat, P. J.,** Selection of winter forage by subarctic browsing vertebrates: the role of plant chemistry, *Annu. Rev. Ecol. Syst.*, 11, 261, 1980.
169. **Milton, K.,** Factors influencing leaf choice by howler monkeys: a test of some hypotheses of food selection by generalist herbivores, *Am. Nat.*, 114, 362, 1979.
170. **Wrangham, R. W. and Waterman, P. G.,** Feeding behaviour of vervet monkeys on *Acacia tortilis* and *Acacia xanthophloea:* with special reference to reproductive strategies and tannin production, *J. Anim. Ecol.*, 50, 715, 1981.
171. **Oates, J. F., Waterman, P. G., and Choo, G. M.,** Food selection by the south Indian leaf monkey, *Presbytis johnii*, in relation to leaf chemistry, *Oecologia*, 45, 45, 1980.
172. **McKey, D. B.,** Soils, vegetation, and seed-eating by black colobus monkeys, *The Ecology of Arboreal Folivores*, Smithsonian Inst. Press, Washington, D.C., 1978, 423.

173. **McKey, D. B., Waterman, P. G., Mbi, C. N., Gartlan, N., and Struhsaker, T. T.,** Phenolic content of vegetation in two African rain forests: ecological implication, *Science*, 202, 61, 1978.
174. **Oates, J. F., Swain, T., and Zantovska, J.,** Secondary compounds and food selection by colobus monkeys, *Biochem. Syst. Ecol.*, 5, 317, 1977.
175. **Glander, K. E.,** Feeding patterns in mantled howling monkeys, in *Foraging Behavior*, Kamil, A. C. and Sargent, T. D., Eds., Garland STPM Press, New York, 1981, 231.
176. **Batzli, G. O. and Jung, H. G.,** Nutritional ecology of microtine rodents: resource utilization near Atkasook, Alaska, *Arc. Alp. Res.*, 12, 483, 1980.
177. **Abaza, R. H., Blake, J. T., and Fisher, E. J.,** Chemical analyses for nutritional elements and toxic oxalate compounds of sedges, rushes and certain broad leafed plants typifying inter-mountain meadowland forage, Utah Agric. Exp. Stn., Mimeo Series 504, 1967.
178. **Johnson, D. R.,** The food habits of rodents on rangelands of southern Idaho, *Ecology*, 42, 407, 1961.
179. **Vorhies, T. and Taylor, W. P.,** Life history and ecology of the white-throated wood rat, *Neotoma albigula albigula* Hurtley, in relation to grazing in Arizona, Tech. Bull. No. 86, Arizona Agric. Exp. Stn., Tucson, 1940, 455.
180. **Schmidt-Nielsen, K.,** *Desert Animals, Physiological Problems of Heat and Water*, Clarendon Press, Oxford, England, 1964.
181. **Green, J. S. and Flinders, J. T.,** Habitat and dietary relationship of the pygmy rabbit, *J. Range Manage.*, 33, 136, 1980.
182. **Braun, C. E., Britt, T., and Wallestad, R. D.,** Guidelines for maintenance of sage grouse habitats, *Wildl. Soc. Bull.*, 5, 99, 1977.
183. **Kufeld, C. R., Wallmo, O. C., and Feddema, C.,** Foods of the Rocky Mountain mule deer, *U.S. Department of Agriculture Forest Service Res. Pap.*, RM-111, 1973.
184. **Sundstrom, C., Hepworth, W. G., and Diem, K. L.,** Abundance, distribution, and food habits of the pronghorn, Wyoming Game and Fish Comm., Bull. No. 15, Cheyenne, 1973.
185. **Farentinos, R. C., Capretta, P. J., Kepner, R. E., and Littlefield, V. M.,** Selective herbivory in tassel-eared squirrels: role of monoterpenes in ponderosa pines, chosen as feeding trees, *Science*, 213, 1273, 1981.
186. **Radwan, M. A., Crouch, G. L., Harrington, C. A., and Ellis, W. D.,** Terpenes of ponderosa pine and feeding preferences by pocket gophers, *J. Chem. Ecol.*, 8, 241, 1982.
187. **Berger, P. J., Sanders, E. H., Gardner, P. D., and Negus, N. C.,** Phenolic plant compounds functioning as reproductive inhibitors in *Microtus montanus*, *Science*, 195, 575, 1977.
188. **Leopold, A. S., Erwin, M., Oh, J., and Browning, B.,** Phytoestrogens: adverse effects on reproduction in California quail, *Science*, 191, 98, 1976.
189. **Kendall, W. A. and Sherwood, R. T.,** Palatability of tall fescue and reed canary grass and some of their alkaloids to meadow voles, *Agron. J.*, 67, 667, 1975.
190. **Kalow, W.,** Pharmacogenetics in animals and man, *Ann. N.Y. Acad. Sci.*, 151, 694, 1968.
191. **Myers, G. T. and Vaughan, T. A.,** Food habits of the plains pocket gopher in eastern Colorado, *J. Mammal.*, 45, 588, 1964.
192. **Cooper-Driver, G., Finch, S., Swain, T., and Bernays, E.,** Seasonal variation in secondary plant compounds in relation to the palatability of *Pteridium acquilium*, *Biochem. Syst. Ecol.*, 5, 177, 1977.
193. **Rothschild, M.,** Some observations on the relationship between plants, toxic insects and birds, in *Phytochemical Ecology*, Harborne, J. B., Ed., Academic Press, London, 1972, 1.
194. **Majak, W., Quinton, D. A., and Broersma, K.,** Cyanogenic glycoside levels in Saskatoon serviceberry, *J. Range Manage.*, 33, 197, 1980.
195. **Schaller, G. B.,** *The Mountain Gorilla*, University of Chicago Press, Chicago, 1963.
196. **Bryant, J. P., Wieland, G. D., Reichardt, P. B., Lewis, V. E., and McCarthy, M. C.,** Pinosylvin methyl ether deters snowshoe hare feeding on green alder, *Science*, 222, 1023, 1983.
197. **Bate-Smith, E. C.,** Attractants and repellents in higher animals, in *Phytochemical Ecology*, Harborne, J. B., Ed., Academic Press, London, 1972, 45.
198. **Swain, T.,** Tannins and lignins, in *Herbivores: Their Interaction with Secondary Plant Metabolites*, Rosenthal, G. A. and Janzen, D. H., Eds., Academic Press, New York, 1979, 657.
199. **Mabry, T. J. and Gill, J. E.,** Sesquiterpene lactones and other terpenoids, in *Herbivores: Their Interaction with Plant Secondary Metabolites*, Rosenthal, G. A. and Janzen, D. H., Eds., Academic Press, New York, 1979, 502.
200. **Rhoades, D. F. and Cates, R. G.,** Toward a general theory of plant anti-herbivore chemistry, *Recent Adv. Phytochem.*, 10, 168, 1976.
201. **Feeny, P. P.,** Seasonal changes in oak leaf tannins and nutrients as a cause of spring feeding by winter moth caterpillars, *Ecology*, 51, 656, 1970.
202. **Nagy, J. G., Steinhoff, H. W., and Ward, G. M.,** Effects of essential oils of sagebrush on deer rumen microbial function, *J. Wildl. Manage.*, 28, 785, 1964.

203. **Oh, H. K., Sakai, T., Jones, M. B., and Longhurst, W. M.,** Effects of various essential oils isolated from Douglas fir needles upon sheep and deer rumen microbial activity, *Appl. Microbiol.*, 15, 777, 1967.
204. **Oh, H. K., Jones, M. B., and Longhurst, W. M.,** Comparison of rumen microbial inhibition resulting from various essential oils isolated from relatively unpalatable plant species, *Appl. Microbiol.*, 16, 39, 1968.
205. **Cluff, L. K., Welch, B. L., Pederson, J. C., and Brotherson, J. D.,** Concentration of monoterpernoid in the rumen ingesta of wild mule deer, *J. Range Manage.*, 35, 192, 1982.
206. **White, S. M., Welch, B. L., and Flinders, J. T.,** Monoterpenoid content of pygmy rabbit stomach ingesta, *J. Range Manage.*, 35, 107, 1982.
207. **Klein, D. R.,** The introduction, increase, and crash of reindeer on St. Matthew Island, *J. Wildl. Manage.*, 32, 350, 1968.
208. **Schwan, H. E. and Swift, L.,** Forage inventory methods with special reference to big game ranges, *Trans. North Am. Wildl. Conf.*, 6, 118, 1941.
209. **Milner, C. and Hughes, R. E.,** Methods for measurement of primary production of grassland, *IBP Handbook No. 6*, Blackwell Scientific, Oxford, England, 1968, 70.
210. **Tadmore, N. H., Breight, A., Noy-Meir, I., Benjamin, R. W., and Eyal, E.,** An evaluation of the calibrated weight estimate method for measuring production in annual vegetation, *J. Range Manage.*, 28, 65, 1975.
211. **McIntyre, G. A.,** Statistical aspects of vegetation sampling, in *Measurements of Grassland Vegetation and Animal Production*, Manntje, L. T., Ed., Commonwealth Agricultural Bureaux, Farnham Royal, England, 1978, 8.
212. **Francis, R. C., Van Dyne, G. M., and Williams, B. K.,** An evaluation of weight estimation double sampling as a method of botanical analysis, *J. Environ. Manage.*, 8, 55, 1979.
213. **Reese, G. A., Bayn, R. L., and West, N. E.,** Evaluation of double sampling estimates of subalpine herbage production, *J. Range Manage.*, 33, 300, 1980.
214. **Pechanec, J. P. and Pickford, G. D.,** A weight estimate method for the determination of range on pasture production, *J. Am. Soc. Agron.*, 29, 894, 1937.
215. **Wilm, H. G., Costello, D. F., and Hipple, G. E.,** Estimating forage yield by the double sampling method, *Am. Soc. Agron. J.*, 36, 194, 1944.
216. **Campbell, N. A. and Arnold, G. W.,** The visual assessment of pasture yield, *Aust. J. Exp. Agric. Anim. Husb.*, 13, 263, 1973.
217. **Haydock, K. P. and Shaw, N. H.,** The comparative yield method for estimating dry matter yield of pasture, *Aust. J. Exp. Agric. Anim. Husb.*, 15, 663, 1975.
218. **Ahmed, J., Bonham, C. D., and Lycok, W. A.,** Comparison of techniques used for reducing biomass estimates by double sampling, *J. Range Manage.*, 36, 217, 1983.
219. **Lyon, L. J.,** Estimating twig production of serviceberry from crown volumes, *J. Wildl. Manage.*, 32, 115, 1968.
220. **Bently, J. R., Seegrist, D. W., and Blakeman, D. A.,** A technique for sampling low shrub vegetation by crown volume classes, *U.S. Department of Agriculture Forest Service Res. Note PSW 215*, Pacific Southwest Forest and Range Exp. Stn., Berkeley, Calif., 1970, 11.
221. **Guy, P. R.,** The estimation of above-ground biomass of the trees and shrubs in the Sengwa Wildlife Research Area, Zimbabwe, S. Afr., *Tydskr. Natuurwet*, 11, 135, 1980.
222. **Dean, S. J., Burkehardt, W., and Meeuwig, R. O.,** Estimating twig and foliage biomass of sagebrush, bitterbrush, and rabbitbrush in the Great Basin, *J. Range Manage.*, 34, 224, 1981.
223. **Shafer, E. L.,** The twig count method for measuring hardwood browse, *J. Wildl. Manage.*, 27, 428, 1963.
224. **Telfer, E. S.,** Twig weight-diameter relationship for browse species, *J. Wildl. Manage.*, 33, 917, 1969.
225. **Lyon, L. J.,** Length- and weight-diameter relations of serviceberry twigs, *J. Wildl. Manage.*, 34, 456, 1970.
226. **Peek, J. M., Krefting, L. W., and Tappeiner, J. C., III,** Variation in twig diameter-weight relationships in northern Minnesota, *J. Wildl. Manage.*, 35, 501, 1971.
227. **Fletcher, J. E. and Robinson, M. E.,** Capacitance meter for estimating forage weight, *J. Range Manage.*, 9, 96, 1956.
228. **Neal, D. L. and Neal, J. L.,** A new electronic meter for measuring herbage yield, *U.S. Department of Agriculture Forest Service Res. Note PSW 210*, 36, 1965.
229. **Johns, G. G.,** The accuracy of a range of capacitance probe methods for estimating pasture yields, *J. Agric. Sci.*, (Cambridge), 79, 273, 1972.
230. **Neal, D. L. and Neal, J. L.,** Uses and capabilities of electronic capacitance instruments for estimating standing herbage. I. History and development, *J. Br. Grassl. Soc.*, 28, 81, 1963.
231. **Alfonso, A., Toledo, J. M., and Burns, J. C.,** Herbage measurement *in-situ* by electronics. I. The multiple-probe-type capacitance meter: a brief review, *Grass Forage Sci.*, 35, 25, 1980.

232. **Alfonso, A., Toledo, J. M., and Burns, J. C.,** Herbage measurements *in situ* by electronics. II. Theory and design of an earth plate capacitance meter for estimating forage dry matter, *Grass Forage Sci.*, 35, 95, 1980.
233. **Toledo, J. M., Burns, J. C., Lucas, H. L., Jr., and Angelone, A.,** Herbage measurement *in situ* by electronics. III. Calibration, characterization and field application of the earth-plate forage capacitance meter: a prototype, *Grass Forage Sci.*, 35, 189, 1980.
234. **Vickery, P. J., Bennett, I. L., and Nichol, G. R.,** An improved electronic capacitance meter for estimating herbage mass, *Grass Forage Sci.*, 35, 247, 1980.
235. **Terry, W. S., Hunter, D. H., and Swindel, B. F.,** Herbage capacitance meter: an evaluation of its accuracy in Florida rangelands, *J. Range Manage.*, 34, 240, 1981.
236. **Teare, I. D., Mott, G. D., and Eaton, J. R.,** Beta attenuation: a technique for estimating forage yield *in situ*, *Radiat. Bot.*, 6, 7, 1966.
237. **Mitchell, J. E.,** An analysis of the beta-attenuation technique for estimating standing crop of prairie range, *J. Range Manage.*, 25, 300, 1972.
238. **Colwell, J. E.,** Vegetation canopy reflectance, *Remote Sensing Environ.*, 3, 175, 1974.
239. **McNaughton, S. J.,** Serengeti migratory wildebeest: facilitation of energy flow by grazing, *Science*, 197, 92, 1976.
240. **Waller, S., Brown, M. A., and Lewis, J. K.,** Factors involved in estimating green biomass by canopy Spectroreflectance measurements, *J. Range Manage.*, 34, 105, 1981.
241. **White, R. S. and Currie, P. O.,** Prescribed burning in the northern Great Plains: yield and cover responses of 3 forage species in the mixed grass prairie, *J. Range Manage.*, 36, 179, 1983.
242. **Sneva, F. A. and Hyder, D. N.,** Estimating herbage production on semarid ranges in the intermountain region, *J. Range Manage.*, 15, 88, 1962.
243. **Murphy, A. H.,** Predicted forage yield based on fall precipitation in California annual grasslands, *J. Range Manage.*, 23, 363, 1970.
244. **Mueggler, W. F.,** Plant development and yield on mountain grasslands in southwestern Montana, U.S. Department of Agriculture Forest Servies Res. Pap. INT-24, 1972, 20.
245. **Shrimal, R. L. and Vyas, L. N.,** Net primary production in grasslands at Udaipur, India, in *Tropical Ecological Systems*, Golley, F. B. and Medina, E., Eds., Springer-Verlag, New York, 1975, 265.
246. **McNaughton, S. J.,** Grassland herbivore dynamics, in *Serengeti: Dynamics of an Ecosysem*, Sinclair, A. R. E. and Norton-Griffiths, M., Eds., University of Chicago Press, Chicago, 1979, 46.
247. **Sala, O., Deregibus, V. A., Schlichter, T., and Alippe, H.,** Productivity dynamics of a native temperature grassland in Argentina, *J. Range Manage.*, 34, 48, 1981.
248. **Samuel, M. J., Ravzi, F., and Hart, R. H.,** Nitrogen fertilization of range: yield, protein content, and cattle behavior, *J. Range Manage.*, 33, 119, 1983.
249. **Leege, T. A., Herman, D. J., and Zamora, B.,** Effects of cattle grazing on mountain meadows in Idaho, *J. Range Manage.*, 34, 324, 1981.
250. **Steuter, A. A. and Wright, H. A.,** Spring burning effects on redberry juniper-mixed grass habitats, *J. Range Manage.*, 36, 161, 1983.
251. **Reiners, R. J. and Urness, P. J.,** Effects of grazing horses managed as manipulators of big game winter range, *J. Range Manage.*, 35, 567, 1982.
252. **Herchel, C. H., Gould, W. I., Leifeste, W. F., and Gribbens, R. P.,** Herbicide treatment and vegetation response to treatment of mesquites in Southern New Mexico, *J. Range Manage.*, 36, 149, 1982.
253. **Trudell, J. and White, R. G.,** The effects of forage structure and availability on food intake, biting rate, bite size and daily eating time of reindeer, *J. Appl. Ecol.*, 18, 63, 1981.
254. **Dodd, J. L., Lavenroth, W. K., and Heitschmidt, R. K.,** Effects of controlled SO_2 exposure on net primary production and plant biomass dynamics, *J. Range Manage.*, 35, 572, 1982.
255. **Merrill, E. H., Mayland, H. F., and Peck, J. M.,** Effects of a fall wildfire on herbaceous vegetation on xerie sites in the Selway-Bitterroot Wilderness, Idaho, *J. Range Manage.*, 33, 363, 1980.
256. **Whittaker, R. H. and Woodwell, G. M.,** Structure, production and diversity of the oak-pine forest at Brookhaven, New York, *J. Ecol.*, 155, 1969.
257. **Wolters, G. L.,** Longleaf and slash pine decreases herbage production and atlers herbage composition, *J. Range Manage.*, 35, 761, 1982.
258. **Wolters, G. L., Martin, A., and Pearson, H. A.,** Forage response to overstory reduction on loblolly-shortleaf pine-hardwood forest range, *J. Range Manage.*, 35, 443, 1982.
259. **Pyke, D. A. and Zamora, B. A.,** Relationships between overstory structure and understory production in the grand fir/myrtle boxwood habitat type of northcentral Idaho, *J. Range Manage.*, 35, 769, 1982.
260. **Clary, W. P. and Jameson, D. A.,** Herbage production following tree and shrub removal in the pinyon-juniper type of Arizona, *J. Range Manage.*, 34, 109, 1981.
261. **Thill, R. E., Ffolliott, P. F., and Patton, D. R.,** Deer and elk forage production in Arizona, mixed conifer forests, *U.S. Department of Agriculture Forest Service Res. Pap.* RM-248, 1983, 1.

262. **Stout, D. G., McLean, A., Brooke, B., and Hall, J.,** Influence of simulated grazing (clipping) on pinegrass growth, *J. Range Manage.*, 33, 286, 1980.
263. **Oswald, B. P. and Covington, W. W.,** Changes in understory production following a wildfire in southwestern ponderosa pine, *J. Range Manage.*, 36, 507, 1983.
264. **Pase, C. P.,** Herbage production and composition under immature ponderosa pine stands in the Black Hills, *J. Range Manage.*, 11, 238, 1958.
265. **Anderwartha, H. G. and Birch, L. C.,** *The Distribution and Abundance of Animals*, University of Chicago Press, Chicago, 1954.
266. **Wallmo, O. C., Carpenter, L. H., Regelin, W. L., Gill, R. B., and Baker, D. L.,** Evaluation of deer habitat on a nutritional basis, *J. Range Manage.*, 230, 122, 1977.
267. **Moen, A. N. and Evans, K. E.,** The distribution of energy in relation to snow cover in wildlife habitat, in *Snow and Ice in Relation to Wildlife and Recreation Symposium Proceedings*, Haugen, A. O., Ed., Iowa State University, Ames, 1971, 1.
268. **Bergerud, A. T.,** Relative abundance of food in winter for Newfoundland caribou, *Oikos*, 25, 379, 1974.
269. **Miller, D. R.,** Seasonal changes in the feeding behaviour of barren-ground caribou on the taiga winter range, *The Behaviour of Ungulates and its Relation to Management*, Geist, V. and Walther, F., Eds., IUCN Publ. NS 44, 1974, 744.
270. **Kelsall, J. P.,** Continual barren-ground caribou studies, *Can. Wildl. Serv. Wildl. Manage. Bull.*, Ser. 1, 12, 1957, 1.
271. **Kelsall, J. P.,** Cooperative studies of barren-ground caribou, 1957—58, *Can. Wildl. Serv. Wild. Manage. Bull. Ser. 1*, 15, 1, 1960.
272. **Kelsall, J. P.,** The Migratory Barren-Ground Caribou of Canada, Canadian Wildlife Service, Ottawa, 1968, 1.
273. **Pruitt, W. O.,** Snow as a factor in the winter ecology of barren-ground caribou, *Arctic*, 12, 158, 1959.
274. **Coady, J. W.,** Influence on snow on behavior of moose, *Nat. Can.*, 101, 417, 1974.
275. **Nasimovitch, A. A.,** The role of the regime of snow cover in the life of ungulates in the USSR, Akad. Nauk. SSSR, Moskva, Transl. Russina, Canadian Wildlife Service, Ottawa, 1955.
276. **Peek, J. M.,** On the nature of winter habitats of Shiras moose, *Nat. Can.*, 101, 131, 1974.
277. **Smith, B. L.,** Influence of snow conditions on winter distribution, habitat use and group size of mountain goat, in Proc. 1st Int. Mountain Goat Symp. Kalispell, Montana, Feb. 19, Samuel, W. and Macgregor, W. G., Eds., Province of British Columbia, Ministry of Recreation and Conservation, Fish and Wildlife Branch, 1977.
278. **Bunnell, F. L.,** Deer-forest relationships on northern Vancouver Island, in Sitka Black-tailed Deer: Proc. Conf. in Juneau, Alaska, Wallmo, D. C. and Schoen, J. W., Eds., U.S. Department of Agriculture, Forest Service, Alaska Region, 1970, 86.
279. **Regelin, W. L.,** Nutritional interactions of black-tailed deer with their habitat in southeast Alaska, in Sitka Black-tailed Deer: Proc. Conf. in Juneau, Alaska, Wallmo, O. C. and Schoen, J. W., Eds., U.S. Department of Agriculture, Forest Service, Alaska Region, 1979, 60.
280. **Schoen, J. W. and Wallmo, O. C.,** Timber management and deer in southeast Alaska: current problems and research direction, in Sitka Black-tailed Deer: Proc. Conf. in Juneau, Alaska, Wallmo, O. C. and Schoen, J. W., Eds., U.S. Department of Agriculture, Forest Service, Alaska Region, 1979, 69.
281. **Stoddart, L. A. and Smith, A. D.,** *Range Management*, McGraw-Hill, New York, 1955.
282. **Jardine, J. T. and Anderson, M.,** Range Management on the National Forests, U.S. Department of Agriculture Bull. 790, Washington, D.C., 1919.
283. **Cassady, J. T.,** General review of methods and techniques for measuring production and utilization, in, Technique and Methods of Measuring Understory Vegetation, U.S. Forest Service, South and Southwest For. Exp. Stn. Proc., 1959, 5.
284. **Stewart, G. and Hutchings, S. S.,** The point-observation-plot (square-foot density) method of vegetation survey, *Am. Soc. Agron. J.*, 28, 714, 1936.
285. **Stoddart, L. A.,** Problem in estimating grazing capacity of range, in *Proc. 6th Int. Grass Congress, Vol. II*, Penn State College, University Park, Pa., 1952, 1367.
286. **Humphrey, R. R.,** Field comment on the range condition methods of forage survey, *J. Range Manage.*, 2, 1, 1947.
287. **Moen, A. N.,** *Wildlife Ecology: An Analytical Approach*, W. H. Freeman, San Francisco, 1973.
288. **Robbins, C. T.,** The Biological Basis for the Determination of Carrying Capacity, Ph.D. thesis, Cornell University, Ithaca, N.Y., 1973.
289. **Hobbs, N. T., Baker, D. L., Ellis, J. E., Swift, D. M., and Green, R. A.,** Energy and nitrogen-based estimates of elk winter range carrying capacity, *J. Wildl. Manage.*, 46, 12, 1982.
290. **Potvin and Huot, J.,** Estimating carrying capacity of a white-tailed deer wintering area in Quebec, *J. Wildl. Manage.*, 47, 463, 1983.
291. **Mentis, M. T. and Duke, R. R.,** Carrying capacity of natural veld in Natal for wild herbivores, *S. Afr. J. Wildl. Res.*, 6, 65, 1976.

Chapter 3

FORAGING BEHAVIOR: DYNAMICS OF DINING OUT

F. L. Bunnell and M. P. Gillingham

TABLE OF CONTENTS

The literature of foraging theory frequently utilizes optimization techniques to predict diet and habitat choice.[1,2] Several authors have criticized the approach for reasons that include objections to the simplicity of many models, mathematical complexity of some models, arguments concerning the stability of optima, or apparent failure to explicitly test hypotheses.[3-6] The basic formulation of many foraging models has been adapted directly from the fields of economics and operations research[7-9] for which state spaces and constraint sets may be well bounded. Yet even in those fields there is discomfort with assumptions underlying optimality criteria. The discomfort has encouraged alternative approaches, one of which is termed "satisficing". Satisficing models investigate decision rules for which the actor's objective is not to maximize or minimize some quantity, but to survive.[10,11] Satisficing while foraging is analogous to the existential game of Slobodkin[12] in which the winnings due to fulfilling the goal consist of staying in the game. An animal adopting that approach acts not to optimize its allocation of time, or choice of habitat or diet, but to stay alive. It may be able to stay alive and reproduce without optimizing any single variable.

We have reservations about optimal foraging theory similar to those noted. We also find that the complexity of most wild herbivores' environments makes it nearly impossible to estimate *a priori* the variable to be optimized. Invoking satisficing rules does not evade the problem. Despite mathematical differences, individuals employing a satisficing decision rule may act "as if" they are optimizing.[10] The direction of effort may be the same, but the target is not a single optimum. Ultimately, the theoretician evaluating foraging behavior faces the same difficulty with satisficing criteria as with optimization: he must specify the variables and units critical to the animal. It may, in fact, be more difficult to specify minimally satisfictory thresholds than specific optima (unless the real but unhelpful threshold of survival is invoked).

Our approach is to synthesize published data, extract apparent generalities, and evaluate those against current theory. We concentrate on wild populations, because domestic species were reviewed recently and comprehensively.[13] Our emphasis is on simple relationships. We consider that when an animal is dining out it has three major kinds of decisions to make: when to dine, where to dine, and what to select from the menu. These three decisions are frequently interrelated (a serious impediment to optimality and satisficing functions, particularly the former). We attempt to acknowledge most interrelationships explicitly in our analyses, and concentrate on the allocation of time because it is the area for which least theory is applicable and which most needs review. Time also is the most easily measured variable.

I. WHEN TO DINE — ALLOCATING TIME

Despite the fact that many studies of foraging document time budgets, there are few theoretical treatments. When treated theoretically, allocation of time is usually considered an optimization process;[14-17] alternative satisficing models[18] are rarely invoked. Each approach is simplified and does not readily accommodate ruminants. For example, it is unclear how handling time should be defined for a ruminant. We focus on broad patterns and simple expectations that do not invoke optimization. The foraging herbivore either makes or has imposed upon it decisions that determine: (1) how much time to spend foraging (total time), (2) how to partition that time (bout lengths), and (3) how intensively to utilize the time spent foraging (the proportions of time spent eating within a foraging bout).

A. Total Time Foraging

The total time an animal spends feeding can be considered a function of two groups of factors: intrinsic energy demands or constraints, and environmental constraints. The first group includes differences in digestive systems, age, sex, weight, and physiological state.

The second includes the amount of daylight available, potential predation, and factors influencing the profitability of foraging (biomass, tiller length, availability, nutrient status).

1. Intrinsic Demands and Constraints

a. Digestive System and Feeding Habit

We recognize two digestive systems: hindgut fermenters and ruminants. We expect hindgut fermenters (HGF) to spend more time foraging for two reasons: (1) less of the forage ingested is digested, and (2) time available is not limited by rumination. We recognize three broad feeding habits: bulk and roughage eaters or grazers, intermediate or mixed feeders, and concentrate selectors or browsers. On the basis of energy readily available for digestion, we expect the ranking of time allocated to foraging to decrease in the order bulk feeders > mixed feeders > concentrate feeders. The decline should be more obvious among ruminants because rumen turnover times will be affected by readily digestible energy.[19,20]

These expectations are met (Table 1). For any class of feeding habits, HGF allocate more time to foraging than do ruminants, especially over a 24-hr period. Among ruminants there is a marked decrease in time spent foraging from bulk to concentrate feeders. Bulk feeders also spend more of their active period foraging; conversely, concentrate feeders spend more time walking and seeking areas in which to forage. The largest number of individual budgets is available for ruminants (n = 112). The time foraging per unit metabolic weight ($F/W^{0.75}$ or F/MW) allocated by ruminants to foraging declines dramatically from bulk to concentrate feeders (Table 1), despite an opposing association with weight (*vide infra*).

b. Weight

Although larger animals might be expected to feed longer to meet their energy requirements, there are several reasons why the effect should not be pronounced. Within taxa, younger and lighter animals have the additional energy demands of growth. Across taxa, energetic demands scale with metabolic weight while bite size scales more directly with weight, and feeding habits influence total time foraging (Table 1). As a result, no clear relationship is evident (Figure 1). Among ruminants, the greater ability of larger animals to utilize low quality forage allows them to be less selective while foraging and reduces the effect of increasing weight. Note that it is the small bulk feeding ruminants, which would profit most from being selective, that also allocate most time to foraging (Figure 1B). Together, these relationships produce a rapid decline in time allocated to foraging per unit body weight (F/W, min/kg) with body weight (W, kg):

Hindgut fermenters:[21-31,60-62]

$$F/W = 0.65 - 0.08\ \ln(W) \quad (1)$$

$$(n = 29,\ r^2 = 0.83)$$

Ruminants:[32-59,63-74]

$$F/W = 4.16 - 0.74\ \ln(W) \quad (2)$$

$$(n = 141,\ r^2 = 0.65)$$

The difference in coefficients between Equations 1 and 2 primarily reflects the larger range in weights of hindgut fermenters (Figure 1).

Within ruminants, we expect the relationships to vary among feeding styles. To evaluate that prediction, effects of age are excluded by treating only adult animals. Both sexes are included but courting or territorial males and lactating females are excluded. Empirical relationships using the references of Equation 2 are

Table 1
MEAN VALUES OF TIME ALLOCATED TO FORAGING PER ACTIVE PERIOD AND PER METABOLIC WEIGHT FOR DOMESTIC AND WILD HERBIVORES GROUPED BY DIGESTIVE SYSTEM AND FEEDING HABITS

Digestive system and feeding habits[a]	Number of species	Number of budgets	Time foraging (%)		Foraging/active		Active/inactive		Time foraging (min/kg$^{0.75}$)	
			12 hr[b]	24 hr	12 hr	24 hr	12 hr	24 hr	12 hr	24 hr
Hindgut fermenters										
Bulk	5	18	59.0	57.8	0.90	0.84	3.26	2.18	7.98	10.29
Intermediate	2[c]	4	76.0	74.2	0.83	0.86	—	6.68	1.58	2.69
Concentrate	2	6	43.8	39.6	0.69	0.61	2.35	1.77	2.99	5.44
Ruminants										
Bulk	12	55	49.1	41.8	0.82	0.96	1.47	0.73	15.26	18.65
Intermediate	6	46	45.9	41.4	0.76	0.77	1.69	1.25	7.83	13.60
Concentrate	5	11	38.5	28.4	0.71	0.72	2.63	0.65	7.21	6.30

[a] Intermediate feeders that were primarily grazing or browsing when activity budgets were acquired are considered to be bulk or concentrate feeders, respectively.

[b] 0700 to 1900 or daylight hours; 24-hr means are usually based on fewer, larger species.

[c] Largest species, *Elephas maximus* and *Loxodonta africana*.

Bulk feeders:

$$F/W = 5.00 - 0.91\ ln(W) \quad (3)$$
$$(n = 78,\ r^2 = 0.63)$$

Intermediate feeders:

$$F/W = 3.45 - 0.62\ ln(W) \quad (4)$$
$$(n = 43,\ r^2 = 0.81)$$

Concentrate feeders:

$$F/W = 2.17 - 0.32\ ln(W) \quad (5)$$
$$(n = 20,\ r^2 = 0.81)$$

The intercepts follow the pattern anticipated from Table 1. The differences in slope reflect both the broad decrease in size from bulk to concentrate feeders (*Giraffa* is an exception) and the fact that rumination confers greater advantage with increasing size among bulk than among concentrate feeders. Whereas large bulk feeders can cope well with relatively low quality forage, concentrate feeders (by definition) avoid such forage and attain consistently higher rates of passage. Equations 1 through 5 follow similar forms when expressed per unit metabolic weight ($W^{0.75}$). In each case, they suggest that increasing size decreases the amount of time allocated to foraging per unit body weight or metabolic weight. Basal metabolic rate is constant per unit metabolic weight, but both costs of thermoregulation and locomotion per unit weight decline with increasing weight.[75,76] Available data suggest that the decline in total time foraging per unit weight is greater than would be predicted from differential costs of thermoregulation and locomotion. We suggest it is also a product of increasing relative bite size and decreasing selectivity with increasing body weight. The empirical observations also suggest that if "time minimizers"[15] are to be found among ruminant herbivores, it will only be among concentrate feeders (Equation 5). Among other ruminants there is sufficient scope for tradeoffs between time digesting and time searching for forages of higher digestibility that neither variable may be optimized and time allocation is more likely to be explained by satisficing criteria.

c. *Age, Sex, and Physiological Condition*

Age, sex, and physiological condition (e.g., lactation) each may be covariate with weight and potential seasonal effects. Seasonal effects were eliminated by restricting comparisons to animals on the same range in the same season (Table 2). Differences between species are functions of weight, digestive system, and feeding habits (Table 1, Equations 3 to 5); comparisons are more informative within species.

Young animals that are still suckling have less need to allocate time to foraging and may be constrained by greater requirements for rest and play. Opposing tendencies that could increase the time allocated to foraging are energy requirements for growth and inexperience at foraging. Very young ruminants are characterized by active/inactive ratios of less than 1.0, a small portion of active time allocated to foraging (45 to 50%), and apparent inefficiencies during foraging (values of 48 to 80 min/$kg^{0.75}$, Table 2). The former two characteristics are likely the results of energy acquisition through suckling, and the advantages of allocating time to play during active periods, thereby developing mechanical and social skills, and encouraging growth. The latter characteristic may reflect inexperience at grazing or requirements to be more selective than are adults.

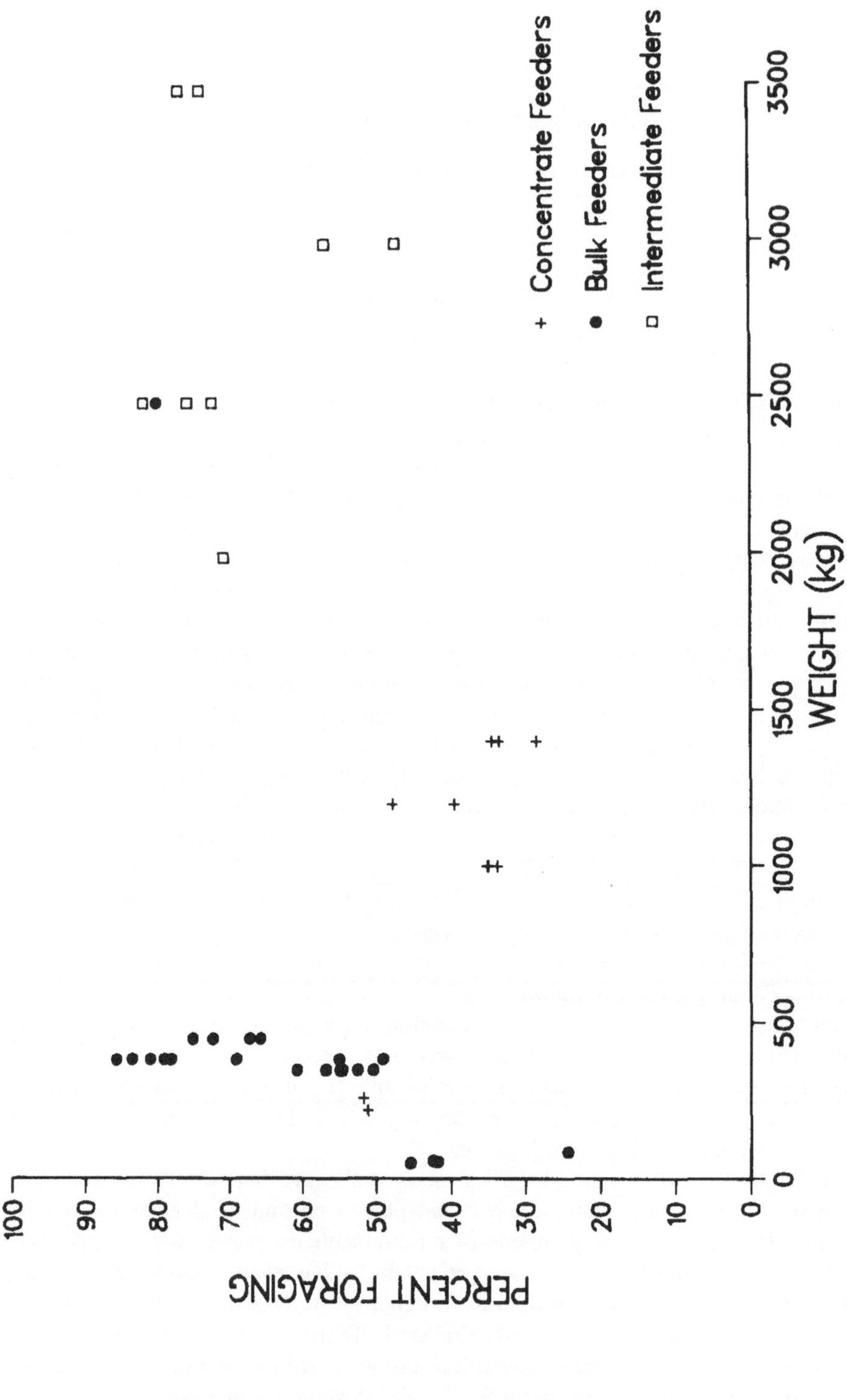

FIGURE 1. Percent of daylight allocated to foraging as a function of body weight and feeding habits. (A) hindgut fermenters, (B) ruminants. Data from References 21 to 74.

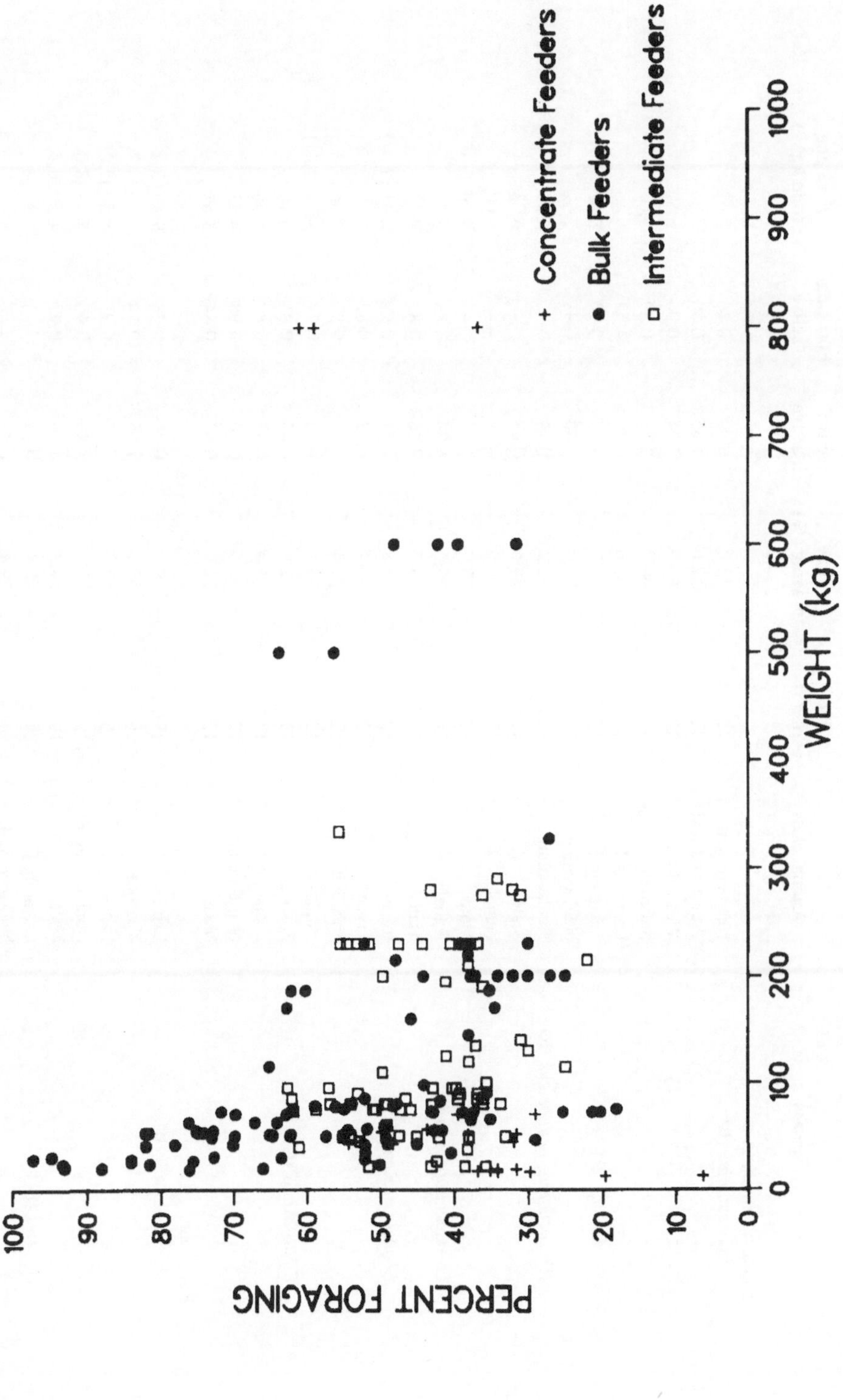
PERCENT FORAGING
100
90
80
70
60
50
40
30
20
10
0
100
200
300
400
500
600
700
800
900
1000
WEIGHT (kg)
+ Concentrate Feeders
● Bulk Feeders
□ Intermediate Feeders

FIGURE 1B

Table 2
DIFFERENCES IN TIME ALLOCATED TO FORAGING BY VARIOUS HERBIVORES OF DIFFERENT AGES OR PHYSIOLOGICAL STATUS BUT EXPERIENCING THE SAME RANGE CONDITONS

Species	Animal class	Diel period (hr)	Time foraging (%)	Active/ inactive	Foraging/ active	Foraging (min/kg$^{0.75}$)	Ref.
Hindgut fermenters							
Equus caballus (B)[a]	Newborn foal	12[b]	13.0	0.39	0.46	3.67	60
	Preweaned foal	12	52.0	1.85	0.80	5.96	
	Weaned foal	12	62.0	2.24	0.90	6.52	
	Subadult	12	60.7	2.13	0.89	5.03	
	Lactating F	12	62.0	2.30	4.53		
E. caballus (B)	Lactating F	17.5	74.7	4.03	0.93	8.02	26
	Nonlactating F	17.5	70.0	3.30	0.91	7.52	
Phacochoerus aethiopicus (B)	Nonlactating F	12	42.2	0.74[c]	—	14.53	21,77
	Male	12	24.6	0.33	—	6.33	
	Female	12	76.0	—	—	26.00	
	Male	12	55.0	—	—	14.19	
Loxodonta africana (I)	Female	24	75.2	6.09	0.88	3.06	29
	Male	24	73.2	7.26	0.83	2.32	
Elephus maximus (I)	Infant	12	75.8	3.27	0.99	2.41 (4.82)[d]	28
	Juvenile	12	70.4	6.12	0.82	1.69 (3.38)	
	Subadult	12	81.4	∞[c]	0.81	1.66 (3.32)	
	Adult	12	76.1	120.90	0.77	1.21 (2.41)	
Equus asinus (C)	Foals	12	22.9	0.43	0.85	6.16	30
	Lactating F	12	44.0	0.93	0.87	5.55	
	Nonlactating F	12	37.5	0.60	0.98	4.73	
	Male	12	32.4	0.63	0.84	3.61	
Ruminants							
Ovis dalli dalli (B)	Lambs	19	19.8	0.71	0.48	79.8 (63.2)[d]	32
	Yearling F	29	52.1	1.95	0.79	48.8 (38.7)	
	Lactating F	19	62.5	2.17	0.91	38.5 (37.0)	
	Nonlactating F	19	57.3	1.49	0.96	35.3 (28.0)	
	Male	19	54.8	1.30	0.97	24.5 (20.8)	
O. aries (B)	Lactating F	24	40.8	0.69	—	49.8	44
	Nonlactating F	24	29.4	0.42	—	35.9	

Kobus ellipsiprymus (B)	6-Month calf	12	60.0	—	—	48.2	66
	Lactating F	12	68.4	—	—	8.08	
	Nonlactating F	12	62.0	—	—	7.32	
	Male	12	44.3	—	—	5.23	
Cervus elaphus (B)	Lactating F	24	49.0	—	—	27.0	38
	Nonlactating F	24	41.0	—	—	21.5	
	Male	24	43.3	—	—	17.7	
Aepyceros melampus (I)	Nonlactating F	12	54.7	5.80	0.64	21.3	46
	Male	12	51.7	4.88	0.62	17.1	
	Nonlactating F	24	44.5	2.02	0.67	17.3	
	Male	24	40.1	1.58	0.66	13.7	
Ovibos moschatus (I)	Calf	24	20.0	0.79	0.45	25.8	45
	Yearling	24	41.0	1.50	0.68	16.3	
	Subadults	24	42.0	1.33	0.72	13.4	
	Adult F	24	44.0	1.22	0.80	11.9	
	Adult M	24	43.0	1.22	0.77	9.9	
Giraffa camelopardalis (C)	Female	12	60.5	6.27	0.70	2.90	56
	Male	12	26.0	1.35	0.45	1.46	
Litocranius walleri (C)	Lactating F	12	60.9	3.59	0.78	33.4	56
	Male	12	52.6	3.59	0.67	22.2	
Capreolus capreolus (C)	Pregnant F	24	32.0	0.58	0.89	57.5	58
	Male	24	36.0	0.67	0.90	61.9	
	Lactating F	24	33.5	0.64	0.85	63.2	
	Male	24	32.9	0.79	0.79	54.1	

[a] Letters in parentheses indicate feeding habits: B = bulk, I = intermediate, C = concentrate feeders.
[b] Numbers less than 24 are daylight hours.
[c] "Time in Hole" considered "inactive".
[d] Numbers in parentheses are estimates for 24-hr period.
[e] Small sample, animals never observed inactive.

The high values of F/MW among young ruminants are much greater than those predicted by weight (Equation 2); observed differences between very young and adult HGF approximate those predicted by weight alone (Equation 1). Young HGF are usually heavier than ruminant young and exhibit lower, weight-related differences in F/MW (Equations 1 and 2). They also allocate more of their active time to foraging, as is expected given their digestive system. Apparent inefficiencies at foraging are less well expressed among HGF. That observation may be a product of young ruminants having to be highly selective before the rumen has developed fully.

Yearlings and subadults typically rank next to suckling young in time allocated to foraging (min/$kg^{0.75}$, Table 2), presumably a product of energy demands for growth. Among horses, weaning increased the time commitment to foraging above that allocated just prior to weaning.[60] Differences associated with lactation and sex were remarkably consistent across taxa. Time allocated to foraging by nonlactating adult females was 85.9 ± 2.9% (SE) of that allocated by lactating females on the same range. This difference also occurs among domestic livestock.[74,78] The increased allocation of 16.4% associated with lactation is less than would be expected if total time foraging scaled directly with energy requirements. The observation suggests that eating rates during foraging (or foraging intensities) also increase in healthy animals as energy demands increase.

During periods when many females are nursing, males allocate less time to foraging than do nonlactating females (Table 2). The difference is marked during daylight (70.8 ± 7.6%) but is less over a 24-hr period (96.3 ± 2.3%). Because males are frequently larger than females, the difference is more evident when considered in terms of min/MW. The allocation of time per MW during daylight by males is 62.7 ± 6.1% of that for nonlactating females on the same range; over 24 hr it is 78.8 ± 1.7%. Together, these observations suggest: (1) females, which are generally smaller and potentially more vulnerable to predation, are encouraged to pursue more of their foraging during daylight than are males, and (2) males may exhibit greater eating rates than females or be more efficient at converting food energy to meet their requirements.

Other sex-related differences can alter the males' allocation of time to foraging. During the breeding season, adult males frequently allocate less time to foraging than do females (Table 3). Adult Asian elephants (*Elephas maximus*) normally spend about 76% of the day feeding,[28] but when accompanied by females in estrus, that value declines to 45.9%.[79] In Soay sheep rams, it declines from 67.7% before the rut to 27.6% during the rut.[65] The resultant differences between foraging times of males and females are especially dramatic among those species forming harems (Table 3). After the breeding season, the participating males are in depleted condition. They then allocate more time to foraging than do females. The reversal is dramatic. During breeding, male values of F/MW are typically less than 50% of female values; after breeding values of F/MW are typically greater for males despite their greater body weight (Table 3, Equation 2).

2. *Environmental Constraints*

Treatment of potential environmental constraints on the time allocated to foraging is often ambiguous. There are three major reasons: (1) most potential environmental constraints vary seasonally and are, therefore, covariate with each other; (2) they are also covariate with intrinsic constraints such as weight, age, and physiological status; and (3) in temperate regions, where daylight can vary dramatically, activity budgets are often available only for daylight hours. Figure 2 summarizes seasonal patterns in time allocated to foraging during daylight by six species.

Some points documented as interspecific generalities are illustrated in Figure 2. HGF spend more time foraging than do ruminants (Figure 2b). Larger species spend less daylight foraging than do smaller species with similar food habits (Figure 2a). Females may increase

Table 3
SEX-RELATED DIFFERENCES IN TIME ALLOCATED TO FORAGING DURING DAYLIGHT BY VARIOUS HERBIVORES DURING AND SHORTLY AFTER THE BREEDING SEASON

	% Daylight		Min/kg$^{0.75}$		
Species	Female	Male	Female	Male	Ref.
Breeding season					
Aepyceros melampus	53.5	26.7	20.80	9.15	46
Cervus elaphus elaphus	58.8	4.0	16.13	0.82	38
C. elaphus canadensis	45.7	17.2	7.31	1.61	39
Kobus ellipsiprymnus	64.1	41.9	9.16	4.95	66
Equus asinus	67.9	40.8	8.56	4.54	30
Ovibos moschatus	41.0	31.0	8.56	8.29	45
Post-breeding season					
Capreolus capreolus	32.0[a]	36.0[a]	57.5[a]	61.9[a]	58
Ovis aries	93.0	97.2	41.3	31.8	65
Rangifer tarandus	36.1	73.5	6.32	11.4	52
Connochaetes taurinus	53.0	60.0	7.18	7.31	43
Kobus ellipsiprymnus	35.1	42.6	5.02	5.03	66
Ovibos moschatus	32.0[b]	38.0[b]	—[c]	—[c]	45

[a] 24-hr period.
[b] Late winter, all others shortly after the end of breeding.
[c] Observation period unknown.

their time foraging after the season of births (Figures 2a, b, and c); males decrease time foraging during breeding but increase it afterwards (Figures 2b, c, and d). Despite these covariate factors, three generalities associated with season can be extracted. These are associated with use of daylight, abiotic conditions, and foraging profitability.

a. Use of Daylight

Data of Figure 2 are all for diurnal species from north temperate regions, but from various latitudes. All species but muskoxen show a depression in the proportion of daylight used to forage during summer (data for muskoxen are for daylight which encompasses 24 hr at midsummer). Although the proportion of daylight used declines with increasing day length, the total amount of time increases absolutely. As a result, the fraction of foraging occurring during daylight peaks during summer. The trend exists among both solitary (Figure 2c) and herd-forming species,[32] and is most pronounced among smaller species. The ratios of day to night foraging in winter and summer were about 0.4 and 3.0 in roe deer,[80] 1.8 and 9.5 in Dall's sheep,[32] and 2.0 and 1.2 in wapiti.[47] The general trend for diurnal species is thus to exploit daylight to the fullest extent possible. If one estimates the time requirements for rumination, some of these species[32,36,65] approximate complete utilization of daylight hours for foraging.

Advantages gained by full use of daylight, particularly for smaller, herd-forming species, include maintenance of herd integrity and increased predator avoidance while foraging, and greater opportunities for selective foraging. Among larger species, the incentive to avoid predators is relaxed, and the larger energy requirements reduce flexibility in concentrating foraging during one portion of the 24-hr period. The seasonal differences in use of daylight are thus a function of underlying patterns across taxa (Table 4) that become more strongly expressed as day length changes (Figure 2). The patterns reflect digestive systems, feeding habits, and broad differences in weight associated with feeding habits.

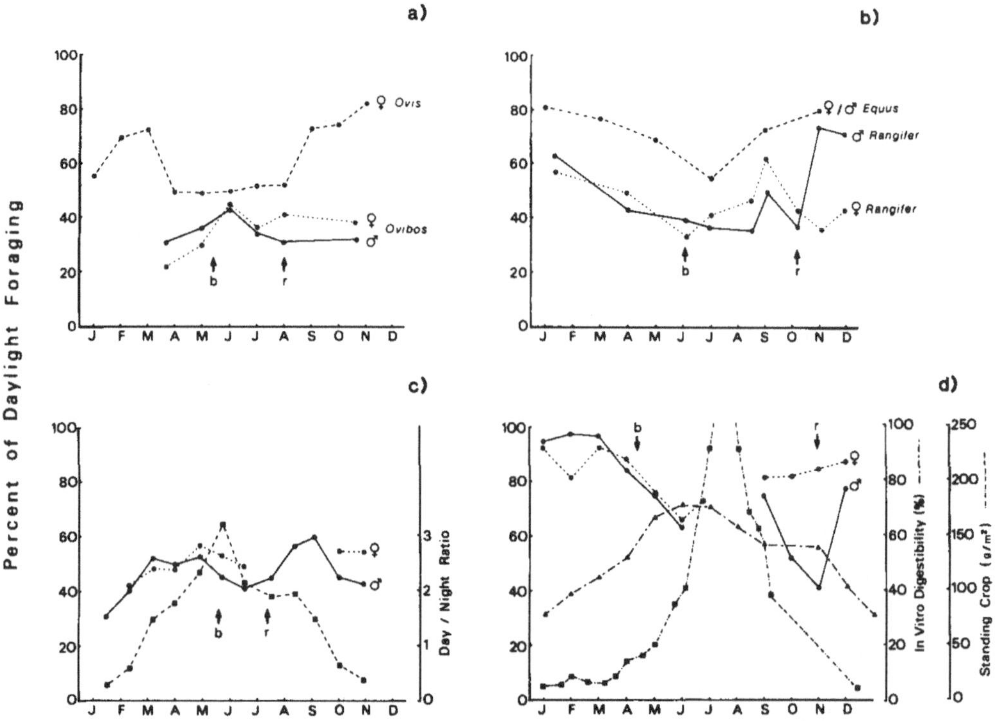

FIGURE 2. Seasonal patterns of the percent of daylight allocated to foraging. (a) California bighorn sheep (*Ovis canadensis californiana*);[36] captive female herd, there was no rut, and muskoxen (*Ovibos moschatus*).[45] (b) Horses (*Equus caballus*)[24] and caribou (*Rangifer tarandus*).[52] (c) Roe deer (*Capreolus capreolus*); the finely dashed line is the ratio of time foraging in daylight to time foraging during night by males.[80] (d) Soay sheep (*Ovis aries*); activity budgets are somewhat inflated as they use the percent of sheep grazing, not time; from Grubb and Jewell.[65] In vitro dry matter digestibilities and standing crop of forage are data of Milner and Gwynne.[81] Legend: b = season of births; r = rut.

Using only their relative abilities to digest their usual forage, one would predict that HGF would show less difference in the ratio of day to night foraging than would ruminants, and that bulk feeders would show smaller differences than concentrate feeders. Those predictions are met (Table 4), despite a bias among available data to sample nocturnal activity budgets from larger taxa within any class of feeding habits. Few data on highly crepuscular or nocturnal species are included in Table 4.

b. Abiotic Conditions

Changing abiotic conditions naturally influence forage production, but can have additional effects as well. For example, abiotic factors can cause sufficient thermal stress to modify the animal's activity patterns. Shade-seeking in response to increasing radiation or temperature is well documented for diverse species including horses,[27] lesser kudu,[82] impala,[46] and Indian blackbuck.[55] For some species, primarily larger ones, African buffalo,[41] Hartmann's zebra,[22] black rhino,[31] and male giraffes,[56] increasing temperature does not necessarily encourage shade-seeking but does encourage inactivity. In some cases, shade-seeking and inactivity significantly reduce the time spent foraging during daylight as temperature or radiation increases; examples of this include horses,[26] gerenuk,[56] female giraffes,[56] wildebeest,[43] and domestic sheep.[83] However, when daylight foraging is decreased, nocturnal foraging may increase so that the total time foraging remains unchanged.[83] Thus, heat stress may not always affect food acquisition adversely.

Table 4
RATIOS OF DIURNAL TO NOCTURNAL FORAGING BY DOMESTIC AND WILD HERBIVORES AS A FUNCTION OF DIGESTIVE SYSTEM AND FEEDING HABITS[a]

	Feeding habits		
Digestive system	Bulk	Intermediate	Concentrate
Hindgut fermenters	1.04	1.05	1.24
Ruminants	1.42	1.24	2.10

[a] Sources of Table 1.

Potential effects of cold stress, including wind and rain, on food acquisition are similarly unclear. Warthogs tend to stay in their dens during rainstorms[21] and other species become less active during ephemeral cool, windy, and wet conditions.[22,25,27,46,54,55] Klingel[25] suggested that steppe zebras may spend less time foraging on rainy days, but statistical evaluation showed no differences between sunny and rainy days for domestic sheep.[83]

Effects of harassment are less equivocal and the degree of harassment is a function of abiotic variables.[45,49,53] The influence of insect harassment on activity budgets has been quantified for several species.[22,27,45,49,52,53,60] As well as reducing the time spent foraging, harassment can increase energy costs substantially by encouraging movement. The result may be a significant reduction in net energy available to the animal. Reimers[84] argued that differential harassment by insects, hikers, and hunters was sufficient to account for the difference in growth of 79 g/day between Hardangervidda and Svalbard reindeer.

c. *Foraging Profitability*

Total time spent foraging should decline as it becomes more profitable to forage (see discussion of Table 6). Profitability of foraging can be influenced by forage biomass density, tiller length, or digestibility. These factors are often covariate and thus difficult to separate, but they allow a herbivore an array of tactics to achieve the same digestible intake. Interpretation of published activity budgets is further hampered because many are expressed as percentages of a daylight period of unknown length. For example, Figure 2d suggests that time foraging declines as biomass density and in vitro digestibility increase. In fact, we calculate that total time foraging during daylight was about 11 hr in March and 11.3 hr in June. Amount of foraging during darkness is unknown. There is, however, evidence that total time foraging is influenced by pasture conditions for many species including: horses,[26,27] black rhinos,[31] impala,[46] waterbuck,[66] reindeer,[85,86] African buffalo,[41] red deer,[38] wildebeest,[43] elephants,[61] bighorn sheep,[36] domestic sheep,[87-90] and cattle.[91-93] Similarly, supplemental feeding reduces total foraging time of sheep, cattle, and horses, especially the latter.[13,94]

Evidence for tropical species above involved direct comparison between activity budgets derived in the wet and in the hot, dry seasons. Digestibility and availability are confounded. For example, the time allocated to foraging by elephants appears dependent on the relative availability of browse and grass.[28,29,61] In most tropical species, the increase in time foraging during the dry season is associated with an increase in time walking, suggesting that searching for appropriate forage is important. Male black rhinos expanded their seasonal home ranges considerably.[31]

We expect somewhat different tactics and controls on the profitability of foraging among bulk and concentrate feeders. Concentrate feeders and HGF should allocate more time to movement as pasture conditions decline and they search for patches of more digestible forage. Generally, they do.[24,27,29,31,56] Bulk feeders confront a more homogeneous distribution of forage and are more likely to be influenced by structural features such as tiller length. Indeed, studies on these species indicate that tiller length, through its influence on bite size, is the dominant factor controlling total foraging time.[13,38,78,93,95]

There is equivocal evidence under experimental conditions that digestibility also modifies total foraging time of bulk feeders.[74,96] Indirect field evidence supports the suggestion. In African buffalo, the time foraging per 24 hr declined from 595 to 580 min from the wet to the late dry season, but the ruminating to foraging ratio increased from 0.76 to 0.93.[41] More time ruminating was required per unit time foraging in the dry season. Conversely, the Uganda waterbuck increased its time foraging from 440 to 595 min, but the ruminating to foraging ratio declined from 1.15 to 0.90.[66] There is apparently a tradeoff depending on the dietary flexibility; either submit to control by rumen passage rate thereby reducing the time foraging, or extend the foraging time by searching for more digestible forage to reduce rates of passage. We expect concentrate or mixed feeders to select that latter option. Data from impala[46] suggest they do because total time foraging increased from 477 to 640 min, while the ratio declined dramatically from 1.52 to 0.73. Although rumination can control total time allocated to foraging, it is more likely to influence the partitioning of the time allocated.

B. Partitioning Foraging Time

The simplest rule is to feed until full, but what is "full" could differ among feeding habits. For example, bulk feeders might be more often limited by physical distention and concentrate feeders by production of volatile fatty acids. Among ruminants (particularly bulk feeders), bout lengths should be related broadly to rumenoreticular volumes. Rumenoreticular volume (RV, liters) increases linearly with metabolic weight (MW, $kg^{0.75}$):[97-102]

$$RV = -3.49 + 0.77\ MW \qquad (6)$$
$$(n = 33,\ r^2 = 0.96)$$

Parra[103] and others have used weight of the gastrointestinal tract but that measurement may have the disadvantage of being dependent on what was eaten. We have used volume as the measurement of capacity and relied largely on the values obtained by Hoffmann[97] to ensure consistency of technique. RV per metabolic weight follows an asymptotic relationship ($n = 33$, $r^2 = 0.48$):

$$RV/MW = 0.80\left(\frac{MW}{8.47 + MW}\right) \qquad (7)$$

The tendency for concentrate and bulk feeders to segregate into different classes is more evident when considered per unit MW (Figure 3). Concentrate feeders are usually on or below the fitted line, bulk feeders are commonly above it (Figure 3B). Apparent adherence to the simplest rule can be evaluated by replacing RV with mean bout lengths (BL, min) in Equations 6 and 7. Empirical relationships are

Daylight:[32,38,45,54,65,71]

$$BL = 33.91 + 2.08\ MW \qquad (8)$$
$$(n = 16,\ r^2 = 0.48)$$

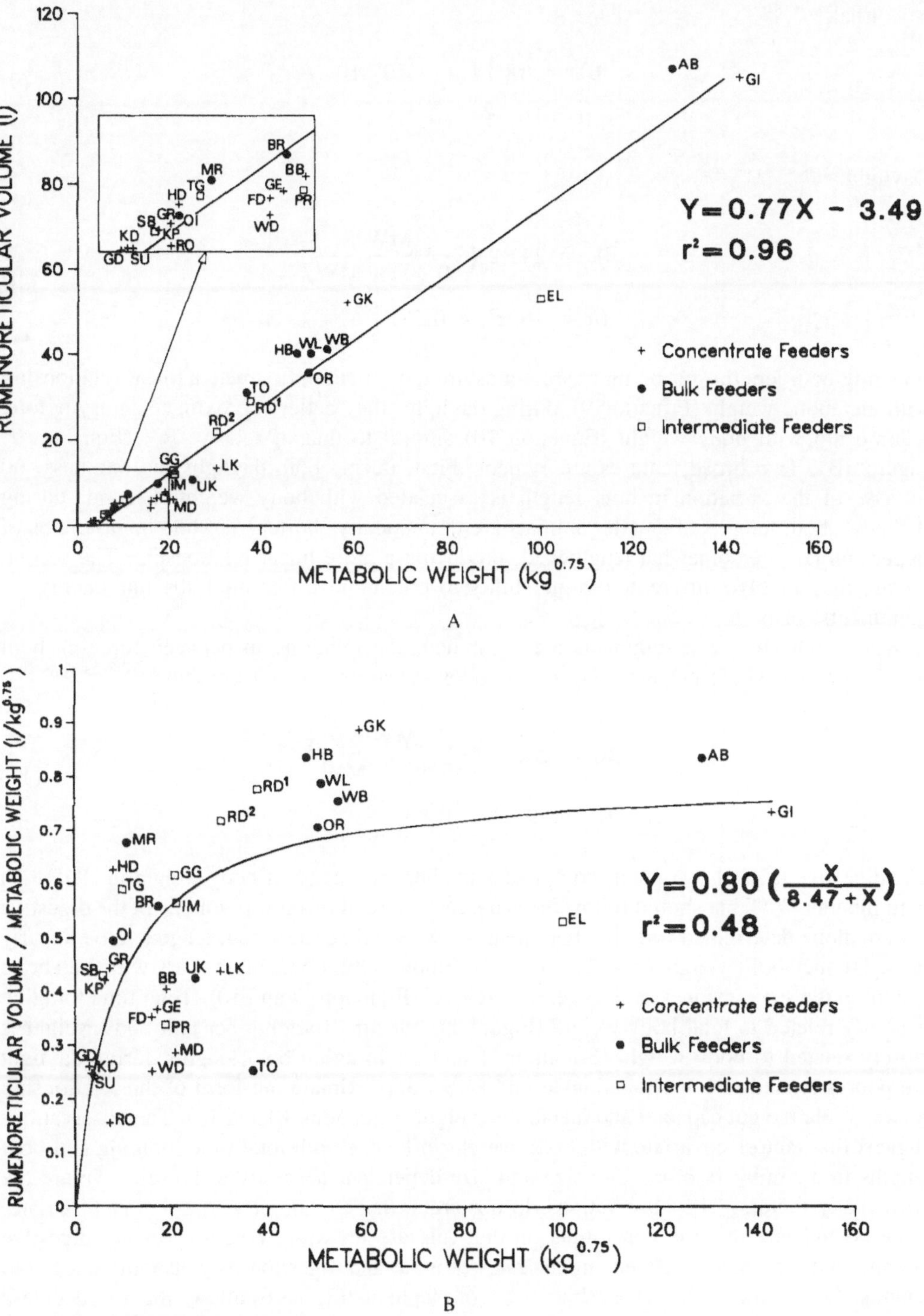

FIGURE 3. (A) Ruminoreticular volume (ℓ) as a function of metabolic weight. (B) Relative ruminoreticular volume (volume/metabolic weight) as a function of metabolic weight. Legend: AB = African buffalo, BB = bushbuck, BR = bohor reedbuck, EL = eland, FD = fallow deer,[98] GD = Gunther's dik-dik, GE = gerenuk, GG = Grant's gazelle, GI = giraffe, GK = greater kudu, GR = grey duiker, HB = hartebeest, HD = Harvey's duiker, IM = impala, KD = Kirk's dik-dik, KP = klipspringer, LK = lesser kudu, MD = mule deer,[99] MR = mountain reedbuck, OI = oribi, OR = oryx, PR = pronghorn,[100] RD[1] = red deer,[101] RD[2] = red deer,[98] RO = roe deer,[98] SB = steenbock, SU = suni, TG = Thomson's gazelle, TO = topi, UK = Uganda kob, WB = waterbuck, WD = white-tailed deer,[102] WI = wildebeest (from Reference 97 unless indicated otherwise.)

Nocturnal:[38,45,54]

$$BL = 18.14 + 1.50\ MW \tag{9}$$
$$(n = 9,\ r^2 = 0.71)$$

Daylight:[32,38,45,54,65,71]

$$BL = 241.2\left(\frac{MW}{40.30 + MW}\right) \tag{10}$$
$$(n = 16,\ r^2 = 0.57)$$

Foraging bout lengths during night are not asymptotic, but approximate a linear relationship with metabolic weight (Equation 9); during daylight, they better approximate an asymptotic relationship with body weight (Equation 10) similar to that of relative RV (Equation 7, Figure 3B). Two broad features are evident. First, during both daylight and darkness, 60 to 70% of the variation in bout length is associated with body weight. Second, during darkness, bout lengths of smaller animals are dramatically shortened, while bout lengths of larger animals are somewhat lengthened, producing a more linear relationship. The second feature may involve differential vulnerability to predation, differential absolute energy requirements, or both.

When both HGF and ruminants are combined, the relationship between foraging bout lengths during daylight and metabolic weight follows an asymptotic relationship:[26,28,32,45,54,65,71]

$$BL = 208.2\left(\frac{MW}{85.55 + MW}\right) \tag{11}$$
$$(n = 20,\ r^2 = 0.83)$$

The digestive tract plus contents comprise a smaller proportion of body weight in HGF than in ruminants,[103-104] but should follow the same form of relationship. If volume of the digestive system alone determined foraging bout lengths, we would expect bout lengths to be linearly related to metabolic weight (e.g., Equation 6). Empirically, bout lengths vary with metabolic weight in the same manner as does relative RV (cf. Equations 7 and 10). Total time foraging is poorly related to total body weight (Figure 1), but time foraging per unit body weight is closely related to body weight (Equations 1 and 2); foraging bout lengths during daylight are poorly related to RV (Equations 6 and 8) but approximate the form of the relationship between relative gut capacity and metabolic weight (Equations 10 and 11). The observations suggest that factors covariate with body weight influence both total time foraging and bout lengths in a similar fashion. Bite size and size-dependent necessity to be selective are the most likely factors. The observations do not argue that the simplest rule is inappropriate. They do indicate that the constraints on that rule are not simply volume of the digestive system, and that factors influencing the rates of input and digestion may also influence bout lengths. For example, the large "bite size" of elephants (Table 6) allows them to decrease both total time foraging and bout lengths below that predicted from linear relationships with metabolic rate.

C. Foraging Intensity

Herbivores facing large energy demands need not necessarily feed longer; they could increase the rate of intake by feeding more intensively. Many indexes of foraging intensity have been used. The most direct measurements are bites per minute and weight per bite.

The most crude indexes are based on the proportion of the active period spent foraging; that is, the portion of time an animal is not resting that is allocated to foraging. Between these extremes lie measurements of the proportion of a foraging period devoted to actual ingestion. Regardless of the index, we should find that animals experiencing greater energy demands feed more intensively. Predictable differences also should exist between digestive systems and feeding habits.

For example, most HGF should be more active than ruminants. Among both HGF and ruminants, the proportion of the active period foraging should decline from bulk to concentrate feeders, reflecting the decreasing effect of plant structure and increasing influence of digestibility on foraging profitability. Differences between ruminants and HGF should be greater among bulk feeders than among concentrate feeders. These expectations are met (Table 1). One effect is that time foraging per MW also declines from bulk to concentrate feeders. Similarly, lactating females should show higher rates of foraging intensity than nonlactating females on the same range. The crude index, foraging/active period, satisfies that expectation (Table 2); lactating reindeer also show higher foraging to active ratios than other herd members.[105] The herbivore may have considerable flexibility which obscures all but the dominant trends. For example, when forage is abundant, the animal may increase the foraging to active ratio, but decrease total time active.[86,105]

Studies using focal animal sampling can document the time within a foraging period spent eating. This index of foraging intensity also should increase with increasing energy demands. The few data available suggest that it does; lactating females forage more intensively than nonlactating females (Table 5). Young animals forage less intensively than adults, and adult males generally forage less intensively than adult females. The male ass, when undisturbed on its territory, is the exception to this latter trend (Table 5). Available data suggest that, despite their larger size, males not only forage less during daylight than do females, but also feed less intensively while foraging. Intraspecific comparisons suggest that males are more likely to forage at night than females, but those data also are sparse.

The two most direct measurements of foraging intensity (bite size and biting rate) are themselves covariate with the structure of the forage. When animals are grazing, the bite size increases and the biting rate decreases with increasing tiller length (Table 6, also[78,87,88,93,106]). Herbivores prehend different forage types differently,[86,111] but generally, when feeding on forbs or browse, biting rates again decrease and bite sizes increase as biomass or leaf surface area increases.[48,86,105,107,112] Thus, there is a broad inverse relationship between biting rate and bite size across forage types (Table 6). Actual intake is further modified by the percent of time foraging. Of these three variables influencing intake, the animal has most control over time foraging and biting rate; bite size is largely determined by range conditions. As energy demands increase, the herbivore often increases total time foraging (Table 2), foraging intensity (Table 5), and biting rate to produce a greater intake of forage.[78,79,113] If forage biomass or structure is such that bite size is much reduced, the variables over which the animal has some control can no longer produce a high rate of intake. There are two broad consequences of the control of forage structure on food intake. First, larger mixed feeders can often attain a higher rate of intake when feeding on browse or forbs than when grazing (Table 6, also References 86, 114, and 115). Second, forage structure and bite size may be the dominant controls on where an animal should feed.

II. WHERE TO DINE

Optimality functions evaluating where an animal should feed, or patch selection, are usually some modification of Charnov's[116] "marginal value theorem". Briefly, the "theorem" states that a forager should leave a particular patch when its net gain from staying drops to the expected gain from traveling to and starting to search in the next patch. The

Table 5
MEASUREMENTS OF FORAGING INTENSITY[a] AMONG DIFFERENT SEX-AGE CLASSES OF SELECTED HERBIVORES DURING DAYLIGHT

	Sex/age class					
	Adult males	Lactating females	Nonlactating females	Yearling females	Young	Ref.
Rangifer tarandus	0.60	0.80	0.68	0.68	—	49
Ovis dalli dalli	0.85	0.85	0.83	0.74	0.28	32
Equus asinus	0.97[b]	0.98	0.81	0.70[d]	0.57	30
	0.57[c]					
Antilope cervicapra	0.90	0.94	0.94	—	—	42

[a] Eating/(eating + searching + alert).
[b] Territorial.
[c] Solitary.
[d] Yearling male.

model addresses when to choose one location over another, and has been applied effectively to relatively simple systems such as great tits foraging in an aviary,[117] hummingbirds foraging,[118] and copulation times for dung beetles.[119] Major difficulties with the model are its use of expected rates of food acquisition and static motivational state in the animal. Gillingham[120] has documented that black-tailed deer develop an expected rate of gain from past foraging experience that produces departures from optimality when average patch quality suddenly changes. Similarly, he[120] demonstrated that hunger modified the animals' foraging tactics. His data were from controlled conditions; in the wild, risk to predation could also modify the animals' motivational state. Additional difficulties with simple optimality functions are that herbivores are frequently catholic feeders and have a variety of means to modify forage intake (e.g., Table 6). Nevertheless, the dominating controls of digestibility on rates of passage and forage biomass or of structure on intake rates suggest that broad patterns of patch selection should be evident. Specifically, we expect herbivores to choose patches or habitats where foraging profitability is greatest (higher digestibility and biomass, larger bite sizes) and to switch when another patch is at least equally profitable.

Herbivore studies are more often concerned with relative use and availability than with comparable states when the animal switches patches. Direct evidence is therefore sparse, but indirect evidence for broad habitat or patch selection on the basis of forage digestibility and biomass exists for several species: horses,[27] African buffalo,[41] red deer,[38] wapiti,[48] reindeer,[121] and caribou.[49,122] Kuropat and Bryant[123] suggested that female caribou maximized their fitness by effective patch selection, and there is wide-ranging evidence that, at the broadest level, herbivore biomass is related to forage biomass.[121,124,125]

Despite our difficulties in defining patches in a manner meaningful to the herbivore, available data suggest that simple rules may explain an animal's decision to leave one patch for another. At a broad scale, Harestad[126] documented that the decision of black-tailed deer to change seasonal home ranges was related to the food energy available to the animal. Seip[34] found that Stone's sheep maintained a constant biting rate (and by implication a constant bite size), by following phenological stages up a mountain so that they were continuously foraging on tiller lengths of about 12 cm. Data of Collins et al.[48] dealt primarily with relative use of habitat rather than switching, but still reveal elements influencing the animal's decision. Using Spearman's rank correlation (r_s) we find that relative use of habitat correlates poorly with biting rate or bite size, in part because these are themselves inversely correlated ($r_s = -0.90$). Relative use and intake rate (g/min) were highly correlated ($r_s = 1.0$). The analysis reveals both the compensatory mechanisms within a herbivore's foraging

Table 6
MEAN BITE SIZES, EATING RATES, PERCENT OF TIME FORAGING, AND RATES OF INTAKE PER METABOLIC WEIGHT OF SELECTED HERBIVORES

Species	Forage	Bite size (g)	Bites/min	Intake (g/min)	Time foraging(%)	Intake ($g/kg^{0.75}/hr$)	Ref.
Elephas maximus	Grass	160	1.3	214	76.1	19.4	28
	Browse	160	1.6	259	76.1	23.6	
	Grass and browse	160	1.2	192	76.1	17.5	
	Grass	140	1.6	230	76.1	22.9	
Bos taurus	Ryegrass	0.60	62.3	37.4	29.2	6.4	
		1.15	34.2	39.3	32.9	7.6	
		1.57	19.7	30.9	29.6	5.3	
		0.14	37.8	5.3	31.7	3.4	
		0.24	24.5	5.0	39.2	4.0	
		0.31	21.0	6.5	36.7	3.8	
Cervus elaphus	Wet meadow	0.45	30.0	13.5	49.5	13.7	48,106
	Dry meadow	0.30	34.0	10.6	49.5	10.8	
	Clearcut	0.40	29.0	9.5	49.5	9.7	
	Mature forest	0.46	10.0	6.2	49.5	6.3	
	Stagnant forest	0.50	8.0	3.7	49.5	3.8	
	Grass	0.75	18.0	13.5	50.0[a]	9.0	
	Browse	0.92	15.0	13.8	50.0[a]	9.2	
Alces alces	Deciduous leaves	—	—	16.7	48.0	5.8	107
	Forbs	—	—	8.3	48.0	2.9	
Ovis aries	Ryegrass (5 cm)	0.035	73.0	2.6	42.0	3.8	95
	Ryegrass (37 cm)	0.420	18.0	7.6	28.0	7.6	
Odocoileus hemionus hemionus	Grass	0.048	45.5	2.18	55.0[a]	5.0	106
	Grass	0.077	27.4	2.11	55.0[a]	4.9	
	Browse	0.268	18.0	13.5	40.0[a]	8.9	
Aepyceros melampus	Browse	0.275	16.0	4.5	45.0[a]	6.5	108
Rangifer tarandus	Vascular plants	0.021	177.0	3.7	53.0	4.9	86
	Lichens	0.031	205.0	6.6	53.0	8.6	
Odocoileus virginianus	Winter browse	0.008	—	0.4	50.0[a]	0.7	109,110
	Forbs	0.034	—	1.2	69.0[a]	3.1	

[a] From other studies on the same or closely related species.

tactics and a simple rule for patch selection. It also suggests that incorporation of specific components of foraging behavior into current theory would assist wildlife managers.

III. MENUS — WHAT TO EAT

What to eat can be considered at two levels of detail: (1) the gross level of feeding habits (bulk, intermediate, or concentrate), and (2) the finer level of specific plant taxa and plant parts. Optimal foraging theory usually addresses questions at the second level.[15,127]

A. Feeding Habits

The actual process of food fermentation appears identical in perrisodactyls and ruminant artiodactyls.[128] Controls on food intake by ruminants and HGF differ primarily in the rates of passage. Intake in ruminants is largely dependent on fiber content and rumen fill which determine rates of passage.[19] In HGF rates of intake and passage may decrease with increasing dietary fiber, but proportionately much less so than in ruminants.[129-131] Because of the greater ability of HGF to increase rates of intake and passage as forage quality declines, differences between foraging tactics of bulk and concentrate feeders should be similar but less marked in HGF than in ruminants.

Relationships between feeding habits and body size of herbivores are usually expressed as broad expectations rather than specific predictions and most have been expressed previously.[132] Briefly, log rumenoreticular volume increases approximately linearly with log metabolic weight[103] or with a slope of 77% of metabolic weight (Equation 6). Nonruminants exhibit a similar linear log-log relationship between gastrointestinal weight and body weight.[103] As a result, in both HGF and ruminants gut capacity per unit metabolic weight is less in smaller species (Figure 3B); larger herbivores have a relatively larger digestive storage capacity than do smaller ones and can afford a longer passage time. Structural constraints thus leave the small herbivore with three broad options: (1) increase total intake per unit time, (2) increase fermentation rate, or (3) select food of a lower fiber to protein ratio (be more selective). The last option is the easiest to attain, particularly among ruminants which have less control over rates of passage.

Foraging time per unit weight is actually higher among smaller species of both HGF and ruminants (Equations 1 and 2), indicating that the first option is not attained. There is a tendency for smaller herbivores to exhibit more rapid rates of fermentation. Among ruminants, Hoppe[133] reported that rates of production of volatile fatty acids in large grazers (120 to 200 kg) were scattered around 200 μmol/g DM · hr; intermediate and concentrate feeders exhibited rates of 300 to 500μmol and 400 to 700 μmol, respectively. Higher rates of fermentation suggest a faster rate of passage. From data collated by Parra[103] and Warner,[134] we calculated mean residency times in the gastrointestinal tract: HGF — 31.6 hr; bulk feeding ruminants — 67.8 hr; concentrate feeding ruminants — 35.2 hr. However, high rates of fermentation can be attained only through selection of rapidly fermenting plant parts. In an extremely dry region where plant quality was uniformly low, there was no difference in fermentation rates among species spanning a range in weights of 32 to 600 kg.[135] Apparently, the second option is attained by exercising the third (feed selectively).

These broad structural constraints suggest only that small herbivores must be selective, not that large species must be bulk feeders. However, highly digestible forage is generally more contagiously distributed, and the high absolute energy requirement of large species suggests that few will be concentrate feeders. This observation is reflected in the smaller proportions of active period spent foraging by concentrate feeders (Table 1); they must spend more time searching for clumps of highly digestible forage. The giraffe is an exception in terms of its size (Figure 3A), but not in the amount of movement during the day. In fact, giraffes exhibit uncommonly low foraging to active ratios: 0.61 for females, 0.47 for males during daylight. The expectations between body weight and food habits are broadly met (Figures 1B and 3B).

B. Selectivity

That portion of optimal foraging theory dealing specifically with particular food items has proven insightful when applied to monogastrics, including granivores, insectivores, carnivores, and nectivores.[2] Attempts have been made to apply it to ruminants,[127,136] but some workers are skeptical of such efforts.[3,7,136-138] We suggest that there are two broad reasons why optimizing formulations may be inappropriate to describe herbivore feeding. The first reason considers simple structural constraints across herbivore taxa, the second considers plant chemistry.

Small ruminants and likely HGF must be selective because of the structural constraints imposed by relative gut capacity (e.g., Figure 3B). However, related size-dependent structural constraints almost impose selectivity on smaller herbivores and certainly facilitate it. One structural constraint is mouth shape. A much narrower muzzle will facilitate a higher degree of selectivity. Bell[132] proposed a measure of incisor insertion angle and muzzle width to relate cranial morphology to diet selectivity. Using Bell's measurement technique for shape of face,[140] Figure 4A expresses the percent reduction from post-alveolar dental width (PADW) to lower incisor row width for 24 species (47 skulls). This measurement incorporates body size as PADW is highly correlated ($r = 0.93$) with basal length of the skull. Concentrate feeders show a higher amount of taper, regardless of size, than do intermediate and bulk feeders. The finding suggests that much of the potential choice in food habits is physically imposed on the animal.

Regulation of food intake differs among broad classes of feeding habits; not all are physically limited by cranial features or rumen volume. Rumen fill which reflects fiber to protein ratios and rate of passage shows almost complete separation between bulk and concentrate feeders as a function of metabolic weight (Figure 4B). Intermediate and bulk feeders overlap in the percent rumen fill by volume, particularly at lower metabolic weights.

The influence of plant chemistry can be summarized briefly: for a species to exhibit the consistent rankings among plants predicted by optimal foraging theory, the gradient of its fitness vs. plant chemistry must be at least as steep as the gradient between plants, and any chemical differences among plants must be consistent. Neither restriction appears to be true. Comparison of reproduction or growth rates of captive ungulates fed different rations[141-145] suggests that the relationship between such indexes of fitness and crude fiber, crude protein, or digestible energy are less steep than those exhibited among or within plants on a range.[146-149] Further, differences within plants are by no means consistent.[148,150,151] A sufficient rule for forage selection would be as follows: ingest the less common, highly digestible forages whenever encountered, regardless of their relative abundance; be indiscriminate with regard to the bulk of plant species in the range of moderate digestibility; and avoid the poorly digestible or toxic species. Such a simple rule would produce the catholic food habits commonly observed among ungulates.[146,152-154] It also invokes satisficing rather than optimizing criteria.

ACKNOWLEDGMENTS

The Natural Sciences and Engineering Research Council of Canada supported F. L. Bunnell's research on foraging theory in ruminants and provided scholarship support to M. P. Gillingham. Dr. D. M. Shackleton provided a helpful critique of the manuscript.

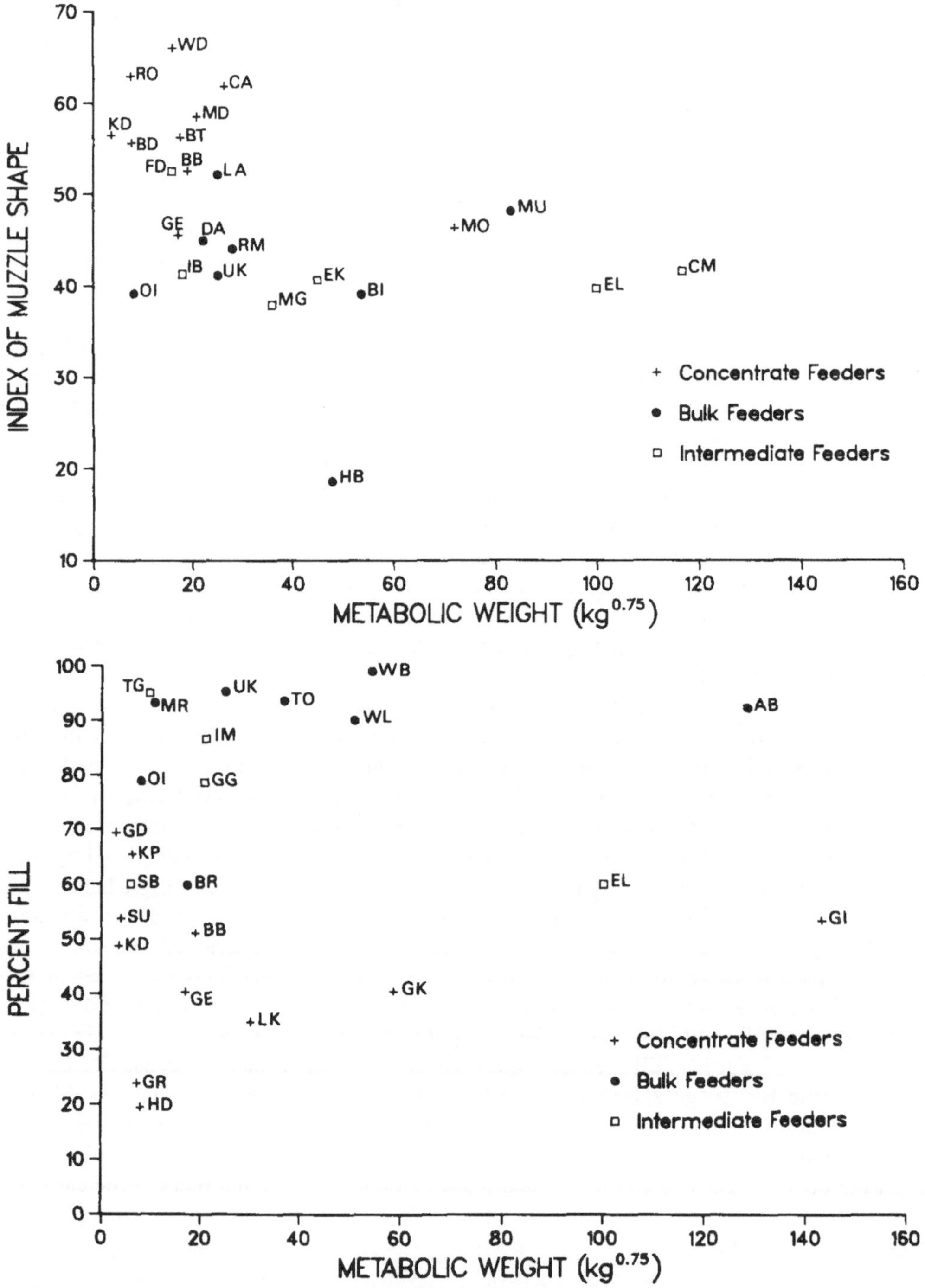

FIGURE 4. (Top) Index of shape of face (see text) as a function of metabolic weight. (Bottom) Percent rumen fill by volume as a function of metabolic weight. Based on data from Hoffman.[97] Legend: AB = African buffalo, BB = bushbuck, BD = blue duiker, BR = bohor reedbuck, BI = bison, BT = black-tailed deer, CA = caribou, CM = camel, DA = Dall's sheep, EK = elk, EL = eland, FD = fallow deer, GD = Gunther's dik-dik, GE = gerenuk, GG = Grant's gazelle, GI = giraffe, GK = greater kudu, GR = grey duiker, HB = hartebeest, HD = Harvey's duiker, IB = ibex, IM = impala, KD = Kirk's dik-dik, KP = klipspringer, LA = llama, LK = lesser kudu, MD = mule deer, MG = mountain goat, MO = moose, MR = mountain reedbuck, MU = muskoxen, OI = oribi, RO = roe deer, RM = Rocky Mountain bighorn sheep, SB = steenbok, SU = suni, TG = Thomson's gazelle, TO = topi, UK = Uganda kob, WB = waterbuck, WD = white-tailed deer, WL = wildebeest.

REFERENCES

1. **Krebs, J. R.,** Optimal foraging: decision rules for predators, in *Behavioural Ecology, an Evolutionary Approach,* Krebs, J. R. and Davies, N. B., Eds., Sinauer, Sunderland, Mass., 1978, chap. 2.
2. **Kamil, A. C. and Sargent, T. D., Eds.,** *Foraging Behavior,* Garland STPM Press, New York, 1981.
3. **Schulter, D.,** Does the theory of optimal diets apply in complex environments?, *Am. Nat.,* 118, 139, 1981.
4. **Jenkins, S. H.,** Management implications of optimal foraging theory: a critique, *J. Wildl. Manage.,* 46, 255, 1982.
5. **Zach, R. and Smith, J. N. M.,** Optimal foraging in wild birds?, in *Foraging Behavior,* Kamil, A. C. and Sargent, T. D., Eds., Garland STMP Press, New York, 1981, chap. 5.
6. **Lewontin, R.,** Fitness, survival, and optimality, in *Analysis of Ecological Systems,* Horne, D., Mitchell, R., and Stairs, G. B., Eds., Ohio State University Press, Columbus, 1978, chap. 1.
7. **Rapport, D. J.,** An optimization model of food selection, *Am. Nat.,* 105, 575, 1971.
8. **Rapport, D. J. and Turner, J. E.,** Feeding rates and population growth, *Ecology,* 56, 942, 1975.
9. **Rapport, D. J. and Turner, J. E.,** Economic models in ecology, *Science,* 195, 367, 1977.
10. **Winter, S. G.,** Satisficing, selection, and the innovating remnant, *Q. J. Econ.,* 85, 237, 1971.
11. **Radner, R.,** Satisficing, *J. Math. Econ.,* 2, 253, 1975.
12. **Slobodkin, L.,** The strategy of evolution, *Am. Sci.,* 52, 342, 1964.
13. **Arnold, G. W. and Dudzinski, M. L.,** *Ethology of Free-ranging Domestic Animals,* Elsevier, Amsterdam, 1978, chap. 1.
14. **Schoener, T. W.,** Optimal size and specialization in constant and fluctuating environments: an energy-time approach, *Brookhaven Symp. Biol.,* 22, 103, 1969.
15. **Schoener, T. W.,** Theory of feeding strategies, *Annu. Rev. Ecol. Syst.,* 2, 369, 1971.
16. **Schoener, T. W.,** The compression hypothesis and temporal resource partitioning, *Proc. Natl. Acad. Sci. U.S.A.,* 71, 4169, 1974.
17. **Norberg, R. A.,** An ecological theory on foraging time and energetics and choice of optimal food-searching method, *J. Anim. Ecol.,* 46, 511, 1977.
18. **Bergers, J. M.,** Time resources and laziness in animals, *Oecologia,* 49, 252, 1981.
19. **Baile, C. A. and Forbes, J. M.,** Control of feed intake and regulation of energy balance in ruminants, *Physiol. Rev.,* 54, 160, 1974.
20. **Robbins, C. T.,** *Wildlife Feeding and Nutrition,* Academic Press, New York, 1983, chap. 14.
21. **Clough, G. and Hassam, A. G.,** A quantitative study of the daily activity of the warthog in the Queen Elizabeth National Park, Uganda, *East Afr. Wildl. J.,* 8, 19, 1970.
22. **Joubert, E.,** Activity patterns shown by Hartmann zebra, *Equus zebra hartmannae* in South West Africa with reference to climatic factors, *Madoqua,* Ser. 1, 5, 33, 1972.
23. **Keiper, R. R.,** The behavior of EIA-positive horses, *Equine Pract.,* 3, 6, 1981.
24. **Keiper, R. R., Moss, M. B., and Zervanos, S. M.,** Daily and seasonal patterns of feral ponies on Assateague Island, *2nd Conf. on Sci. Res. Natl. Parks,* 8, 1980, 369.
25. **Klingel, H.,** Soziale Organisation und Verhalten freilebender Steppenzebras, *Z. Tierpsychol.,* 24, 580, 1967.
26. **Martin-Rosset, W., Doreau, M., and Cloix, J.,** Étude des activitiés d'un troupeau de poulinières de trait et de leurs poulains au paturage, *Ann. Zootech.,* 27, 33, 1978.
27. **Tyler, S. J.,** The behaviour and social organization of the New Forest Ponies, *Anim. Behav. Monogr.,* 5, 85, 1972.
28. **McKay, G. M.,** Behavior and ecology of the Asiatic elephant in southeastern Ceylon, *Smithsonian Contribution to Zoology,* No. 125, Smithsonian Inst. Press, Washington, D.C., 1973.
29. **Wyatt, J. R. and Eltringham, S. K.,** The daily activity of the elephant in the Rwenzori National Park, Uganda, *East Afr. Wildl. J.,* 12, 273, 1974.
30. **Moehlman, P. D. R.,** Behavior and Ecology of Feral Asses *(Equus asinus),* Ph.D. thesis, University of Wisconsin, Madison, 1974.
31. **Goddard, J.,** Home range, behaviour, and recruitment rates of two black rhinoceros populations, *East Afr. Wildl. J.,* 5, 133, 1967.
32. **Bunnell, F. L. and Farr, A. C. M.,** unpublished data for *Ovis dalli dalli,* 1977.
33. **Geist, V.,** *Mountain Sheep. A Study in Behavior and Evolution,* University of Chicago Press, Chicago, 1971.
34. **Seip, D. R.,** Foraging Ecology and Nutrition of Stone's Sheep, Ph.D. thesis, University of British Columbia, Vancouver, 1983.
35. **Horejsi, B. L.,** Behavioral differences in bighorn lambs (*Ovis canadensis canadensis* Shaw) during years of high and low survival, *North. Wild Sheep Counc. Proc.,* 1972, 51.
36. **Eccles, T. R.,** Aspects of Social Organization and Diurnal Acrtivity Patterns of Californian Bighorn Sheep *(Ovis canadensis californiana* Douglas, 1829), M.Sc. thesis, University of British Columbia, Vancouver, 1981.

37. **Eccles, T. R.,** Activity patterns of captive California bighorn sheep (*Ovis canadensis californiana*) at Penticton, B.C., *Proc. North. Wild Sheep and Goat Conf.*, 1978, 274.
38. **Clutton-Brock, T. H., Guinness, F. E., and Albon, S. D.,** *Red Deer. Behavior and Ecology of Two Sexes*, University of Chicago Press, Chicago, 1982.
39. **Struhsaker, T. T.,** Behavior of elk (*Cervus canadensis*) during the rut, *Z. Tierpsychol.*, 24, 80, 1967.
40. **Segal, A. N.,** The periodicity of pasture and physiological functions of reindeer, in *Reindeer in the Karelian ASSR*, Petrozavodsk: Academie Nauka, transl. by Dept. Secretary of State Bureau of Translation, Ottawa, Canada, 1962, 130.
41. **Sinclair, A. R. E.,** *The African Buffalo. A Study of Resource Limitation of Populations*, University of Chicago Press, Chicago, 1977.
42. **Schaller, G. B.,** *The Deer and the Tiger. A Study of Wildlife in India*, University of Chicago Press, Chicago, 1967.
43. **Berry, H. L., Siegfried, W. R., and Crowe, T. M.,** Activity patterns in a population of free-ranging wildebeest *Connochaetes taurinus* at Etosha National Park, *Z. Tierpsychol.*, 59, 229, 1982.
44. **Béchet, G., Fléchet, J., and Diez, E.,** Enregistrement des activités alimentaires et Merycigues des ovins au paturâge, *Ann. Zootech.*, 27, 107, 1978.
45. **Jingfors, K. T.,** Seasonal activity budgets of a reintroduced Alaskan muskoxen herd, *J. Wildl. Manage.*, 46, 344, 1982.
46. **Jarman, M. V. and Jarman, P. J.,** Daily activity of impala, *East Afr. Wildl. J.*, 11, 75, 1973.
47. **Craighead, J. J., Craighead, F. C., Jr., Ruff, R. L., and O'Gara, B. W.,** Home ranges and activity patterns of non-migratory elk of the Madison drainage herd as determined by radiotelemetry, *Wildl. Monogr.*, 33, 1, 1973.
48. **Collins, W. B., Urness, P. J., and Austin, D. D.,** Elk diets and activities on different lodgepole pine habitat segments, *J. Wildl. Manage.*, 42, 799, 1978.
49. **White, R. G., Thomson, B. R., Skogland, T. S., Persson, S. J., Russell, D. E., Holleman, D. F., and Luick, J. R.,** Ecology of Caribou at Prudhoe Bay, Alaska, in *Ecological Investigations of the Tundra Biome in the Prudhoe Bay Region, Alaska*, Brown, J., Ed., Biol. Pap. Univ. Alaska, Spec. Rep. 2, 151, 1975.
50. **Thomson, B. R.,** *Wild Reindeer Activity, Hardangervidda: 1971*, Report from the grazing project of the Norwegian IBP Committee, Statens Viltundersøkelser, Trondheim, 1973.
51. **Kastnes, K.,** Svalbardreinens activitetsmonster gjennom året (*Rangifer tarandus platyrhynchus*), Hovedfagsoppgave, Universitet i Oslo, (cited from Reimers, 1980), 1970.
52. **Roby, D. D.,** Behavioral Patterns of Barren-Ground Caribou of the Central Arctic Herd Adjacent to the Trans-Alaska Pipeline, M.Sc. thesis, University of Alaska, Fairbanks, 1978.
53. **Thomson, B. R.,** The Behaviour of Wild Reindeer in Norway, Ph.D. thesis, University of Edinburgh, Edinburgh, 1977.
54. **Walther, F. R.,** Round-the-clock activity of Thomsons gazelle (*Gazella thomsoni* Günther 1884) in the Serengeti National Park, *Z. Tierpsychol.*, 32, 75, 1973.
55. **Mungall, E. C.,** *The Indian Blackbuck Antelope: A Texas View*, Kleberg Studies in Natural Resources, 1978.
56. **Leuthold, B. and Leuthold, W.,** Daytime activity patterns of gerenuk and giraffe in Tsavo National Park, Kenya, *East Afr. Wildl. J.*, 16, 231, 1978.
57. **Bunnell, F. L.,** Energetics of ungulates — with particular reference to black-tailed deer, Proj. Rep. to B. C. Fish and Wildlife Branch, Victoria, British Columbia, 1978.
58. **Ellenberg, H.,** Beiträge zur Ökologie des Rhes (*Capreolus capreolus* L. 1758), Daten aus den Stammhamer Versuchsgehen, Ph.D. thesis, Christian Albrechts Univeristy, Kiel, 1974.
59. **Yahner, R. H.,** Time budgets in captive Reeve's muntjacs (*Muntiacus reevesi*), *Appl. Anim. Ethol.*, 6, 277, 1980.
60. **Boy, V. and Duncan, P.,** Time-budgets of Camarque horses. I. Developmental changes in the time budgets of foals, *Behaviour*, 71, 187, 1979.
61. **Guy, P. R.,** Diurnal activity patterns of elephant in the Sengwa Area, Rhodesia, *East Afr. Wildl. J.*, 14, 285, 1976.
62. **Joubert, E. and Eloff, F. C.,** Notes on the ecology and behaviour of the black rhinoceros *Diceros bicornis* Linn. 1758 in South West Africa, *Madoqua*, Ser. No. 3, 5, 53, 1971.
63. **Foster, B. R.,** Observability and Habitat Characteristics of the Mountain Goat (*Oreamnos americanus* Blainville, 1816) in West-central British Columbia, M.Sc. thesis, University of British Columbia, Vancouver, 1982.
64. **Hoefs, M.,** Ecological Investigation of Dall Sheep (*Ovis dalli dalli*, Nelson) and their Habitat on Sheep Mountain, Kluane National Park, Yukon Territory, Canada, Ph.D. thesis, University of British Columbia, Vancouver, 1975.

65. **Grubb, P. and Jewell, P. A.,** Movement, daily activity, and home range of Soay sheep, in *Island Survivors: the Ecology of the Soay Sheep of St. Kilda*, Jewell, P. A., Milner, C., and Morton Boyd, J., Eds., Athlone Press, London, 1974, 160.
66. **Spinage, C. A.,** A quantitative study of the daily activity of the Uganda Defassa waterbuck, *East. Afr. Wildl. J.*, 6, 89, 1968.
67. **Tomlinson, D. N. S.,** The feeding behaviour of waterbuck in the Lake McIlwaine Game Enclosure, *Rhod. Sci. News*, 13, 11, 1979.
68. **Waser, P.,** Diurnal and nocturnal strategies of the bushbuck *Tragelaphus scriptus* (Pallas), *East Afr. Wildl. J.*, 13, 49, 1975.
69. **Hanley, T. A.,** Cervid activity patterns in relation to foraging constraints, *Northwest Sci.*, 56, 208, 1982.
70. **Hubert, B.,** Estimated Productivity of Muskoxen (*Ovibos moschatus*) on Northwestern Devon Island, N.W.T., M.Sc. thesis, University of Manitoba, Winnipeg, 1974.
71. **Jungius, H.,** The biology and behaviour of the reedbuck (*Redunca arundinum* Boddaert 1785) in the Kruger National Park, *Mammalia Depicta*, Verlag Paul Parey, Hamburg, 1971.
72. **Bannikov, A. G., Zhirnov, L. V., Lebedeva, L. S., and Fandeev, A. A.,** *Biology of the Saiga*, Israel Program for Scientific Translations, 1967.
73. **Leuthold, W.,** Variations in territorial behaviour of Uganda kob *Adenota kob thomasi* (Neumann 1896), *Behaviour*, 27, 215, 1966.
74. **Hodgson, J. and Jamieson, W. S.,** Variations in herbage mass and digestibility, and the grazing behaviour and herbage intake of adult cattle and weaned calves, *Grass Forage Sci.*, 36, 39, 1981.
75. **Tracy, C. R.,** Newton's law: its application for expressing heat losses from homeotherms, *BioScience*, 22, 656, 1972.
76. **Tucker, V. A.,** The energetics of moving about, *Am. Sci.*, 220, 70, 1975.
77. **Bradley, R. M.,** Some Aspects of the Ecology of the Warthog (*Phacochoerus aethiopicus*, Pallas) in Nairobi National Park, M.Sc. thesis, University of East Africa, Nairobi, 1968.
78. **Arnold, G. W. and Dudzinski, M. L.,** Studies on the diet of the grazing animal. III. The effect of pasture species and pasture structure on the herbage intake of sheep, *Aust. J. Agric. Res.*, 18, 657, 1967.
79. **Eisenberg, J. F., McKay, G. M., and Jainudeen, M. R.,** Reproductive behaviour of the Asiatic elephant (*Elephas maximus maximus* L.), *Behaviour*, 38, 193, 1971.
80. **Cederlund, G.,** Daily and seasonal activity pattern of roe deer in a boreal habitat, *Viltrevy*, 11, 315, 1981.
81. **Milner, C. and Gwynne, D.,** The Soay sheep and their food supply, in *Island Survivors: the Ecology of the Soay Sheep of St. Kilda*, Jewell, P. A., Milner, C., and Morton Boyd, J., Eds., Athlone Press, London, 1974, chap. 11.
82. **Mitchell, A. W.,** Preliminary observations on the daytime activity patterns of lesser kudu in Tsavo National Park, Kenya, *East Afr. Wildl. J.*, 15, 199, 1977.
83. **Bueno, L. and Rucklebusch, Y.,** Ingestive behaviour in sheep under field conditions, *Appl. Anim. Ethol.*, 5, 179, 1979.
84. **Reimers, E.,** Activity pattern: the major determinant for growth and fattening in *Rangifer?*, in *Proc. 2nd Int. Reindeer/Caribou Symp.*, Reimers, E., Gaare, E., and Skjenneberg, S., Eds., Directoratet for vilt og ferskvannsfisk, Trondheim, Norway, 1980, 466.
85. **White, R. G.,** Nutrient acquisition and utilization in arctic herbivores, in *Comparative Mechanisms of Cold Adaptations*, Folk, E. and Underwood, L., Eds., Academic Press, New York, 1979, chap. 2.
86. **Trudell, J. and White, R. G.,** The effect of forage structure and availability on food intake, biting rate, bite size and daily eating time of reindeer, *J. Appl. Ecol.*, 18, 63, 1981.
87. **Allden, W. G.,** Rate of herbage intake and grazing time in relation to herbage availability, in *Proc. Aust. Soc. Anim. Prod.*, IV, 163, 1962.
88. **Arnold, G. W. and Dudzinski, M. L.,** The effect of pasture density and structure on what the grazing animal eats and animal productivity, in *Intensive Utilization of Pastures*, James, B. J. F., Ed., Angus and Robertson, Sydney, 1969, 42.
89. **Young, B. A. and Corbett, J. L.,** Maintenance energy requirement of grazing sheep in relation to herbage availability. I. Calorimetric estimates, *Aust. J. Agric. Res.*, 23, 57, 1972.
90. **Lynch, J. J. and Hedges, D. A.,** The influence of paddock, type of fence and stocking rate on grazing behaviour and social facilitation in Merino sheep, *Appl. Anim. Ethol.*, 5, 321, 1979.
91. **Harker, K. W., Taylor, J. I., and Rollinson, D. H. L.,** Studies on the habits of zebu cattle. I. Preliminary observations on grazing habits, *J. Agric. Sci.* (Cambridge), 44, 193, 1954.
92. **Joblin, A. D. H.,** The influence of night grazing on the growth rates of zebu cattle in East Africa, *J. Br. Grassl. Soc.*, 15, 212, 1960.
93. **Chacon, E. and Stobbs, T. H.,** Influence of progressive defoliation of a grass sward on the eating behaviour of cattle, *Aust. J. Agric. Res.*, 27, 709, 1976.
94. **Holder, J. M.,** Supplementary feeding of grazing sheep — its effect on pasture intake, in *Proc. Aust. Soc. Anim. Prod.*, IV, 154, 1962.

95. **Allden, W. G. and Whittaker, I. A. McD.,** The determinants of herbage intake by grazing sheep. The interrelationships of factors influencing herbage intake and availability, *Aust. J. Agric. Res.*, 21, 755, 1970.
96. **Hodgson, J., Rodriguez Capriles, J. M., and Fenlon, J. S.,** The influence of sward characteristics on the herbage intake of grazing calves, *J. Agric. Sci.* (Cambridge), 89, 743, 1977.
97. **Hoffmann, R. R.,** *The Ruminant Stomach*, East African Monographs in Biology, Vol. 2, East African Literature Bureau, Nairobi, 1973, chaps. 4 to 6.
98. **Prins, J. and Geelen, M. J. H.,** Rumen characteristics of red deer, fallow deer, and roe deer, *J. Wildl. Manage.*, 35, 673, 1971.
99. **Short, H. L., Medin, D. E., and Anderson, A. E.,** Ruminoreticular characteristics of mule deer, *J. Mammal.*, 46, 196, 1963.
100. **O'Gara, B. W.,** personal communication, in **Hanley, T. A.,** Nutritional Constraints of Food and Habitat Selection by Sympatric Ungulates, Ph.D. thesis, University of Washington, Seattle, 23, 1980.
101. **Gill, J. and Jaczewski, Z.,** The capacity of various parts of the digestive tract of Red Deer (*Cervus elaphus* L.) (in German), *Z. Jagdwiss.*, 4, 168, 1958.
102. **Short, H. L.,** Rumen fermentations and energy relationships in white-tailed deer, *J. Wildl. Manage.*, 27, 184, 1963.
103. **Parra, R.,** Comparison of foregut and hindgut fermentation in herbivores, in *The Ecology of Arboreal Folivores*, Montgomery, G. G., Ed., Smithsonian Inst. Press, Washington, D.C., 208, 1978.
104. **Benedict, F. G.,** *The Physiology of the Elephant*, Carnegie Inst. Wash., Washington, D.C., 1936.
105. **Skogland, T.,** Wild reindeer range use and selectivity in southern Norway, in *Proc. 1st Int. Reindeer and Caribou Symposium*, Luick J. R., Lent, P. C., Klein, D. R., and White, R. G., Eds., Biol. Pap. University of Alaska, Spec. Rep. No. 1, 1975, 342.
106. **Wickstrom, M. L.,** Food Intake and Foraging Energetics of Elk (*Cervus elaphus nelsoni*) and Mule Deer (*Odocoileus hemionus hemionus*), M.Sc. thesis, Washington State University, Pullman, 1983.
107. **Belovsky, G. E. and Jordan, P. A.,** The time-energy budget of a moose, *Theor. Pop. Biol.*, 14, 76, 1978.
108. **Dunham, K. M.,** The feeding behavior of a tame impala *(Aepyceros melampus), Afr. J. Ecol.*, 18, 253, 1980.
109. **Crawford, H. S. and Whelan, J. B.,** Estimating food intake by observing mastications by tractable deer, *J. Range Manage.*, 26, 372, 1973.
110. **Crawford, H. S.,** Seasonal food selection and digestibility by tame white-tailed deer in central Maine, *J. Wildl. Manage.*, 46, 974, 1982.
111. **Willms, W.,** Foraging strategy of ruminants, *Rangeman's J.*, 5, 72, 1978.
112. **Owen-Smith, N. and Novellie, P.,** What should a clever ungulate eat?, *Am. Nat.*, 119, 151, 1982.
113. **Arnold, G. W. and Birrell, H. A.,** Food intake and grazing behaviour of sheep varying in body condition, *Anim. Prod.*, 24, 243, 1977.
114. **Hendricksen, R. and Minson, D. J.,** The feed intake and grazing behavior of cattle grazing a crop of lablab — *Purpereus cultivar* Rongai, *J. Agric. Sci.*, 95, 547, 1980.
115. **Dunham, K. M.,** The foraging behavior of impala *Aepyceros melampus, S. Afr. J. Wildl. Res.*, 12, 36, 1982.
116. **Charnov, E. L.,** Optimal foraging: the marginal value theorem, *Theor. Pop. Biol.*, 9, 129, 1976.
117. **Cowie, R. J.,** Optimal foraging in great tits *(Parus major), Nature (London)*, 268, 137, 1977.
118. **Pyke, G. H.,** Optimal foraging in hummingbirds: testing the marginal value theorem, *Am. Zool.*, 18, 739, 1978.
119. **Parker, G. A.,** Searching for mates, in *Behavioural Ecology: an Evolutionary Approach*, Krebs, J. R. and Davies, N. B., Eds., Blackwell Scientific, Oxford, 1978, chap. 8.
120. **Gillingham, M. P.,** unpublished data on *Odocoileus hemionus columbianus*, 1984.
121. **Skogland, T.,** Comparative summer feeding strategies of arctic and alpine *Rangifer, J. Anim. Ecol.*, 49, 81, 1980.
122. **White, R. G. and Trudell, J.,** Patterns of herbivory and nutrient intake of reindeer grazing tundra vegetation, in *Proc. 2nd Int. Reindeer/Caribou Symp.*, Reimers, E., Gaare, E., and Skjenneberg, S., Eds., Direktoratet for vilt og ferskvannsfisk, Trondheim, Norway, 1980, 180.
123. **Kuropat, P. and Bryant, J. P.,** Foraging behavior of cow caribou on the Utukok calving grounds in northwestern Alaska, in *Proc. 2nd Int. Reindeer/Caribou Symp.*, Reimers, E., Gaare, E., and Skjenneberg, S., Eds., Direktoratet for vilt og ferskvannsfisk, Trondheim, Norway, 1980, 64.
124. **Coe, N. J., Cumming, D. H., and Phillipson, J.,** Biomass and production of large African herbivores in relation to rainfall and primary production, *Oecologia*, 22, 341, 1976.
125. **Bobek, B.,** Summer food as a factor limiting roe deer population size, *Nature (London)*, 268, 47, 1977.
126. **Harestad, A. S.,** Seasonal Movements of Black-tailed Deer on Northern Vancouver Island, Ph.D. thesis, University of British Columbia, Vancouver, 1979.
127. **Belovsky, G. E.,** Diet optimization in a generalist herbivore: the moose, *Theor. Pop. Biol.*, 14, 105, 1978.

128. **Janis, C.,** The evolutionary strategy of the *Equidae* and the origins of rumen and cecal digestion, *Evolution*, 30, 757, 1976.
129. **Heinlein, E. F. W., Holdren, R. D., and Yoon, R. M.,** Comparative responses of horses and sheep to different physical forms of alfalfa hay, *J. Anim. Sci.*, 25, 740, 1966.
130. **Darlington, J. M. and Hershberger, T. V.,** Effect of forage maturity, digestibility, intake and nutritive value of alfalfa, timothy and orchard grass by equine, *J. Anim. Sci.*, 27, 1572, 1968.
131. **Vandernoot, G. W. and Gilbreath, E. C.,** Comparative digestibility of components of forage by geldings and steers, *J. Anim. Sci.*, 31, 351, 1970.
132. **Bell, R. H. V.,** The Use of the Herb Layer by Grazing Ungulates in the Serengeti National Park, Tanzania, Ph.D. thesis, University of Manchester, 1969.
133. **Hoppe, P. P.,** Rumen fermentation and body weights in African ruminants, *Proc. XIII Int. Congr. Game Biologists*, 1977, 141.
134. **Warner, A. C. I.,** Rate of passage of digesta through the gut of mammals and birds, *Nutr. Abstr. Rev. Ser. B*, 51, 789, 1981.
135. **Giesecke, D. and Gylswyk, N. O.,** A study of feeding types and certain rumen functions in six species of South African wild ruminants, *J. Agric. Sci.* (Cambridge), 85, 75, 1975.
136. **Belovsky, G. E.,** Food selection by a generalist herbivore: the moose, *Ecology*, 62, 1020, 1981.
137. **Pulliam, H. R.,** Diet optimization with nutrient constraints, *Am. Nat.*, 109, 765, 1975.
138. **Westoby, M.,** An analysis of diet selection by large generalist herbivores, *Am. Nat.*, 108, 290, 1974.
139. **Westoby, M.,** What are the biological bases of varied diets?, *Am. Nat.*, 112, 627, 1978.
140. **Bell, R. H. V.,** personal communication, 1983.
141. **Murphy, D. A. and Coates, J. A.,** Effects of dietary protein on deer, *Trans. N. Am. Wildl. Nat. Resour. Conf.*, 31, 129, 1966.
142. **Ullrey, D. E., Youatt, W. G., Johnson, H.E., Fay, L. D., and Bradley, D. L.,** Protein requirements of white-tailed deer fawns, *J. Wildl. Manage.*, 31, 679, 1967.
143. **Thorne, E. T., Dean. R. E., and Hapworth, W. G.,** Nutrition during gestation in relation to successful reproduction in elk, *J. Wildl. Manage.*, 40, 330, 1976.
144. **Bahnak, B. R., Johnson, D. E., Carpenter, L. H., Wallmo, O. C., and Gill, R. B.,** Seasonal and nutritional effects on serum nitrogen constituents in white-tailed deer, *J. Wildl. Manage.*, 43, 454, 1979.
145. **Verme, L. J. and Ozoga, J. J.,** Influence of protein-energy intake on deer fawns in autumn, *J. Wildl. Manage.*, 44, 305, 1980.
146. **Mautz, W. W.,** Nutrition and carrying capacity, in *Big Game of North America*, Schmidt, J. L. and Gilbert, D. L., Eds., Stackpole Books, Harrisburg, Pa., 1978, 321.
147. **Milchunas, D. G., Dyer, M. I., Wallmo, O. C., and Johnson, D. E.,** *In vivo/in vitro* relationships of Colorado mule deer forages, Spec. Rep., No. 3, Colorado Division of Wildlife, 1978.
148. **Rochelle, J. A.,** Mature Forests, Litterfall and Patterns of Forage Quality as Factors in the Nutrition of Black-tailed Deer on Northern Vancouver Island, Ph.D. thesis, University of British Columbia, Vancouver, 1980.
149. **Seip, D. R.,** Energy Intake in Relation to Puberty Attainment in Female Black-tailed Deer Fawns, M. Sc. thesis, Simon Fraser University, Vancouver, British Columbia, 1980.
150. **Wolters, G. L.,** Southern pine overstories influence herbage quality, *J. Range Manage.*, 26, 423, 1973.
151. **Blair, R. M., Alcaniz, R., and Harrell, A.,** Shade intensity influences the nutrient quality and digestibility of southern deer browse leaves, *J. Range Manage.*, 36, 257, 1983.
152. **Welsh, D.,** Population, Behaviour, and Grazing Ecology of the Horses of Sable Island, Nova Scotia, Ph.D. thesis, Dalhousie University, Nova Scotia, 1975.
153. **Boyd, R. J.,** American elk, in *Big Game of North America*, Schmidt, D. L. and Gilbert, D. L., Eds., Stackpole Books, Harrisburg, Pa., 1978, 67.
154. **Peek, J. M.,** A review of moose food habits studies in North America, *Nat. Can.*, 101, 195, 1974.

Chapter 4

REGULATION OF FORAGE INTAKE

G. W. Arnold

TABLE OF CONTENTS

Food intake in herbivores is determined by a range of factors that relate to the animal's genotype, its physiological state, diet, and environment. Some of the factors involved in the control of intake of ruminants are indicated in Figure 1. The same sets of factors will be involved in nonruminant herbivores; however, those of the gastrointestinal tract will vary with the type of digestive system. In some wild herbivores, seasonally varying set points may be the dominant factor. The vast bulk of information about intake regulation is for domestic herbivores.[1-4] However, there is little reason to suppose that the mechanisms are different in wild herbivores.

I. MECHANISMS OF INTAKE CONTROL

This section is based primarily on the review by Freer[1] who claims that the ultimate regulation of food intake comes from the brain, but it is not yet certain whether the hypothalamus plays the dominant role in balancing and integrating signals from various parts of the body.[5] It has been suggested that the lateral hypothalamus controls the pattern of eating behavior by responding to sensory characteristics of the food and by monitoring the onset and decay of satiety after eating. In this model, the activity of the lateral areas is regulated by feed-back from the ventro-medial hypothalamus. The function is to meter signals that indicate the level of nutrient stores in the body, and to adjust the satiety controls in the lateral areas so as to maintain a set point for the stores and, thereby, for the long term energy balance of the animal. However, Morrison[5] considers that there are many situations where intake control is extra-hypothalamic and it is likely that these centers represent just one junction in a hierarchy of control systems stretching from peripheral control in the gut to higher centers in the brain (Figure 2).[6]

The nature of the signals to the control centers is not well understood. Those indicating short term satiety include nervous impulses from stretch receptors in the lining of the gut, changes in body temperature, intestinal or circulating hormones, indicators of fatigue, and changes during a meal in the plasma concentration of metabolites arising from digestion and metabolism. Long term stability in energy balance is thought to be controlled by the size of the fat reserves but no known pathway exists for transmitting this information to the hypothalamus. The settings for satiety control or the capacity of the alimentary tract need to be adjusted to allow increased intake during lactation, for example. Here, endocrine factors are involved.

There are problems with applying this theory to wild herbivores. Many species in temperate and arctic areas appear not to have stability in energy balance even in a constant nutritional environment. This may be because in the real world the nutritional environment is never constant for long enough to find out whether energy balance is maintained at a certain level of body fatness, or whether photoperiodic effects on intake intervene.

II. ANIMAL FACTORS

There have been many reviews of animal factors affecting intake regulation in herbivores, but the most comprehensive have been by Weston,[2,3] and these are drawn on extensively here.

The effects of genotype and physiological state on voluntary feed consumption are indicated via the animal's metabolism and generally, but not always, are independent of diet quality.

A. Genotype

It has been widely assumed that the interspecific metabolic base $W^{0.75}$ should be used to make interspecific comparisons of feed intake. However, comparisons between adult cattle

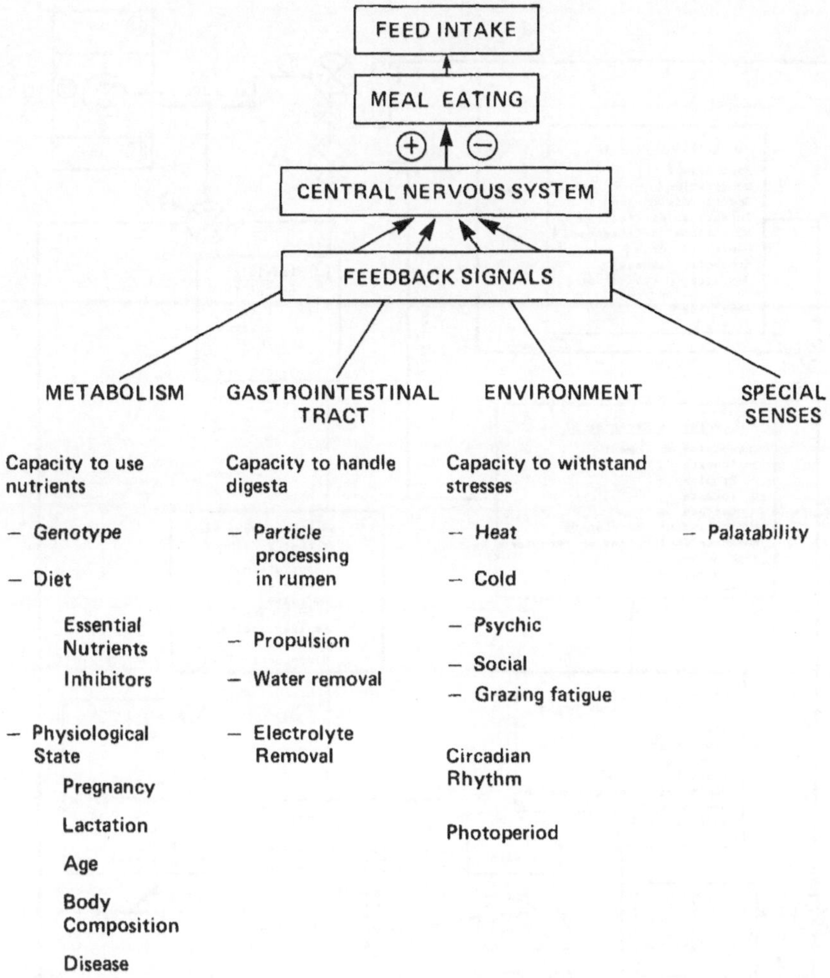

FIGURE 1. Factors involved in feed intake by free-ranging herbivores. (From Weston, R. H., Proc. 1981 Sheep and Wool Refresher School, Queensland Dept. Primary Industries and the Australian Wool Corporation, Queensland, Australia, 1981.)

and sheep are better made using $W^{0.9}$ (References 8 and 9), and probably $W^{1.0}$ is appropriate with comparisons of adults.[2] Weston[2] states that, under optimal conditions, feed intake should be determined by the animal's genetic potential to use energy, and that there appears to be no convincing evidence of a general difference in voluntary consumption per unit body size among sheep, goats, cattle, and buffalo. There is, however, evidence that sheep differ from the larger macropods. Hume[10] found that on three feeds of different quality (lucerne, oaten chaff, and wheat straw), sheep ate significantly more digestible energy than either euros or red kangaroos. However, two smaller macropods, the pademelon and red-necked wallaby, ate the same amount of digestible energy as sheep when offered lucerne hay.[11] Other species comparisons are given in Table 1. The problem in comparing species is that they can differ in the degree to which they digest fiber on high fiber or on low fiber diets. Thus, true comparisons are possible only over a range of feed quality. These are described later.

B. Physiological State

1. Age

The food intake of 8-month-old sheep per unit $W^{1.0}$ was found by Langlands[14] and

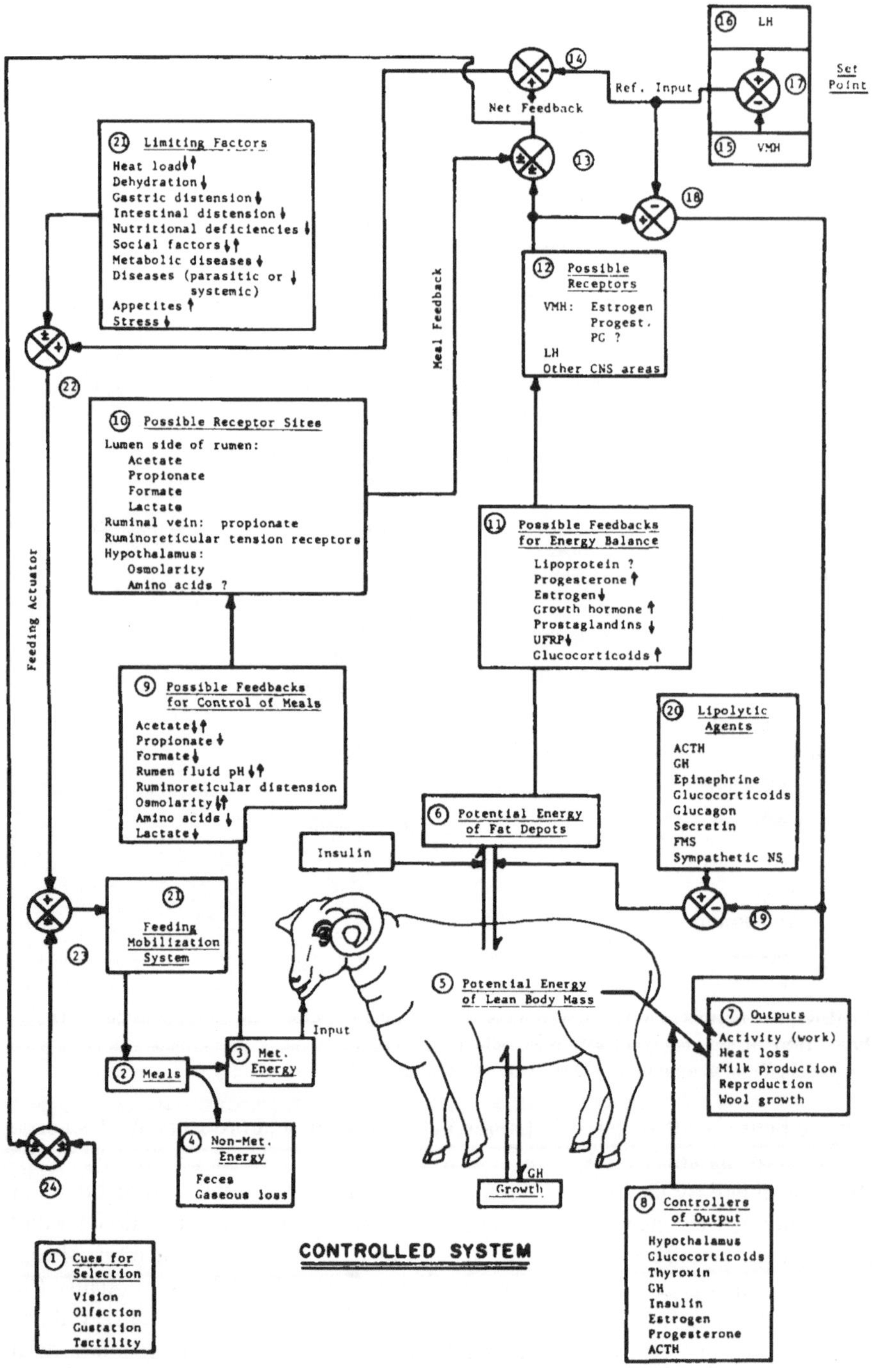

FIGURE 2. Scheme illustrating some interrelationships of a regulatory system of energy balance with special emphasis on control of feed intake. Factors for hunger-satiety system probably differ greatly between species and within species under various environmental, physiological, and nutritional conditions. Arrows in boxes 9, 11, and 21 indicate each factor's probable effect on feeding behavior. (From Baile, C. A. and Forbes, C. A., *Physiol. Rev.*, 54, 160, 1974. With permission.)

Table 1
COMPARISONS OF VOLUNTARY INTAKE AND DIGESTIBILITY BY VARIOUS WILD HERBIVORES

Animal species	Body weight (kg)	Organic matter intake (kg/$BW^{0.75}$)	Organic matter digestibility (%)	Ref.
Sheep	36.0	91.6[a]	60.9[a]	10
Euro	27.0	52.7[b]	61.5[a]	10
Red kangaroo	30.0	53.4[b]	54.8[b]	10
Sheep	37.8	62.0[a]	61.2[a]	11
Red-necked wallaby	11.3	54.6[b]	55.0[a]	11
Red-necked pademelon	5.5	69.2[a]	61.6[a]	11
Sheep	48.2	54.9[a]	59.0[a]	12
Red kangaroo	20.8	51.5[a,b]	55.9[b]	12
Red-necked pademelon	5.0	47.9[b]	55.8[b]	12
Tammar wallaby	4.8	26.8[c]	60.9[c]	12
Sheep	51.4	88.2[a]	58.5[a]	13
Red deer	74.5	82.3[b]	57.3[a]	13

Note: In each study, values within columns that have no superscript in common are statistically significantly different from one another (e.g., in the study from Reference 10, organic matter intake for sheep differs from that of euro and red kangaroo).

Langlands and Hamilton[15] to be 22% higher than at 20 months, and 27% higher than at 32 months; higher intakes in younger animals are accompanied by larger quantities of digesta in the rumen. Although the decrease in voluntary consumption with advancing age is consistent with changes in fasting metabolism, a strict proportionality may not apply.[2]

2. *Body Composition*

Although evidence with sheep and cattle is not unequivocal, the majority of studies shows that with increasing fat content in the body, feed intake decreases, and this has been shown to be associated with a reduction in the quantity of digesta.[16-18] The mechanism involved is not elucidated. Increasing abdominal fat may physically reduce rumen capacity and hence food intake. However, metabolic changes may accompany increase in body fat content.[19] As adipocytes increase in size, they may have a reduced capacity to synthesize triglycerides, and fatty acids released during normal turnover of triglycerides may escape more readily from the cell. These fatty acids could influence metabolic receptors associated with feeding, i.e., act as a lipostatic control mechanism.

3. *Pregnancy and Lactation*

Feed intake in monogastric animals generally increases during pregnancy to match the increased nutrient demand of the conceptus. However, with cattle and sheep, a decline is sometimes seen in late pregnancy.[19,20] It is not known why this occurs. Although rumen volume is reduced because of pressure from the amniotic sac, food intake can be sustained.[2] Metabolic changes, perhaps the increase in estradiol, may reduce the capacity to use energy. In wapiti (elk), feed intake was found to be relatively constant during the last half of pregnancy.[21]

Both ruminant and monogastric animals achieve their highest voluntary food intake during lactation, which for dairy cows may be 70% above that of the nonpregnant animal. Sadlier[22] found that the food intake of black-tailed deer does rearing twins reached a peak between 20 and 40 days of lactation and was more than double that prior to parturition (Figure 3).

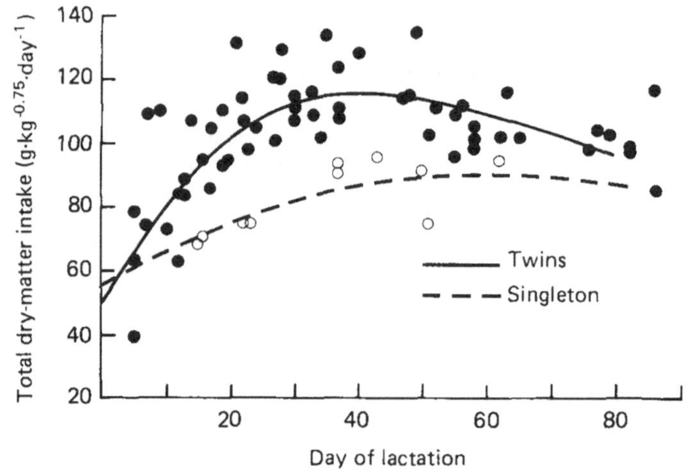

FIGURE 3. Changes in total dry matter intake with stage of lactation. Each point represents a 24-hr determination based on a 3-day average. (From Sadlier, R. M. F. S., *Can. J. Zool.*, 60, 382, 1982. With permission.)

Does rearing singletons had much smaller increases in intake. Similar increases occur in elk and red deer.[21,23] More limited data for white-tailed deer were obtained by Short et al.,[24] which showed an increase of 30% in the first 14 days of lactation in a doe rearing twins.

Increased intake in lactation is associated with increased rumen digesta and rumen tissue weight, and enhanced rumen propulsive activity.[12] The increases in food intake in pregnancy and lactation are achieved by the animal increasing grazing time in lactation.[25,26]

4. Hormonal Status

Many species of wild herbivores are responsive to the gradual variation in the duration of the photoperiod and scotoperiod, the direction of the effect being photostimulatory or photoinhibitory to a particular system. The effect of photostimulation on the reproductive system is well known. Perhaps less well known is the effect on metabolic activity in general. Both effects are mediated through hormonal status. Recent studies have shown that prolactin stimulates food intake in young male reindeer.[27] Detailed studies have been made in several species of deer. Nichol[28] was the first to show that black-tailed deer reduce their voluntary food intake to maintenance levels in mid-autumn and increase it again with the approach of spring. Subsequently, other workers[23,29] have detailed the yearly cycles in food intake in this species and in white-tailed deer and red deer. Adult males show larger fluctuations than adult females (Figure 4), and this can involve more than a twofold difference in food intake between winter and spring. Pollock[35] found that, in red deer, appetite reached its peak, about twice that of maintenance, as daylength approached its longest and remained high through the declining phase. It declined suddenly as the stags came into sexual activity. Appetite recovered gradually after the rut. This general pattern, which is much more marked than in domestic sheep and cattle, holds for other species of deer.

There is considerable variation in the time of peak and minimum levels, which may reflect the influence of two different factors; time of rut, and the rate at which animals gain weight during spring. On poor quality feed, deer reach maximum weight later in the year than on good quality food. Since food intake declines as animals get fatter, it is likely that some early declines in food intake seen in Figure 4 are due to either body composition (fatness) or body-weight "targets" being reached early in the year. Usually, food intake declines before maximum body weight is achieved, but this was not the case in a study made by

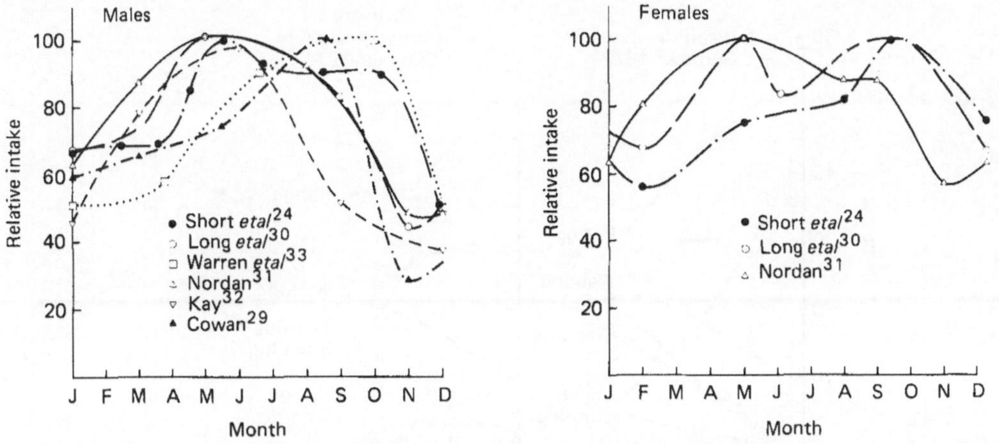

FIGURE 4. Seasonal changes in voluntary food intake of deer.

Warren et al.,[33] in which the increase in intake was delayed usually until late in the spring, perhaps because the study animals were slow to adjust to the feed regimes.

C. Environmental Influences

Besides the influence of daylength on hormonal status and hence food intake, temperature, radiation, and humidity have physiological effects on herbivores that in turn can influence food intake. In many parts of the world, insects cause considerable disruption to feeding of herbivores, and so influence food intake. However, behavioral adaptation may partially compensate depressive effects.

Food intake is reduced when herbivores consume dry forage and water is restricted. The distribution and availability of water is thus important and will be discussed later. In species that have evolved in environments that are normally short of free water, physiological and behavioral adaptations reduce or eliminate the need for water.

1. Thermal Environment

Solar radiation contributes to heat input but depends on latitude, cloud cover, available shade, and animal's pelage. Free-ranging cattle increasingly use shade as temperatures rise, and stand rather than lie when resting,[36] and this behavioral adaptation applies to other herbivores with similar pelages. Nevertheless, grazing time may decrease linearly with temperatures above 21°C.[37] Voluntary food intake in the wombat decreases linearly when temperatures exceed 10°C.[38]

Metabolic heat input depends mainly on the diet of the animal, increasing with the digestibility of the diet so that voluntary food intake is liable to be affected at lower levels of radiation when herbivores are grazing on green rather than on dry forage.

Ambient temperature markedly affects heat dissipation, but because many factors affect thermal balance, the onset of change in voluntary consumption of feed does not occur within narrow ranges of temperatures. For example, in *Bos taurus* cattle, the onset of decline in consumption ranges between 27°C and >35°C.[2] Likewise, the increase in food intake with low temperatures is not clearly related to temperature. The increase in intake by cattle can be 40% when temperatures decrease from 10 to − 15°C.[39,40]

Humidity, by affecting rate of heat loss, can modify food intake. Humidity has little effect on cattle when ambient temperatures are less than 27°C. Above 27°C consumption decreases as humidity increases above 40%.[41,42] In sheep, high humidity affects food intake when temperatures exceed 30°C.[43]

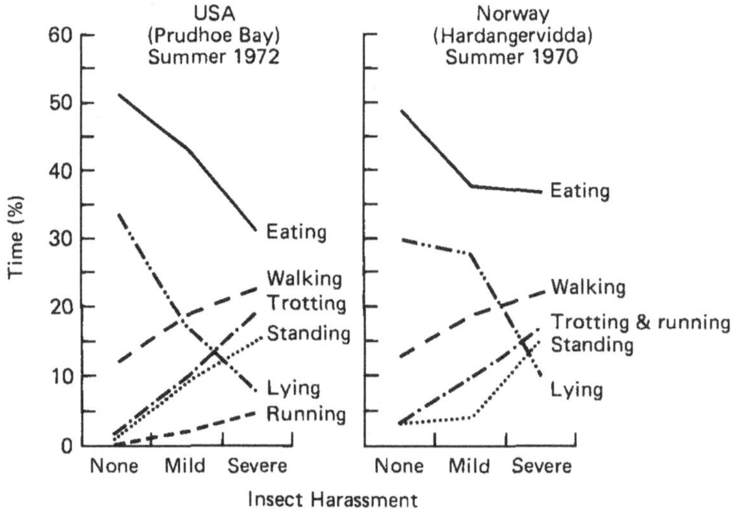

FIGURE 5. Effect of insect harassment on daily activity patterns of caribou at Prudhoe Bay and reindeer at Hardangervidda, Norway. (From White, R. G., Thomson, B. R., Skogland, T., Person, S. J., Russell, D. E., Hollerman, D. F., and Luick, J. R., Ecology of Caribou at Prudhoe Bay, Alaska Region, Alaska, Brown, J., Ed., Biological papers of the University of Alaska, Special Rep. No. 2, 150, 1975.)

2. Insect Harassment

It is difficult to quantify the effects of insect irritation on time spent feeding, and thus on food intake. As insect harassment of caribou in Alaska increased, groups of caribou moved more quickly and to areas with least insect activity.[44] In general, insect harassment causes animals to disturb their grazing frequently and quite often reduces grazing time and so food intake. This is shown in Figure 5 for reindeer in Alaska and Norway.

III. PLANT FACTORS

Both the physical and chemical composition of the diet influence food intake, the former by affecting the rate at which particles can be broken down by mastication and microbial digestion, and the latter by affecting nutrient supply to the microbial populations involved in fiber digestion.

A. Physical Characteristics

It is generally considered that the accumulation of digesta solids in the rumen imposes an important limitation on forage intake, and thus the rate of removal of digesta from the rumen is important in intake regulation. Factors involved are shown in Figure 6. Similar factors are involved in nonruminants but little is known about them.

Absorption and eructation remove products of digestion quite rapidly. But transfer of residual material to the omasum is a comparatively slow process. Propulsive activity and the physical characteristics of the diet primarily determine the transfer rate. Particle size is particularly important. Fiber digestion requires the attachment of microorganisms and the penetration of their enzymes. The availability of substrate for these organisms is likely to be the rate-limiting step. Thus, factors such as feed particle size and mastication by influencing substrate availability are important determinants of fiber digestion rate. Rumination, by affecting particle size, influences both rate of digestion and transfer rate to the omasum.[2]

Physical differences exist between leaf and stem with more leaf than stem being eaten

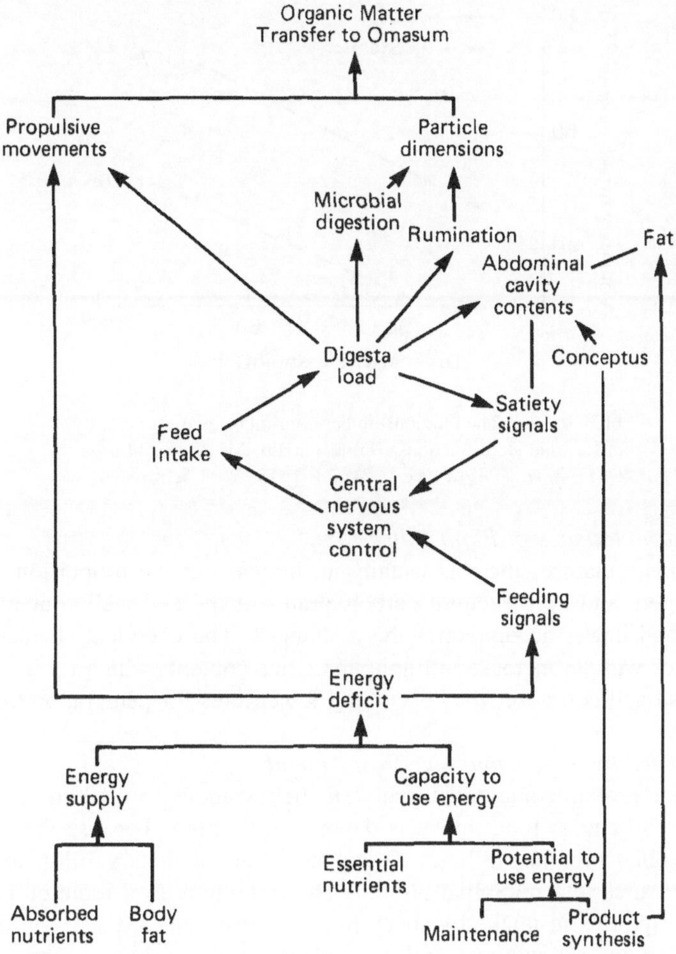

FIGURE 6. Factors contributing to the transfer of organic matter from the rumen to the omasum. (From Weston, R. H., *Nutritional Limits to Animal Production from Pasture*, Hacker, J. B., Ed., Commonwealth Agricultural Bureaux, Farnham Royal, 1981, 183. With permission.)

even when these two fractions have the same digestibility (Figure 7). Several studies have shown that the weight of dry matter in the rumen of sheep fed leaf or stem is the same.[45-47] However, the leaf stays in the rumen a shorter time and more leaf than stem can be consumed. Leaf has a larger surface area per unit weight than stem and this might allow more rapid digestion, but this is not proven. The most probable reason for the difference is the higher proportion of large particles after mastication of stem than of leaf.[48]

For similar reasons the intake of forage legumes and herbs is higher, at the same digestibility, than that of grasses.[49]

B. Chemical Composition

Forage digestion depends predominantly, in ruminants, on microbial enzymes and the metabolic activity of the microbiota depends on the molecular environment of the rumen, this is turn being influenced by the diet, secretions into the rumen, and end-product absorption. Deficiencies of essential nutrients such as nitrogen and various minerals impair microbiota metabolic activity and accordingly enzyme production; low rumen pH and the presence of toxic substances have similar consequences.[2]

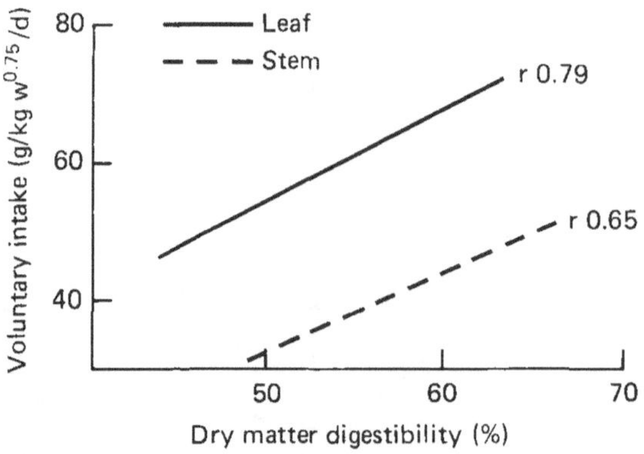

FIGURE 7. The relationship between intake and digestibility of leaf and stem fractions. (From Laredo, M. A. and Minson, D. I., *Aust. J. Agric. Res.*, 24, 33, 1975. With permission.)

1. Chemical Characteristics of Plant Fiber

As pasture plants mature, there is usually an increase in the proportion of fiber and a reduction in protein and nonstructural carbohydrates of the cell-wall contents. A decrease in digestibility and intake accompanies these changes. The chemical characteristics of the fiber change also, with an increase in lignin and silica contents which form a sheath around the hemicellulose and cellulose, thus providing fewer sites for penetration by microbiota.

2. Nutrients Needed by Microbiota and Host Animal

When the crude protein content of forage falls below about 13 g/100 g digestible organic matter in sheep and cattle, food intake is depressed.[50] This is because there is inadequate protein for microbial protein synthesis. Adequate levels of dietary sulfur are required for the conversion of urea into microbial protein. The minimum level required is about 0.10% sulfur,[51-53] but responses in intake by sheep have occurred up to 0.17% sulfur.[54]

Insufficient sodium, phosphorus, and trace elements like cobalt and selenium, and imbalances in trace elements cause reductions in food intake in sheep and cattle, and although the critical levels will vary with species, all herbivore species will be subject to reduced food intake at low levels of these minerals.

3. Toxic Factors

If forage contains elements toxic to either the animal or its microbiota, food intake will be depressed. Often substances that are secondary plant metabolites cause toxic effects. Examples include alkaloids, some amino acids, cyanogenic glycosides, and tannins. Their effects on herbivorous marsupials like the koala and bushtail possum have been reviewed by Hume.[34]

Essential oils of sagebrush (*Artemesia* spp.) foliage[55] and Douglas fir (*Pseudotsuga menziesii*) needles[56] have an inhibiting effect on rumen microorganisms from black-tailed deer not recently exposed to them. The inhibiting effects vary with type of oil and concentration.[57] However, continuous exposure appears to produce a microbial population tolerant of them and which can metabolize them.[57] This was substantiated by Cluff et al.[58] who found that sagebrush in winter was a highly digestible forage for mule deer. Adaptation of the gut flora of koalas and possums to eucalypt essential oils seems likely.[59]

Alkaloids and tannins can have drastic effects on food intake in sheep.[60] For example, 8% tannic acid in the diet reduced food intake by 20%, while 1% of the alkaloid gramine reduced it by 50%. Both effects were due to reduced digestibility.

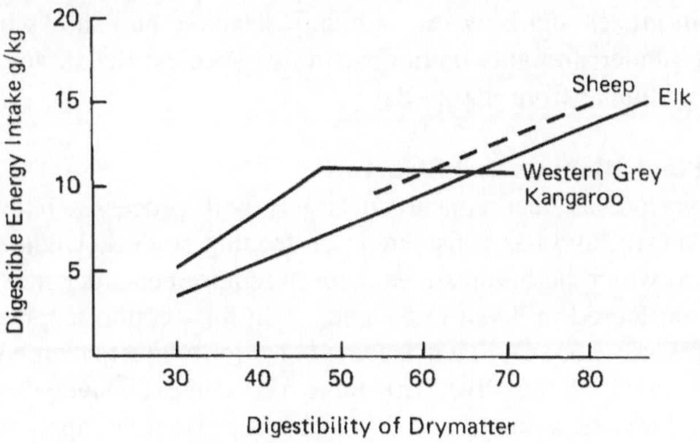

FIGURE 8. Relationship between voluntary feed intake and digestibility for sheep, elk, and Western Grey kangaroos. (Information on elk from Mould, E. D. and Robbins, C. T., *J. Wildl. Manage.*, 37, 195, 1973. With permission. Information on sheep from Minson, D. J., *Nutritional Limits to Animal Production from Pastures*, Hacker, J. B., Ed., Commonwealth Agricultural Bureaux, Farnham Royal, 1981, 167. With permission. Information on kangaroos from Arnold, G. W., unpublished.)

Condensed tannins can reduce the availability of proteins by complexing with them,[61] and at low levels of protein could consequently affect intake.

IV. PLANT-ANIMAL INTERACTIONS

A. Genotype x Forage Quality

It was mentioned earlier that species of herbivores differ in the efficiency with which they digest different quality forages. This may be because of differences in digestive systems, but also occurs among ruminant species. For example, sheep and euros have equal ability to digest cell-wall constituents, but both are superior in this regard to red kangaroos.[10] However, all three species digest hemicellulose and soluble carbohydrates equally well. This difference in ability to digest cell-wall constituents is associated with a difference in digestible energy intake per unit body weight.

Sheep and red deer ate similar quantities of medium quality (c. 65% digestibility) grass. But on low quality forage, deer ate twice as much as sheep even though their digestive efficiencies were similar.[13] Other studies have shown that deer digest roughage less well than sheep or cattle[62,63] probably because the mean retention time in the digestive system is lower in deer.[61] Thus, deer are able to eat more low quality forage than sheep and cattle by having a faster throughput.[64]

It seems that species differ in their response to changes in food quality in the short term (daily), and this is illustrated in Figure 8 for Western Grey kangaroos, sheep, and wapiti (elk). The data represent the relationships between voluntary intake of different forages and digestibility of these forages. Western Grey kangaroos maintain a constant digestible energy intake over a wide range of forage digestibilities, but sheep and wapiti show linear increases in digestible energy intake with increasing digestibility. Presumably, differences reflect adaptations to the seasonal variations in forage quality in different environments, and also to foraging strategies. The strategy for grazers like sheep and wapiti that face a long period each year of low quality feed and a short period of high quality feed is likely to be to maximize energy intake when the opportunity exists, until a control such as body fat content sets in. These species show marked fluctuations in body weight over the year. The Western

Grey kangaroo both grazes and browses, is highly selective and shows little fluctuation in body weight except under drought conditions. In this species, the strategy is to maintain a near constant energy intake from day to day.

B. Influences of Plant Biomass and Structure

The discussion so far has been concerned largely with processes internal to the animal that regulate food intake, and has considered pen feeding studies. Under free-ranging conditions, the extent to which herbivores meet energy requirements depends on their foraging behavior. This is considered in detail in Chapter 3. In this section emphasis is given to the ways in which herbivores can adapt their feeding behavior to meet changing feed availability in the environments in which they live. This topic has been extensively reviewed by Arnold and Dudzinski[67] for domestic animals, and the same basic principles apply to wild herbivores.

Figure 9 illustrates how a grass pasture is grazed, and Table 2 lists factors influencing the whole process of food collection, termed grazing in this discussion. The time spent grazing is determined by the individual's food requirement and the rate at which it can consume food in the current environment. The rate of consumption depends on the time spent searching for areas in which to feed, in selecting each bite, in the mechanical process of biting, in manipulating the food in the mouth, and then in swallowing. Constraints on the animal are the size of bite it can take and, in old animals, teeth condition may limit choice or rate of biting, or both. There will also be upper physical limits to rate of biting and the total time that can be spent grazing because of the need to spend time in other activities. Ruminants must allocate time for rumination, which may require as much as 8 hr/day. Finally, feedback from internal stimuli such as gut distention may constrain time spent grazing.

1. Feeding Behavior

Table 3 illustrates how time spent grazing and browsing varies between and within species. Data are limited because of the difficulty in observing wild herbivores through the hours of darkness. Most species studied spend less than 12 hr a day on grazing and browsing, but under some circumstances have to spend a much longer period of time. The values are similar to those for domestic sheep and cattle. Horses, however, spend longer grazing than any other species so far studied.

There are few measures of the time spent searching on the total time spent grazing, but Trudell and White[76] found that in caribou it was about 50% (Figure 10). Animals have to adopt their method of feeding to the type of material being grazed or browsed.

Selection of areas for feeding will be influenced by weather conditions, availability of water, and type of terrain as well as by abundance and type of food present. One of the major problems in analyzing feeding behavior is the understanding of the animal's perception of what food items are present. This is particularly so when many plant species are present and selective grazing or browsing is very pronounced. Nevertheless, the animal can adjust the time it spends feeding, the degree to which it selects, the rate of biting, and size of bites in order to obtain its food requirements under varying plant structure and biomass conditions. Thus, it is natural to find a wide diversity in the time spent grazing or browsing by a species, and in the different components of grazing and browsing.

Trudell and White[76] found that caribou feeding on shrub willows use a combination of biting and pulling motions, the latter serving to strip leaves, new stem growth, and some bark from branches. When lichens were ingested, lip movements are used to tease the lichens from the moss layer, while graminoids are eaten in a different manner again. As a consequence, the rate of prehension (gathering and biting) and rate of consumption (intake per minute) varied greatly even at the same biomass.

Bite rate varied from 150 to 230 bites per minute, and intake per bite from 5 to 50 mg.

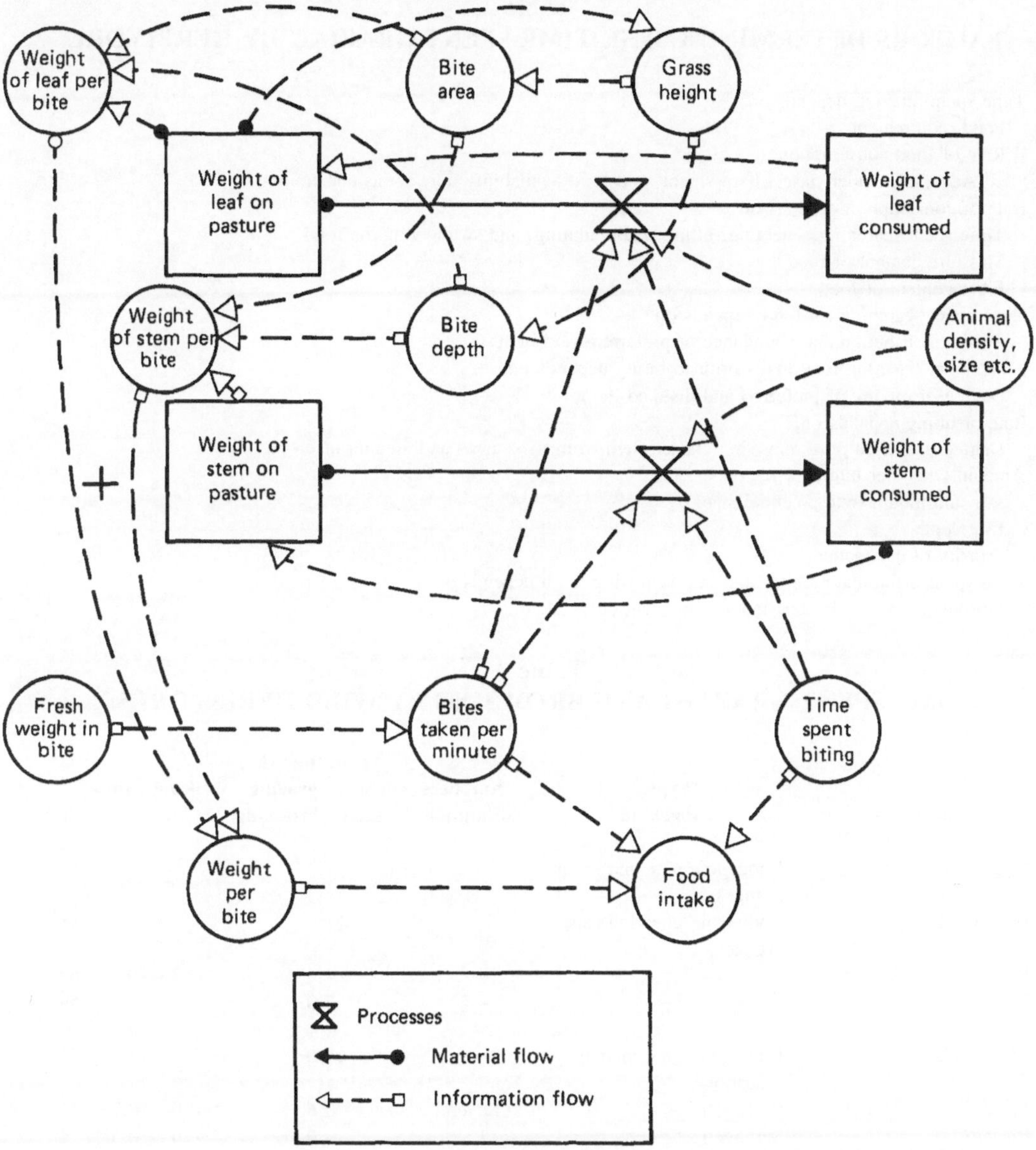

FIGURE 9. Systems diagram of a model of grazing.

Rates of consumption thus varied accordingly to plant growth form and biomass (Figure 11), because while bite rate increased with plant biomass, intake per bite was constant over a wide range of biomass. Collins and Urness[68] reported a wide range in intake rate in wapiti (3.7 to 17.2 g/min) but not in mule deer (1.8 to 3.3 g/min) on aspen and lodgepole pine range in Utah. Wickstrom et al.[71] also compared these two species and found that intake rate plateaued at a much lower biomass in deer than in wapiti. This is to be expected because of the smaller size of the deer. In contrast, red deer grazing on grass pastures in Scotland maintained the same biting rate and grazing time when biomass increased from 1350 to 2350 kg/ha (plant height also increased from 3.2 cm to 7.6 cm) and intake per bite was much increased.[78] Thus, changes in bite rate and intake per bite with biomass depend on plant structure. The net result is that the biomass that allows the maximum rate of intake varies with both plant structure and animal species (Figure 12).

Table 2
FACTORS DETERMINING THE TIME SPENT GRAZING BY HERBIVORES

Time spend grazing depends on:
- Food requirement
- Rate of food consumption
  - (Adverse weather, insect harassment, and food availability may act as constraints)

Rate of consumption depends on:
- Time spent searching, selecting, biting, manipulating, and swallowing the food
- Moisture content of food
- Fiber content of food

Time spent searching for food depends on:
- Spatial distribution and abundance of preferred food items

Time spent selecting food items prior to biting depends on:
- Degree of mixing of preferred and disliked items

Rate of biting depends on:
- Degree of difficulty in removing chosen item from its sources and amount taken per bite

Amount taken per bite depends on:
- Jaw and mouth (bite potential size)
- Bite depth
- Density of dry matter

Time spent manipulating bites and in swallowing them depends on:
- Fibrousness of food eaten and size of plant parts taken.

Table 3
TIME SPENT GRAZING AND BROWSING BY WILD HERBIVORES

Species	Types rangeland	Numbers of animals	Mean time (hr) spent grazing and/or browsing	Range in values	Ref.
Mule deer	Lodgepole pine and aspen ranges	1—2	11.9	—	68
Wapiti (elk)	Lodgepole pine and aspen ranges	5—7	11.0	—	68
Red deer	—	—	—	8—10	69
Roe deer	—	—	7.0	—	69
	—	—	—	12—16	70
Wapiti (elk)	Lodgepole pine, marsh meadows	2—7	10.6	7.7—14.3	72
Impala	Serengeti	Several	9.6	8.0—10.7	73
Buffalo	Serengeti	1	10.0	9.0—10.5	74
Black rhino	Etosha NP	2	6.0	6.0	75
Caribou	Tundra	—	12.7	9.2—20.2	76
Wild reindeer	Tundra	—	11.8	—	76[a]
Western Grey kangaroos	Herbage and browse	6—10	12.9	11.5—14.0	—[b]
Horse	Pasture	12	14.2	10.8—16.0	77
Cattle	Pasture	12	10.9	5.4—12.7	77
Sheep	Pasture	20	9.4	8.4—10.6	77

[a] Cited.
[b] Unpublished.

2. *Daily Food Intake*

It is impossible to measure food intake in truly wild herbivores, so where estimates have been made they are based on indirect methods, or by the use of tame animals.

Alldredge et al.[79] estimated forage intake of mule deer in Colorado from fallout cesium-137 and found a 30% difference between summer and winter. Trudell and White[76] predicted

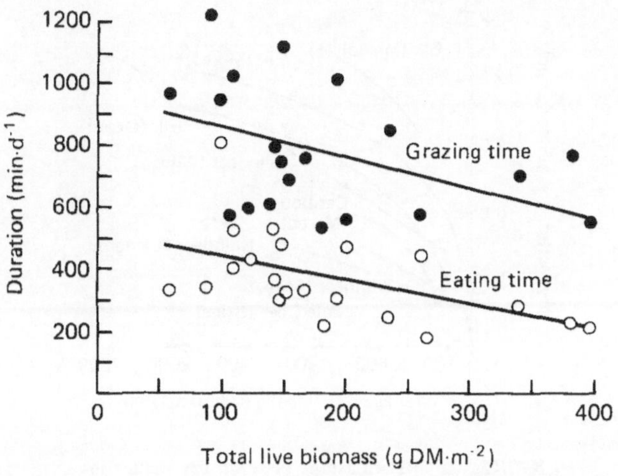

FIGURE 10. Daily grazing (foraging) time (●) and daily eating time (○) in relation to total live biomass (vascular plants + lichens) for reindeer tethered on tundra vegetation types throughout the summer. (From Trudell, J. and White, R. G., *J. Appl. Ecol.*, 18, 63, 1981. With permission.)

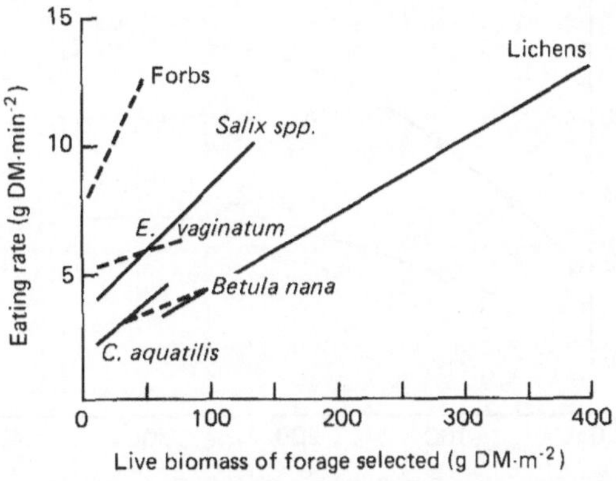

FIGURE 11. Comparison of plant growth form in the relation of eating rate (g/min) and available selected live biomass (g/m^2). Solid lines indicate significant relationships. (From Trudell, J. and White, R. G., *J. Appl. Ecol.*, 18, 63, 1981. With permission.)

a quite different food intake for caribou grazing *Salix* spp. and lichens spp., with intake on *Salix* being strongly influenced by plant biomass (Figure 13). Predicted seasonal trends in food intake are different on different plant communities[80] (Figure 14) due to different changes in biomass and feed quality on the different communities. But these estimates assume the animals would feed year long on one plant community, which they do not. So actual food intake is dependent upon structural characteristics of the plant communities grazed at a given time.

3. Effect of Climate

The direct effects of temperature and humidity on the animal and on its food intake have been described earlier. But there are indirect effects which influence the location of animals

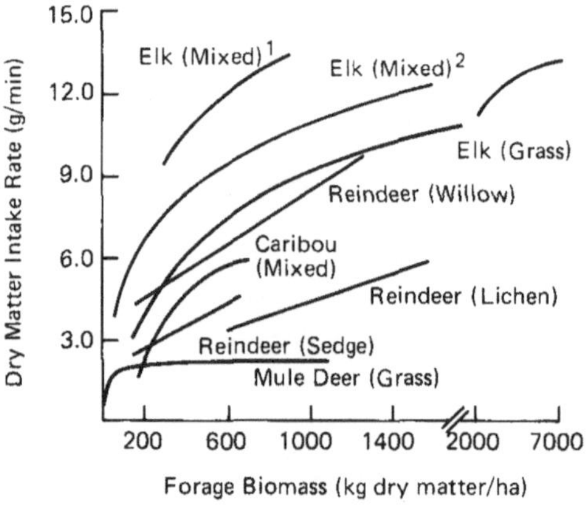

FIGURE 12. Relationships between dry matter intake and forage biomass from various studies with wild herbivores. (From Wickstrom, M. L., Robbins, C. T., Hanley, T. A., Spalinger, D. E., and Parrish, S. M., *J. Wildl. Manage.*, 48, 1285, 1984. With permission.)

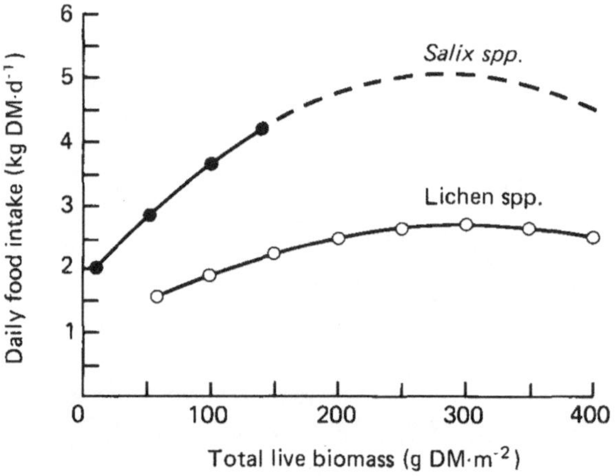

FIGURE 13. Predicted daily forage intake by caribou grazing *Salix* and lichens. (From Trudell, J. and White, R. G., *J. Appl. Ecol.*, 18, 63, 1981. With permission.)

in the environment. Since different locations often have different plant communities or food availabilities, then food intake is going to be affected by location.

Yazan and Glushkov[82] observed that when winter temperatures are greater than $-20°C$, moose feed in open areas of pasture, on hillsides, in clearings, low woods, and riverside osier beds. But they abandon these open areas when temperatures are lower, and especially when there is a wind, and choose dense forest. Since forage is often more abundant in open areas, weather conditions affect food intake.

Conversely, with heat stress animals seek either shaded areas or areas exposed to wind to allow cooling, and feed in these areas.

4. Effect of Plant Distribution in Relation to Water Sources and Home Range Location

Most wild herbivores occupy home ranges that may or may not differ from season to

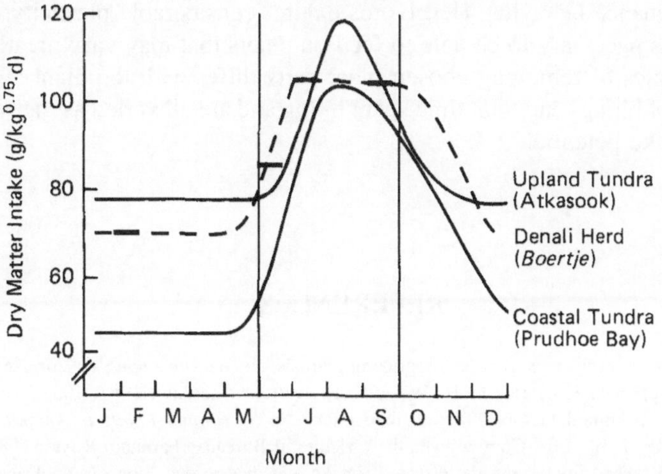

FIGURE 14. Seasonal trends in estimates of daily intake of dry matter in reindeer. Dashes represent periods over which estimates were calculated and dashed lines represent fits by eye to those points for interior Alaska. Solid lines for upland coastal tundra based on literature. (From White, R. G., *Oikos*, 41, 1984. With permission.)

season. What is significant to food intake is the distribution of resources within the home range. Home ranges of different individuals, or groups of individuals, will inevitably vary in size and in the areas within them of the different plant communities, in topography, and in water sources or nearness to water. Each of these factors can influence food intake.

The greater the distance that animals have to travel to water, the longer it takes and the shorter the time that can be spent feeding. This is especially a problem for lactating animals. What often occurs is that the biomass close to water is lower than at a distance from water, but lactating animals may have to feed there because of their greater need for water than in males or dry females. The consequence is a suboptimal food intake. Distribution of favored vegetation types in relation to water will vary between home ranges. If favored types are well removed from water, animals either have to walk longer distances to feed there, and so have increased energy expenditure and food demands, or graze on less preferred types closer to water.

The effects of these interactions on the food intake of animals have not been quantified for wild herbivores but are likely to be as great as has been shown in sheep.[83] Food intake in sheep on a salt-loaded diet decreased linearly with distance to water once the distance was above 1.6 km, and was 20% below normal when the distance was 4.8 km.

The energy costs of locomotion and its effect on "foraging radius" from a fixed point such as water have been calculated.[84] Adaptations to this problem are found in the type of gait the species has, and in migration between two home range areas, as occurs in some African ungulates and in deer and bighorn sheep in North America.

V. SYNOPSIS

The regulation of forage intake by free-ranging wild herbivores is through both internal controls concerned with digestion, rate of passage of digesta through the digestive system, and set points probably including energy balance and body composition. Endocrine influences on the internal controls are stimulated by daylength in some species, pregnancy, and lactation. The extent to which the animal satisfies its food intake potential, as set by these internal controls, depends on its ability to find and eat enough food within the time constraints of its overall maintenance behavior. Herbivores exhibit considerable plasticity in feeding behavior, and this is necessary to be able to feed on plants that may vary greatly in structure. Thus, the mechanics of removing chosen plant parts differ with the plant being eaten, the size of bite, rate of biting, and total time spent biting and are all varied as the animal attempts to achieve its intake potential.

REFERENCES

1. **Freer, M.,** The control of feed intake by grazing animals, in *Grazing Animals,* World Animal Science B, Disciplinary Approach 1, Morley F. H. W., Ed., Elsevier, Amsterdam, 1981, 105.
2. **Weston, R. H.,** Animal factors affecting feed intake, in *Nutritional Limits to Animal Production from Pasture,* Hacker, J. B., Ed., Commonwealth Agricultural Bureaux, Farnham Royal, 1981, 183.
3. **Weston, R. H.,** Feed intake and the effect of intake and diet composition on wool growth, *Proc. 1981 Sheep and Wool Refresher School,* Queensland Department of Primary Industries and the Australian Wool Corporation, Queensland, Australia, 1981.
4. **Minson, D. J.,** Effects of chemical and physical composition of herbage eaten upon intake, in *Nutritional Limits to Animal Production from Pastures,* Hacker, J. B., Ed., Commonwealth Agricultural Bureaux, Farnham Royal, 1981, 167.
5. **Morrison, S. D.,** The hypothalamic syndrome in rats, *Fed. Proc. Fed. Am. Soc. Exp. Biol.,* 36, 139, 1977.
6. **Baile, C. A. and Forbes, C. A.,** Control of feed intake and regulation of energy balance in ruminants, *Physiol. Rev.,* 54, 160, 1974.
7. **Panksepp, J.,** Hypothalamic regulation of energy balance and feeding behaviour, *Fed. Proc. Fed. Am. Soc. Exp. Biol.,* 33, 1150, 1974.
8. **Graham, N. McC.,** Units of metabolic body size for comparisons amongst adult sheep and cattle, *Proc. Aust. Soc. Anim. Prod.,* 9, 352, 1972.
9. **Ternouth, J. H., Poppi, D. P., and Minson, D. J.,** The voluntary food intake, ruminal retention time and digestibility of two tropical grasses fed to cattle and sheep, *Proc. Nutr. Soc. Aust.,* 4, 152, 1979.
10. **Hume, I. D.,** Nitrogen and sulphur retention and fibre digestion by euros, red kangaroos and sheep, *Aust. J. Zool.,* 22, 13, 1974.
11. **Dellar, D. W.,** Production of volatile fatty acids in two species of wallaby and in sheep, *Comp. Biochem. Physiol.,* 56A, 299, 1977.
12. **Dellar, D. W. and Hume, I. D.,** Studies on the nutrition of Macropodine marsupials. I. Intake and digestion of lucerne hay and fresh grass, *Phalaris aquatica, Aust. J. Zool.,* 30, 391, 1982.
13. **Milne, J. A., Macrae, J. C., Spence, A. M., and Wilson, S.,** A comparison of the voluntary intake and digestion of a range of forages at different times of the year by the sheep and the red deer *(Cervus elaphus). Br. J. Nutr.,* 40, 347, 1978.
14. **Langlands, J. P.,** The feed intake of grazing sheep different in age, breed, previous nutrition and liveweight, *J. Agric. Sci.* (Cambridge), 71, 167, 1968.
15. **Langlands, J. P. and Hamilton, B. A.,** Efficiency of wool production of grazing sheep. II. Differences between breeds and strains varying in age, *Aust. Exp. Agric. Anim. Husb.,* 9, 254, 1969.
16. **Taylor, J. C.,** A relationship between weight of internal fat, "fill" and the herbage intake of grazing cattle, *Nature (London),* 1984, 2021, 1959.
17. **Bines, J. A., Suzuki, S., and Balch, C. C.,** The quantitative significance of long-term regulation of feed intake in the cow, *J. Nutr.,* 23, 695, 1969.
18. **Cowan, R. T.,** Grazing time and pattern of grazing of Friesian cows on a tropical grass-legume pasture, *Aust. J. Exp. Agric. Anim. Husb.,* 15, 32, 1975.
19. **Forbes, J. M.,** The voluntary feed intake of pregnant and lactating ruminants: a review, *Br. Vet. J.,* 126, 1, 1970.
20. **Forbes, J. M.,** Physiological changes affecting voluntary food intake in ruminants, *Proc. Nutr. Soc.,* 30, 135, 1971.
21. **Robbins, C. T., Podbielanchik-Norman, R. S., Wilson, D. L., and Mould, E. D.,** Growth and nutrient consumption of elk calves compared to other ungulate species, *J. Wildl. Manage.,* 45, 172, 1981.
22. **Sadlier, R. M. F. S.,** Energy consumption in lactating black-tailed deer, *Can. J. Zool.,* 60, 382, 1982.
23. **Arman, P., Kay, R. N. B., Goodall, E. D., and Sharman, G. A. M.,** The composition and yield of milk from captive red deer, (*Cervus elaphus* L.), *J. Reprod. Fert.,* 37, 67, 1974.

24. **Short, H. L., Newsome, J. D., McCoy, G. L., and Fowler, J. F.,** Effects of nutrition and climate on southern deer, *North Am. Wildl. Nat. Resour. Conf. Trans.*, 34, 137, 1969.
25. **Arnold, G. W. and Dudzinski, M. L.,** Studies on the diet of the grazing animal. II. The effect of physiological status in ewes and pasture availability on herbage intake, *Aust. J. Agric. Res.*, 18, 349, 1967.
26. **Arnold, G. W.,** Herbage intake and grazing behaviour in ewes of four breeds at different physiological states, *Aust. J. Agric. Res.*, 26, 1017, 1975.
27. **Ryg, M. and Jacobsen, E.,** Effect of thyroid hormones and prolactin on food intake and weight change in young male reindeer *(Rangifer tarandus, tarandus), Can. J. Zool.*, 60, 1652, 1982.
28. **Nichol, A. A.,** Experimental feeding of deer, University of Arizona, College of Agriculture, Agric. Exp. Stn. Tech. Bull. No. 75, 1938.
29. **Cowan, I. McT., Wood, A. J., and Kitts, W. D.,** Feed requirements of deer, beaver, bear and mink for growth and maintenance, *Trans. North Am. Wildl. Conf.*, 22, 179, 1957.
30. **Long, T. A., Cowan, R. L., Strawn, G. D., Wetzel, R. S., and Miller, R. C.,** Seasonal fluctuations in feed consumption of the white-tail deer, Pennsylvania State University, College of Agriculture, Prog. Rep. 262, 1965.
31. **Nordan, H. C., Cowan, I. McT., and Wood, A. J.,** Nutritional requirements and growth of black-tailed deer, *Odocoileus hemionus columbianus*, in captivity, *Symp. Zool. Soc. London*, 21, 89, 1968.
32. **Kay, R. N. B.,** Seasonal changes of appetite in deer and sheep, *Agric. Res. Counc. Rev.*, 5, 13, 1979.
33. **Warren, R. J., Kirkpatrick, R. L., Oelschaeger, A., Scanlan, P. F., and Gwazdaeeskas, F. C.,** Dietary and seasonal influences on nutritional indices of adult male white-tailed deer, *J. Wildl. Manage.*, 45, 926, 1981.
34. **Hume, I. D.,** *Digestive Physiology and Nutrition of Marsupials*, Cambridge University Press, Cambridge, 1982.
35. **Pollock, A. M.,** Seasonal changes in appetite and sexual condition in red deer stags maintained on a six-month photoperiod, *J. Physiol.*, 244, 95P, 1975.
36. **Larkin, R. M.,** Observations on the grazing behaviour of beef cattle in tropical Queensland, *Queensland J. Agric. Sci.*, 11, 115, 1954.
37. **Dwyer, D. D.,** Activities and grazing preferences of cows with calves in Northern Osage Country, Oklahoma, Oklahoma Agric. Exp. Stn., Bull. B-588, 1961.
38. **Wells, R. T.,** Physiological and Behavioural Adaptation of the Hairy-Nosed Wombat, *Lasiothimus latifrons* (Owen), to its Arid Environment, Ph.D. thesis, University of Adelaide, Australia.
39. **Ragsdale, A. C., Brody, S. H., Thompson, H. L., and Worstell, D. M.,** Influence of temperature, 50° to 105°F, on milk production and feed consumption in dairy cattle, University of Missouri, Agric. Exp. Stn., Res. Bull., 425, 1948.
40. **Ragsdale, A. C., Thompson, H. J., Worstell, D. M., and Brody, S.,** Influence of increasing temperature, 40° to 105°F, on milk production in Brown Swiss cows, and on feed and water consumption and body weight in Brown Swiss and Brahman cows and heifers, University of Missouri, Agric. Exp. Stn., Res. Bull., 471, 1951.
41. **Ragsdale, A. C., Thompson, H. J., Worstell, D. M., and Brody, S.,** The effect of humidity on milk production and composition, and water consumption, and body weight in cattle, University of Missouri, Agric. Exp. Stn., Res. Bull., 521, 1953.
42. **Johnson, D. H., Ragsdale, A. C., Berry, I. L., and Shanklin, M. D.,** Temperature-humidity effects including influence of acclimatisation on feed and water consumption of Holstein cattle, University of Missouri, Agric. Exp. Stn., Res. Bull., 846, 1963.
43. **Guerrini, V. H.,** Food intake of sheep exposed to hot-humid, hot-dry and cool-humid environments, *Am. J. Vet. Res.*, 42, 658, 1981.
44. **White, R. G., Thomson, B. R., Skogland, T., Person, S. J., Russell, D. E., Hollerman, D. F., and Luick, J. R.,** Ecology of Caribou at Prudhoe Bay, Alaska Region, Alaska, Brown, J., Ed., Biological paper of the University of Alaska, Special Rep. No. 2, 150, 1975.
45. **Laredo, M. A. and Minson, D. J.,** The voluntary intake, digestibility and retention time by sheep of leaf and stem fractions of five grasses, *Aust. Agric. Res.*, 24, 33, 1975.
46. **Laredo, M. A. and Minson, D. J.,** The effect of pelleting on the voluntary intake and digestibility of leaf and stem fractions of three grasses, *Br. J. Nutr.*, 33, 159, 1975.
47. **Poppi, D. P., Minson, D. J., and Ternouth, J. H.,** Studies of cattle and sheep eating leaf and stem fracton of grasses. III. The retention time in the rumen of large feed particles, *Aust. J. Agric. Res.*, 32, 123, 1981.
48. **Hendrickson, R. E., Poppi, D. P., and Minson, D. J.,** The voluntary intake, digestibility and retention time by cattle and sheep of leaf and stem fractions of tropical legume, *(Lablab purpureus), Aust. J. Agric. Res.*, 32, 389, 1981.
49. **Demarquilly, C. and Jarrige, R.,** The comparative nutritive value of grasses and legumes, *Vaxtadling*, 28, 33, 1974.

50. **Blaxter, K. L. and Wilson, C. A.,** The assessment of a crop husbandry technique in terms of animal production, *Anim. Prod.*, 5, 27, 1963.
51. **Playne, M. J.,** Th effect of sodium sulphate and gluten supplements on the intake and digestibility of a mixture of spear grass and Townsville lucerne hay by sheep, *Aust. J. Exp. Agric. Anim. Husb.*, 9, 393, 1969.
52. **Kennedy, P. M. and Siebert, B. D.,** The utilisation of spear grass (*Heteropogon centartus*). III. The effect of sulphur on the utilisation of spear grass by sheep, *Aust. J. Agric. Res.*, 24, 153, 1973.
53. **Rees, M. C., Minson, D. J., and Smith, F. W.,** The effect of supplementary and fertilizer sulphur on voluntary intake, digestibility, retention time in the rumen, and site of digestion of pangola grass in sheep, *J. Agric. Sci.* (Cambridge), 83, 419, 1974.
54. **Rees, M. C. and Minson, D. J.,** Fertilizer sulphur as a factor affecting voluntary intake, digestibility and retention time of pangola grass (*Digitaria decumbens*) by sheep, *Br. J. Nutr.*, 39, 5, 1978.
55. **Nagy, J. A., Steinhoff, H. W., and Ward, G. M.,** Effects of essential oils of sagebrush on deer rumen microbial function, *J. Wildl. Manage.*, 28, 785, 1964.
56. **Oh, H. K., Sakai, J., Jones, M. B., and Longhurst, W. M.,** Effect of various essential oils isolated from Douglas fir needles upon sheep and deer rumen microbial function, *Appl. Microbiol.*, 15, 777, 1967.
57. **Schwartz, C. C., Nagy, J. A., and Regelin, W. Y.,** Juniper oil yield, terpenoid concentration, and antimicrobial effects on deer, *J. Wildl. Manage.*, 44, 107, 1980.
58. **Cluff, L. K., Welch, B. L., Pederson, J. J. C., and Brotherson, J. D.,** Concentration of monoterpenoids in the rumen ingesta of wild mule deer, *J. Range Manage.*, 35, 192, 1982.
59. **Eberhard, I. H., McNamara, J., Pearce, R. H., and Southwell, I. A.,** Ingestion and excretion of *Eucalyptus punctata* D. C. and its essential oils by the Koala, *Phascolaretos cineraus*, (Goldfuss), *Aust. J. Zool.*, 23, 169, 1975.
60. **Arnold, G. W., De Boer, E. S., and Boundy, C. A. P.,** The influence of odour and taste on the food preferences and food intake of sheep, *Aust. J. Agric. Res.*, 31, 571, 1980.
61. **McLeod, M. N.,** Plant tannins — their role in forage quality, *Nutr. Abstr. Rev.*, 44, 803, 1974.
62. **Maloy, G. M. O., Kay, R. N. B., Goodall, E. D., and Topps, J. H.,** Digestion and metabolism in sheep and red deer given large or small amounts of water and protein, *Br. J. Nutr.*, 24, 843, 1970.
63. **Short, H. L.,** Rumen fermentation and energy relationships in white-tailed deer, *J. Wildl. Manage.*, 27, 184, 1963.
64. **Kay, R. N. B. and Staines, B. W.,** The nutrition of the red deer *(Cervus Elaphus)*, *Nutr. Abstr. Rev.*, Ser. B, 51, 601, 1981.
65. **Amman, A. P., Cowan, R. L., Mothershead, C. L., and Baumgardt, B. R.,** Dry matter and energy intake in relation to digestibility in white-tailed deer, *J. Wildl. Manage.*, 37, 195, 1973.
66. **Mould, E. D. and Robbins, C. T.,** Digestive capabilities in elk compared to white-tailed deer, *J. Wildl. Manage.*, 46, 22, 1982.
67. **Arnold, G. W. and Dudzinski, M. L.,** *The Ethology of Free-Ranging Domestic Animals*, Elsevier, Amsterdam, 1978.
68. **Collins, W. B. and Urness, P. J.,** Feeding behaviour and habitat selection of mule deer and elk on northern Utah summer range, *J. Wildl. Manage.*, 47, 646, 1983.
69. **Bubenik, A.,** *Grundlagen der Wildeenäuhreing*, Dectscher Bauern Verlag, 1959.
70. **Mottl, S. T.,** Die Nahrung des Rehwildes, *Z. Jagdwiss.*, 4, 228, 1958.
71. **Wickstrom, M. L., Robbins, C. T., Hanley, T. A., Spalinger, D. E., and Parish, S. M.,** Food intake and foraging energetics of elk and mule deer, *J. Wildl. Manage.*, 48, 1285, 1984.
72. **Craighead, J. J., Craighead, F. C., Ruff, R. L., and O'Gara, B. W.,** Home ranges and activity patterns of nonmigratory elk of the Madison drainage herd as determined by biotelemetry, *Wildl. Monogr.*, 33, 1973.
73. **Jarman, M. V. and Jarman, P. J.,** Daily activity of impala, *East Afr. Wildl. J.*, 11, 75, 1976.
74. **Sinclair, A. R. E.,** The natural regulation of buffalo populations in East Africa. IV. The food supply as a regulating factor, and competition, *East Afr. Wildl. J.*, 12, 291, 1984.
75. **Joubert, E. and Eloff, F. C.,** Notes on the ecology and behaviour of the black rhinoceros *Diceros bicornis* Linn 1758 in south-west Africa, *Madoqua*, (Ser. 1), 3, 5, 1971.
76. **Trudell, J. and White, R. G.,** The effect of forage structure and availability on food intake, biting rate, bite size and daily eating time in reindeer, *J. Appl. Ecol.*, 18, 63, 1981.
77. **Arnold, G. W.,** Comparison of the time budgets and circadian patterns of maintenance activities in sheep, cattle and horses grouped together, *Appl. Anim. Ethol.*, 13, 19, 1985.
78. **Loudon, A. S. I., Darrach, A. D., and Milne, J. A.,** The lactation performance of red deer on hill and improved species pastures, *J. Agric. Sci.* (Cambridge), 102, 149, 1984.
79. **Alldredge, A. W., Lipscombe, J. F., and Whicker, F. W.,** Forage intake rates of mule deer estimated with fallout cesium-137, *J. Wildl. Manage.*, 38, 508, 1974.

80. **White, R. G. and Trudell, J.,** Patterns of herbivory and nutrient intake of reindeer grazing trundra vegetation, in *Proc. Int. Reindeer/Caribou Symp.*, Reimers, E., Gaare, E., and Skjenneberg, S., Eds., Direktoratet for vilt og ferskvannsfisk, Trondheim, Norway, 1980, 180.
81. **White, R. G.,** Foraging patterns and their multiplier effects on productivity of northern ungulates, *Oikos*, 41, 1984.
82. **Yazan, T. P. and Glushkov, V. M.,** The influence of low air temperatures on the distribution and feeding characteristics of elks, *Tr. Kirov. Skh. Inst.*, 21, 125, 1969.
83. **Squires, V. R. and Wilson, A. D.,** Distance between food and water supply and its effect on drinking frequency, food and water intake of Merino and Border Leicester sheep, *Aust. J. Agric. Res.*, 22, 283, 1971.
84. **Pennycuick, C. J.,** Energy costs of locomotion and the concept of "Foraging Radius", in *Serengti: Dynamics of an Ecosystem,* Sinclair, A. R. E. and Norton-Griffiths, M., Eds., Univeristy of Chicago Press, Chicago, 1979, 164.

Chapter 5

DIGESTION

W. van Hoven and E. A. Boomker

TABLE OF CONTENTS

I. ANATOMICAL ADAPTATIONS

Evolutionary development of the mammalian gut must be seen as a reaction to changes that occurred in foraging opportunities and diet. Various types of grasslands and shrublands developed from the tropical forests originally covering a large part of the earth. These new vegetation types correlated with cooler, drier, and more seasonal climates. Throughout evolution, animal tissues retained similar nutrient demands[1] while plants became very diverse in composition, structure, and ecology. Grasslands supplied a new nutritional environment that was inhospitable compared with those previously available. Readily available nutrients became progressively diluted with fiber. Fibrous material tends to be bulky, and digestion is slow. Losses of nutrients also occurred due to shedding of seed, leaching by rain, and other physical and chemical changes. In order to meet their nutritional demands, ungulates evolved enlarged and complicated compartments in the digestive tract which serve to (1) increase the amount of feed digested by microbial action at any point in time, and (2) increase retention time, thus exposing digesta to microbial action for longer periods.

A. Foregut and Hindgut Fermentation

Digestion in mammalian herbivores is generally a two-stage process; namely, fermentation and enzymatic digestion. On the basis of digestive anatomy and function, herbivores are divided into two main groups. Those in which forage is subjected to fermentation before enzymatic digestion are generally referred to as ruminants, although this group also includes certain nonruminants such as the *Camelidae* (camel), *Hippopotamidae* (hippopotamus), and *Tayassuidae* (peccary). In the second group, food is first digested enzymatically and then subjected to fermentation further along the tract. This group consists of monogastric herbivores of the orders *Perissodactyla* and *Proboscidea* as well as the closley related *Hyracoidea* (hyrax).[2,3]

The ruminant stomach is divided into four compartments which separate fermentative and enzymatic regions. The first two compartments, the rumen and reticulum, are largest and harbour the bulk of the fermentative microbial population. The third chamber, the omasum, acts as a filter between the fermentive and the enzymatic region, the abomasum. In the caecum and colon of ruminants, fermentation occurs although on a much smaller scale than in the reticulo-rumen.

Less is known of hindgut fermenters than of foregut fermenters. Nutrition of the domestic horse has been studied intensively but very little information is available on wild representatives such as the zebra and rhino. Van Hoven et al.[4] studied gastrointestinal fermentation in the adult African elephant. In this hindgut fermenter, the principal site for extraction of protein is the glandular or enzymatic stomach. Food enters the stomach directly from the esophagus. Protein digestion takes place without either mechanical maceration such as chewing the cud or initial microbial attack of cell-wall components. Consequently, protein extraction is less efficient than in ruminants but protein is immediately assimilated as amino acids. It is not subjected to the recycling and reconstitution that increase the efficiency of protein utilization in the ruminant. The strategy of rapid rate of passage was developed in the hindgut system. Although digestion is less efficient than in ruminants of similar size, more food can pass through the gut during the same period of time. Hindgut fermenters are more tolerant of dietary cell wall constituents than ruminants but must maintain a high rate of intake to meet nutrient requirements. Generally, they are less selective in their foraging behavior than ruminants.[5]

B. Adaptive Radiation of the Ruminant Stomach

Ruminants underwent particularly spectacular adaptative radiation. Hofmann[6] classifies African wild ruminants into three groups: the concentrate selectors, the intermediate feeders, and the bulk and roughage feeders. This classification also applies to wild ruminants in general.

Differences are based on the total volume and proportions of the digestive tract as well as adaptive changes in delaying structures and absorptive surfaces (Figure 1).[6,7] This adaptive radiation probably arose from a relatively simple form in response to diversification of the forage resources.[8,9]

Anatomical differences have clear functional consequences. A direct relationship is observed between fermentation rate, volatile fatty acid (VFA) production rate, rumen turnover rate, and thus rate of passage through the gut. These factors hold an inverse relationship with body weight.[10,14]

1. Concentrate Selectors

Hofmann[6] divides the concentrate selectors into two groups: the fruit and dicotyledon foliage selectors and the tree and shrub foliage selectors. The first group includes small antelope up to 50 kg body weight. The second group includes large species such as giraffe, kudu, and gerenuk. Typical anatomical features of concentrate selectors are the small capacity of the rumen and reticulum in relation to the peritoneal cavity. Hungate[15] determined the full and the empty weight of the stomachs of several concentrate selectors. The rumen-reticulum in these animals is a simple folded blind-ending tubular structure. The *ostium ruminorecticulare* is a very large opening between these two sections of the stomach, 70% larger than in roughage feeders.[6] This anatomical feature ensures flow of digesta which is facilitated additionally by a wide *ostium intraruminale*, *ostium reticulo-omasicum*, and underdeveloped coronary pillars in the rumen.

The mucosal relief of concentrate selectors is highly papillated to enhance absorption of the high quantities of fermentive end-products. Hofmann[6] found increases in absorptive mucosal surfaces between 10- and 25-fold. There are no unpapillated zones in the rumen and papillation even stretches into a third of the dorsal reticulum surface.

The omasum is small and functionally unimportant in the smallest ruminants and absent in the ancestral tragulids (e.g., lesser mouse-deer).[16]

2. Roughage Feeders

African wild ruminants considered bulk or roughage feeders are divided into three groups by Hofmann:[6] (1) fresh grass grazers, dependent upon water such as the buffalo, oribi, bohor reedbuck, waterbuck, and wildebeest; (2) roughage grazers such as hartebeest, topi, tsessebe, and mountain reedbuck; and (3) dry region grazers such as the oryx and gemsbok.

The relative capacity of the rumen and reticulum in relation to the peritoneal cavity is 60 to 80% larger than in concentrate selectors. When grazing, these animals fill these compartments almost to capacity, while the concentrate selectors rarely utilize more than 50 to 60% of the complex, possibly due to differences in rate of passage. However, there is little information on the role played by distention thresholds.

In bulk feeders, the rate of passage is strongly restricted by anatomical structures. The increase in absorptive mucosal surface is clearly much lower in bulk feeders; papillation is uneven (size and distribution) and the doral ruminal sac is devoid of papillae. In contrast to the concentrate selectors, virtually no ruminal-type papillae are found in the reticulum. The most densely spaced ruminal papillae are found in the ventral region of the *atrium ruminis* and in all the cavities in the rumen where digesta temporarily collect.[6] The production of fermentation end-products serves as a stimulus to papillae growth and formation.

3. Intermediate Feeders

The intermediate feeders or mixed feeders fall between the concentrate selectors and bulk feeders. As a group they are more adaptable to changing habitats and vegetation and are

FIGURE 1. Comparison of concentrate and bulk feeders: (a) rumen position of concentrate feeder; (b) omasum; (c) dorsal wall of rumen with papillae; (d) rumen position of bulk feeders; (e) omasum; (f) dorsal wall of rumen — few papillae.

also more regionally and seasonally adapted. They are opportunistic, utilizing the best and most nutritious substrates available. Seasonal changes in diet are accompanied by a rearrangement of mucosal structures, which reduces or enlarges the absorptive surface.[6]

C. Other Foregut Fermenters

Several other artiodactyls are forestomach fermenters but not true ruminants. Langer[19] deals with this group in general and with the hippopotamus in particular. The adaptational significance of the forestomach of the collared peccary (*Dicotyles tajacu*) is described. The pigs, peccaries, and hippos are grouped in the suborder *Suina*. Only the peccaries and hippos have extensive pregastric fermentation. The *Camelidae* (camels and llamas) are placed in the suborder *Tylopoda*, but are true forestomach fermenters and also chew the cud. Although the ruminant digestive tract is considered the most highly evolved, camels have high digestive efficiency. Forestomach fermenters have part or the whole forestomach covered with cornified epithelium. In most cases, this occurs over the papillae which increases the absorptive surface area.[18]

II. DIGESTIVE PROCESSES

Mammalian herbivores have not developed a digestive enzyme to hydrolyze β-glycosidic structural polysaccharides.[20] Instead, they harbor microorganisms in the digestive tract as symbionts to ferment structural polysaccharides. The relationship is one of true mutualistic symbiosis. The herbivore provides the forage substrate and microecological conditions for the microbes to maintain themselves. In return, the microbes provide various vitamins and fermentation end-products which are essential to the metabolism of the host.

A. The Digestive Symbionts

Rumen microorganisms are predominantly obligate anaerobes. A diverse microbial population is normally present as a result of the variety of substrates for fermentation. Microbes of the gut consist of two main groups, the bacteria and protozoa.

Classification of bacteria is based on metabolic, morphological, and cultural features as well as antigenic characteristics and chemical composition. The most generally accepted system of classification and nomenclature is that of Sherman et al.[21] The classification and functional role of bacteria in the gut of mammalian herbivores and particularly in domestic ruminants have been adequately reviewed.[22-29] Very little information is available on the activity of bacteria in wild herbivores.[30] But, since most of the bacteria found in wild herbivores also occur in domestic ruminants, similar functions can be assumed.

Dogiel[31] compiled an extensive review on rumen protozoa which includes those occurring in wild ruminants. These protozoa are divided into flagellates and ciliates, although very few flagellates are found in the rumen. Since Dogiel's monograph was published in 1927, a number of studies have been undertaken to determine the taxonomic relationships of ciliate protozoa. This has resulted in the description of a number of new species.[25,30] Availability of study material, uncertain taxonomy, and time involved probably act as major deterrents to more in-depth studies.[30]

The two main families of ciliates are the *Isotrichidae* and *Ophryoscolecidae*, averaging about 96% of all protozoa present in the rumen. According to Hungate,[15,22] ciliates in cattle double their numbers each day and in general comprise 2% of the total rumen content. They ferment both structural and soluble carbohydrates. The *Ophryoscolecidae* ingest rumen bacteria and thus slow down the rate of starch digestion. Individual *Entodinium* species engulf between 1,000 and 20,000 bacteria per hour depending on the size of the protozoa. In the larger genera, such as *Epidinium*, *Polyplastron*, and *Diplodinium*, up to 100,000 bacteria are consumed per individual per hour.[32] As in the case with plant material, the bacteria

Table 1
RUMEN GAS COMPOSITION OF SOME AFRICAN WILD HERBIVORES

	CH_4	CO_2	H_2	Ref.
Concentrate selectors				
Nyala	25.80	74.20	0.005	37
Dik-dik	10.80	89.10	0.10	38
Grey duiker	22.40	74.02	0.07	40
Giraffe	15.36	84.44	0.20	37
Intermediate feeder				
Springbok	21.21	74.62	0.09	37
Bulk eaters				
Buffalo	27.25	72.75	0.03	37
Black wildebeest	16.93	76.26	0.06	39
Red hartebeest	20.10	78.52	0.05	37
Oryx	19.95	73.18	0.06	37
Blue wildebeest	26.30	73.00	0.07	37
Elephant (colon)	65.40	34.00	0.60	4

consumed by protozoa are fermented to VFA, H_2, CO_2, and CH_4. Microbial protein is subsequently digested in the abomasum.

Van Hoven and Prins[33] and Prins and van Hoven[34] studied the metabolism of the holotrich protozoa, *Isotricha prostoma* and *Dasytricha ruminantium*. These ruminal holotrichs have the ability to ferment various carbohydrates such as cellobiose, galactose, raffinose, fructose, maltose, pectin, and glucose. Soluble carbohydrates in particular are taken up and transformed into storage amylopectin. When the glucose concentration in the solution falls below a certain minimum, storage amylopectin is again fermented with the release of VFA. This might also be the case in the unusually high numbers of holotrichs in the stomach compartments of the hippopotamus.[35] The hippopotamus grazes on land at night and lies in the water during the day while digesting its feed.[36] These holotrichs may absorb surplus soluble carbohydrates during the night and transform it into storage amylopectin which is fermented during the day providing a more constant supply of VFA.

A variety of factors in the herbivore gut influence the optimal activity of both bacteria and protozoa. Temperature can be regarded as one of the most important since it not only affects cellular reactions but also affects pH, water and ion activity, viscosity, hydration and aggregation of macromolecules, toxic action of metabolic products, and solubility of gases.[24]

Gases in the gut, apart from N_2 and O_2, are of microbial origin. The composition of the gas mixture varies with animal and feeding type (Table 1). The generally low level of H_2 reflects the activity of the methanogenic bacteria which use molecular H_2 to reduce CO_2 and CH_4. The CO_2 originates both directly from fermentation, and from bicarbonate and carbonates.

The protozoa are generally more sensitive to pH than bacteria. In general, both bacteria and protozoa are found in highest numbers at pH 5 to 7.5. In ruminants, most microbes are destroyed and digested in the abomasum where the pH is usually 2.

B. Fermentation Rate

The rate at which fermentation takes place in the gastrointestinal tract can be expressed in terms of the volume of gas produced per gram of dry matter per hour (μmol/g DM·hr). This rate varies according to the quality of the feed and the amount consumed. In various species, the rate of fermentation is lower in winter than in summer and diurnal variations are related to feeding times.[4,8,9,12,37,39,40]

A direct relationship exists between production of gas and VFA (mmol/100 g rumen contents · hr) within a species at a particular point in time. Using data from Hungate,[22] Prins et al.[30] computed a correction factor of 0.6 for converting gas production to VFA production. Giesecke and van Gylswyk[41] used a factor of 0.9 and Van Soest,[26] 1.0. We found this factor to vary with species. The black wildebeest (with a rumen fermentation rate of 202.7 µmol/g DM·hr and a rumen VFA production rate of 1.67 mmol/100 g·hr in summer) gives a factor of 0.8 but in winter this becomes 1.1. During winter, the factor in the elephant hindgut was found to be 0.9, in the buffalo rumen, 0.6, and in dik-dik, 0.3. Although the fermentation rate in small ruminants such as dik-dik is much more rapid per gram DM than in the larger bulk eaters, VFAs are less important in satisfying energy requirements.[26,30,42] Thus, small ruminants must depend on the absorption of appreciable quantities of nutrients from the gastrointestinal tract beyond the rumen to supplement energy that becomes available from VFA production during the fermentation process.

C. Volatile Fatty Acids

Wild herbivores (especially ruminants) utilize the end-products of microbial fermentation to satisfy a large part of their energy requirements. Natural feedstuffs are broken down and converted to substances such as VFAs, methane, amino acids, and ammonia which are then either excreted or used by the herbivore.

Volatile fatty acids in the rumen and hindgut include formic, acetic, propionic, butyric, valeric, isobutyric, isovaleric, and traces of higher acids. Variation in the total VFA concentrations as well as the molar proportions can be found in Tables 2 to 4. The large range can be expected due to various factors affecting rumen fermentation, different analytical and sampling methods, and differences in diets. However, molar proportions are remarkedly similar, even though much of the data includes animals consuming a variety of natural feedstuffs.

In the elephant, concentrations of VFAs are similar to those found in bulk or roughage feeders; the major contribution is from acetic acid. The rhinoceros resembles the elephant in this respect.[44] The hippopotamus shows a pattern of VFA production which is strikingly similar to typical ruminants.

Among ruminants, acetic acid is present in high proportions regardless of feeding type. Therefore, the term "concentrate" feeder does not refer to a high proportion of propionic acid, but rather to food preferences and anatomical variations. On the other hand, no data are available regarding the specific absorption of these acids across the rumen wall and higher rates of absorption might be expected in concentrate selectors because the rumen walls are densely packed with well-developed papillae. The high food intake and rapid rate of passage also imply rapid absorption in order to maintain an adequate supply of energy. However, there may be an additional supply of VFAs from the hindgut of these animals as found in the grey duiker.[46] The amount of VFA produced in the caecum during winter was twice that produced in summer. In mule deer, however, the fermentation rates in the caecum and colon were slightly higher than in the rumen, but the small volume of ingesta in these organs may account for the minor contribution to total VFA production.[53]

The proportion of VFAs originating from protein is not well understood but branched-chain acids (isobutyrate and isovalerate) arise from the fermentation of certain amino acids. All three groups of African ruminants have examples where these acids are found in large quantities. The hindgut fermenters do not produce as much branched-chain acid from which one would deduce limited conversion from protein.

Forages containing an abundance of soluble carbohydrates produce more volatile fatty acids than those high in fiber and insoluble components. This leads to greater proportions of propionic acid relative to acetic acid and less methane production.[46] This implies better energy utilization, particularly in concentrate feeders. Inhibition of methane production

Table 2
TOTAL VOLATILE FATTY ACIDS AND MOLAR PROPORTIONS FOUND IN THE ELEPHANT, HIPPO, AND RHINOCEROS

	TVFA mmol/100 mℓ	C_2	C_3	iC_4	C_4	iC_5	C_5	Ref.
African elephant								4
Caecum	8.6	76	10	0.5	11	0.5	0.8	
Anterior colon	9.6	72	11	0.9	11	1.0	1.6	
Posterior colon	10.1	71	10	1.9	10	2.1	2.8	
Mesocolon	8.3	75	10	1.2	9.3	2.1	2.1	
Hippopotamus								36
Blindsacs								
Dorsal visceral	13.13	62	24	0.7	10	0.5	1.0	
Vental visceral	12.94	62	24	0.4	10	0.5	1.9	
Dorsal parietal	14.64	62	25	0.7	10	0.6	1.7	
Ventral parietal	15.00	63	24	0.8	10	0.6	1.5	
Vestibulum	14.96	63	24	0.7	10	0.4	1.3	
Connecting chamber	12.38	66	23	0.5	8	1.0	0.8	
Diverticula	11—17							43
Vestibulum	6—8							
Posterior chamber	2—5							
Black rhinoceros								44
Caecum	14.4	79	14	6				
Colon	5—8	69	17	10				

Table 3
TOTAL VOLATILE FATTY ACIDS (TVFA, mmol/100 mℓ) AND MOLAR PROPORTIONS (%) FOUND IN AFRICAN RUMINANTS

	TVFA	C_2	C_3	iC_4	C_4	iC_5	C_5	Ref.
Concentrate selectors								
Suni	10.5	77	16		6	0.3	0.4	38
Dik-dik	10.9	70	20		8	0.5	1.3	38
Duiker (Harvey's)	17.3	75	18	0.04	6	0.9	0.8	45
Duiker (common)	11.7	70	19	1.2	7	1.0	1.8	46
	12.4	70	19	1.0	7	0.7	2.0	
Kudu	14.2	74	16	0.3	8	0.7	2.0	41
	15.2	71	15	1.1	10	0.8	1.3	37
Giraffe	15.7	76	12	0.4	9	0.4	0.6	37
Intermediate feeders								
Springbok	12.0	66	17	0.9	15	0.7	0.8	41
	13.3	61	25	1.6	10	0.8	1.5	37
	15.3	66	21	0.9	9.3	0.4	2.5	
Eland	14.9	76.3	13.0		9.4			47
Roughage feeders								
Buffalo	12.0	72	16	0.7	10	0.7	0.5	41
Wildebeest (black)	13.9	70	12	1.4	13	1.7	1.2	39
Wildebeest (blue)	12.6	65	16	1.3	13	1.0	3.4	37
	20.5	71	14	0.6	12	0.3	0.7	
Hartebeest (Coke's)	13.8	76	13	0.8	10	<1	<1	10
		60	20	1.4	12	0.6	1.5	
Hartebeest (red)	11.2—18.8	71	13	1.1	15	0.5	1.4	37
Topi	27.9	65	19	1.6	12	<1	<1	10
Gemsbok	13.7	75	13	0.8	10	0.5	0.4	41
	16.1	67	19	1.5	11	0.7	1.7	37
	18.8	71	17	1.1	8	0.5	1.8	
Waterbuck	15.6	76.4	14.1		8.3			

Table 4
TOTAL VOLATILE FATTY ACIDS AND MOLAR PROPORTIONS FOUND IN NORTHERN RUMINANTS

	TVFA mmol/100 mℓ	C_2	C_3	iC_4	C_4	iC_5	C_5	Ref.
Moose	6.5	73	18		9		4.1	48
Mule deer		66	21		11		2.4	49
Red deer	10.64	67	24	0.4	10	0.2	0.4	50
Fallow deer	11.9	57	29	0.5	11	1.1	1.2	50
Black-tailed deer	6.5	68	16		13		2.0	51
	9.0	56	21		19		4.0	
Reindeer	3.0	76	13		8.3		3.0	52
White-tailed deer	8.9	62	26		9.7			52
Pronghorn	11.7	70	17		9.0		4.0	52

channels carbon and hydrogen into propionate, thus increasing energy available for the host. However, this process could decrease availability of microbial protein, even though this may not be a primary effect. Similar shifts in acid balances and proportions in the rumen are often associated with the species and numbers of protozoa present, which also depends on the host being studied. Such shifts may increase the total VFA concentration and cause a drop in rumen pH. Excess of acetic and propionic acid cause a drop in voluntary food intake, but this phenomenon is seldom seen where animals are fed native forages.

Volatile fatty acid levels and ammonia concentrations in sheep and deer given the same food showed similarities.[53] Small differences in the deer suggested limited buffering capacity. The major factors affecting rumen fermentation patterns would be a choice of different diets and the seasonal variation in food intake. Roughage feeders usually have a long retention time which encourages a diverse ciliate fauna and high degree of cellulolysis with concomitant rises in acetic acid production.

III. ENERGY AND NITROGEN SUPPLIES

A. Energy

The potential energy available to herbivores from gastrointestinal fermentation includes the bodies of microorganisms and their reduced products with the exception of methane. Anaerobic fermentation allows volatile fatty acids to be used by the host, even though the process may not be as energetically efficient as an aerobic one. If an aerobic system were used, or the retention time too long, then no metabolizable products other than the microbial cells would be available to the host. Thus, the "inefficient" anaerobic system optimizes yields of VFAs and microbial cells in order to maximize energy supplied to the host. Energy expended in maintaining a bacterial population is lost to the host. Some substrates of low digestibility may be lost to both the bacteria and host because they cannot be retained long enough to be fermented.

In a study on the African elephant, the amount of energy produced per day from VFAs was 186 MJ,[4] enough to supply 1.6 times the fasting metabolic requirements (298 $W^{0.75}$). However, 73.4 MJ was lost as methane. According to Foose,[54] 601 MJ organic material is consumed per day and the amount of energy lost per day as methane would then be 27% of the digestible energy. If volatile fatty acid production were less, the loss of energy as methane probably would critically limit digestive efficiency. But an additional source of energy (which has not been measured) is the digestion of soluble sugars in the stomach. This could well compensate for energy losses further along the gut. Hindgut fermenters absorb these soluble sugars in the small intestine, leaving the fibrous portions of the diet to be fermented in the hindgut.

In the moose, the energy contained in VFAs is 57% of the digestible energy and 70% of the metabolizable energy.[55] This was calculated by estimating VFA production rates in the rumen, determining molar percentages of fatty acids present, and converting the total to energy. In October, the energy available from VFAs was 69% of the basal metabolic rate. This poses the question of how these animals survive the winter months. Gasaway and Coady[55] found that additional energy was mobilized from adipose tissue stores and could also be obtained from fermentation in the hindgut. In spring, VFAs from the rumen contributed 98% to the basal metabolic rate. This is due to the selection of superior quality foods during the summer and a greater intake of those foods.

Approximately 30% of the 320 KJ daily energy requirement of white-tailed deer is supplied by VFAs[56] if the turnover rate is low. Smaller animals have higher relative maintenance requirements than larger animals. A study of pronghorn antelope indicated that 68 to 73% of the gross energy intake was necessary for maintenance,[57] while the eland and wildebeest needed 59%.[58]

Allo et al.[51] compared energy production in black-tailed deer and sheep. The energy produced by VFAs in the rumen contributed between 45 and 72% of the maintenance requirement of the sheep. Contributions from the hindgut were found to be small. However, Boomker[46] found high VFA contributions from the hindgut of grey duiker. The energy obtained from rumen volatile fatty acid production was 36% in the summer and 75% of the basal metabolic requirement in the winter. The caecum further contributed 6% in summer and 11% in winter. Extrapolation to the colon of the grey duiker showed a total of 55% of basal metabolic energy obtained from volatile fatty acids in summer and 107% in the winter. The grey duiker is a small concentrate selector which ferments the more fibrous parts of the diet in the hindgut, especially in winter.

B. Nitrogen

In larger ruminants, up to 2 kg of microbial protein can be digested in the abomasum each day. During fermentation, plant nitrogenous constituents are transformed to microbial protein. The microbial protein is subsequently digested and incorporated into the tissues of the host. Several biochemical steps are thus obvious during the transition of plant nitrogen to herbivore nitrogen. Present day knowledge of these nitrogen transformations is a result of many investigations on domestic stock. Very little information is available for wild herbivores.

Protozoa and bacteria have proteolytic enzymes which digest plant proteins to amino acids and peptides, and the latter may be further fermented to VFA, CH_4, ATP, NH_3, and CO_2. In addition to plant protein, protozoa also hydrolyze bacterial and other protozoan protein which have been engulfed to the same end-products plus urea. The ATP, to a large extent, is again used in the synthesis of microbial protein. In the rumen, the rate of plant protein hydrolysis depends largely on the solubility of protein. Branched-chain fatty acids such as isobutyric acid, 2-methyl- and 3-methyl-butyric acid, and isovaleric acid are the result of fermentation of branched amino alcids such as valine, isoleucine, and leucine. Since cellulolytic and methanogenic bacteria are not capable of synthesizing these amino acids (and they are important growth factors), most of the branched chain fatty acids are directly used again in the synthesis of their corresponding amino acids (thus, the normally low concentrations in free rumen fluid).

Plants also contain nonproteinous nitrogen (NPN) in the form of nitrates, ammonium salts, fatty acids, chlorine, and betaines. These NPN molecules are rapidly hydrolyzed in the rumen to NH_3 which is also the case with urea from saliva.

Loss of nitrogen through urea in the urine can be minimized by reabsorption in the kidneys, recycling to the saliva glands, and back into the rumen to support microbial protein synthesis. Recycling also occurs directly across the rumen wall. Kay et al.[53] review data available on recycling of endogeous urea in wild herbivores. Changes from a high nitrogen summer diet to a low nitrogen winter diet may be associated with increases in urea recycling of 50 to 60%. Under free-ranging conditions, one would assume that rumen conditions would promote microbial growth.

From work on domestic ruminants, it appears that, under physiological conditions, there is a rate of microbial protein synthesis of around 0.22 g/g FOM (fermented organic matter). As this area has not been studied in wild herbivores, it is difficult to assume that a similar rate would apply. However, if seems that the available plant proteins and carbohydrates affect the amount of microbial protein produced. This would also be coupled to the rate of passage of digesta and types of microbes present. The ammonia concentration in the rumen would not limit microbial protein production unless it fell below 1 m*M*. Normally, pH and osmolarity do not vary enough to affect protein production by rumen microbes.

IV. DIGESTIVE EFFICIENCY

The ability of the herbivore to digest foodstuffs lies in the microbial population, retention times, and whether the animal is a foregut or hindgut fermenter. The tempo at which the food is digested will determine how much energy and protein will become available to the animal. This is closely linked to how fast the food passes through the gut. If digestion rates are low and the rate of passage fairly high, the animal will have a better chance to utilize a greater mass of food to acquire enough energy for daily maintenance. This strategy works especially well for hindgut fermenters.

Where gut capacity is small and highly nutritious food is taken (e.g., concentrate selecting ruminants), rapid digestion would increase the availability of nutrients and rapid passage would increase the mass of food taken in for digestion. Both these factors would be advantageous in terms of extracting the maximum amount of nutrients from the available plant population.

The actual rate of digestion can be measured using in vitro trials and calculating the rate of production, or loss of a specific end-product of digestion. We have looked at gas production and found that specific plant substrates vary in rates of digestion.[37,68] However, very little is known in this field as digestion coefficients are usually used. The coefficient is an estimate of the amount digested in relation to the amount given in an in vitro trial or the amount supplied in an in vitro experiment, and does not reflect the tempo at which the plant material is digested.

A. Digestibility

Various methods for studying digestibility have been employed, among others the in vivo apparent digestibility method using natural feedstuffs and in vitro methods such as the nylon bag and Tilley and Terry[59] techniques. Forage digestibility has also been predicted using summation equations based on chemical analyses and digestibility estimates of various components. In vitro systems are more accurate since the microorganisms and enzymes are sensitive to various factors which may influence the rate of digestion. Chemical analyses are often faster and more repeatable but do not precisely simulate the actual biological process.

Comparative data on digestive capacities of various herbivores have started to accumulate largely from studies on zoo animals.[54,60] Foose[54] estimated that elephants could digest 45% of the dietary matter of hay, but did have a low retention time and could possibly take in more feed than a ruminant. Another very large herbivore, the white rhino, was found to digest hay at the 60% level while in the black rhino the digestibility was only 35%.[60] This is due to the fact that the black rhino is a selective browser and does not digest roughages to the same extent as the white rhino which is a bulk feeder.[60] Foose[54] also found similar values during zoo studies on the two species.

Arman and Field[43] used a hippopotamus in various digestibility studies. The initial values were similar to other ruminants, but lower than would be expected (47%). The digestibility of elephant grass was 68% of the dry matter which compares well with values obtained in sheep on similar diets. These observations were confirmed by Foose[54] in zoo studies and correlated well with the finding in other wild ruminants.[60]

There are great differences in digestibility among the ruminants. Van Soest[26] carefully reviewed available literature and concluded that large grazing species have a higher digestive ability than cattle while smaller concentrate selectors have a low digestive ability. Bison and yak are superior to cattle in their ability to digest hays[61] but this is not necessarily associated with better feed conversion and gain.[62] Prins et al.[30] found that various grazers and intermediate feeders digested more than 80% of available cell-wall matter (buffalo, bison, banteng, blesbok, waterbuck, and others) while two browsers, the kudu and okapi, showed low digestibility coefficients.

Several studies have used ruminal fluids from caribou, mule deer, and white-tailed deer for in vitro digestibility estimates.[63-66] Values ranged from 24 to 68% depending on the plant species. Alfalfa or hay showed high or low digestibilities depending on whether the animal was a concentrate or roughage feeder.

Mehra et al.[67] found two distinct phases during in vitro digestion. The disappearance of nonfibrous dry matter was completed in phase (A) while the disappearance of fiber was confined to phase (B). Production of methane and volatile fatty acids was low during phase (A) and some 40% higher than accepted values during phase (B). Inoculum source may have profound influence on the digestibility of the plant material.[68] This may be related to the time of year and the donors' feed selection. Different donor species resulted in different digestibilities of the same plant substrate which agreed with in vivo zoo studies of Foose.[54] However, similar results were obtained within one species (kudu) when the donor animal was obtained during different months of the year. Thus, variation among and within species must be examined as well as differences resulting from different diets. Variations in experimental techniques also may complicate the matter.

Arman and Hopcraft[69] examined in vivo digestibilities, using inoculum from eland, hartebeest, Thomson's gazelle, and grey duiker. The mean digestibility of dry matter was found to be higher in hartebeest and duiker than in gazelle and eland. The relative efficiencies of the animals fed pelleted diets was not the same when animals were given more natural diets, and showed no relationship to body size.

Boomker and van Hoven[70] studied dry matter disappearance in relation to energy production in terms of methane and volatile fatty acids in vitro. Various plant species were incubated with rumen fluid from kudu. Each plant species differed in terms of methane and fatty acid formed, but plant species preferred by the kudu had high digestibilities and produced large amounts of energy in the form of volatile fatty acids. If the in vivo situation closely approximates the rate of production of fatty acids in vitro, quantitative data regarding the natural situation could easily be obtained.

Factors that would effect both in vivo and in vitro digestibility trials include the plane of nutrition of the animal. Digestibility in vivo decreases as intake increases due to decreased retention time. The amount of dietary fiber and lignin will depress digestibility as fiber increases. Retention time will also be relatively shorter because fiber cannot be completely utilized. Species differences are inherent and can affect in vitro and in vivo trials. The division into bulk and concentrate feeders is clearly shown when one type of feed is used in trials concerning both groups. Nutrient imbalance will depress the digestibility of any feedstuff, especially if there is a protein deficiency. In vivo trials will also be affected by such factors as appetite, frequency of feeding, adaption to ration changes, and feed preparation as well as various associative effects.

B. Rate of Passage

Rate of passage determines how long individual portions of ingesta are subjected to various processes including mixing, digestion, microbial fermentation, and absorption. Passage rate is commonly expressed as retention time. To simplify mathematical calculations, the assumption of a steady-state system is made. This implies that the removal of digesta from the system equals the inflow into the system and that all volumes remain constant. Under these conditions, the mean retention time in any part of the gut is equal to the volume of the contents divided by the volumetric flow rate. In nonsteady-state situations, relations are not this simple.

Measurement of rate of passage is usually accomplished by giving an indigestible substance (marker) with the food. The marker can subsequently be measured in the feces or various gut sections. Markers can be either soluble or particulate. Polyethyleneglycol and Cr-ethylenediaminetetra-acetic acid remain in solution through the gut. Chromic oxide and radiopaque pellets are particulate markers. Some markers such as stains and ruthenium-phenanthroline are particulate markers that associate with food particles, although they are administered to the animal in solution.

Certain markers may be slightly digested and most are difficult to determine accurately. The distribution of solutes between the fluid compartments of the animal body and the solid digesta is generally unknown, and some portion of the marker may disperse to the fluid compartments. Particulate markers differ in size, specific gravity, and surface properties from the insoluble material of the digesta and could thus pass through the gut at a different rate. These factors all have an effect on calculated retention time. How much may be difficult to determine.

Various methods for estimated rate of passage are evaluated by Warner.[71] Two methods eminently suitable for wild herbivores are pulse-dose total collection and sample collection methods. The marker is given in a single dose and, in the first case, all the feces are subsequently collected and analyzed. The marker concentration rises rapidly after a delay and then falls slowly. The curve of the cumulative excretion of marker is consequently sigmoid shaped. In the second case, only samples of feces are collected for determination of marker concentration.

The propulsive movement of the stomach and gut may be the motive force for digesta movement, but the actual movement is also influenced by the volume of material in them and other parts of the gut. Herbivores with fermentation chambers are able to retain the solid fraction of the digesta in the chamber for microbial fermentation, while the more fluid portion can proceed along the gut. Retrograde movement of fluid and fine particles of digesta in the colon and caecum of certain animals also increases the time for microbial fermentation.

Two hindgut fermenters, the Indian elephant and the horse, have been used in rate of passage experiments.[72-74] The elephant has a retention time (approximately 33 hr) which is very similar to that found in the horse (25 to 35 hr, depending on diet). The gut of the elephant is much longer and more voluminous and one would expect a longer retention time of similar foodstuffs. The conditions in these experiments were not identical and could explain these differences. The elephant has very coarse and fibrous feces, whereas the fibrous particles in the horse feces are finer and give a more homogenous appearance. It would thus apear that the relatively longer retention time of the horse aids digestion and would lead to a better efficiency of food utilization. In a study of liquid and solid digesta retention time in several animals including the horse,[75] it was found that the soluble marker retention time was slightly less than that of the particulate marker and both were faster than 33 hr (23 and 18).

Retention times vary widely among ruminants. Using three forms of alfalfa (pelleted, cubed, and baled), Dean et al.[76] measured retention times in wapiti (90% excretion of marker). They found differences between yearling and adult animals as well as among three forms of alfalfa. On the pelleted diet, the retention time was 68 hr for yearlings and 80 hr for adult animals while the baled diet gave retention times of 120 and 140 hr for yearlings and adults, respectively.

Different breeds of cattle, and also the zebra, seem to have similar retention times. The bison and yak have a longer retention time than cattle.[61] These animals were fed a mixture of alfalfa and brome grass. The bison and yak had mean times of 78 hr, 18% longer than the cattle. This would appear to be an inherent animal effect, rather than environment, as the animals were treated similarly for almost one year before the experiment.

Maloiy,[77] in an experiment with Zebu cattle and camels fed poor quality chopped hay, found the two species differed considerably in mean retention time. The mean retention time for camels was 46 hr and that of cattle was 55 hr. The red deer has a retention time 70% that of sheep,[71] but this seems to vary with diet.

The white-tailed deer is one of the few wild animals where retention time has been measured on natural feed.[78] A retention time of 22.5 hr was estimated. When fed a mixture of hay and concentrates, the retention time was 22.7 hr. The mixture closely approximated natural diets.

Two of the smaller African antelope have been used in rate of passage studies, but not on natural feeds. Hoppe and Gwynne[79] studied the suni fed lucerne leaves and lucerne stems and found that the low retention time of 16.8 hr on the leaves increased to 20.5 hr when fed stems. Using Cr-EDTA, the retention time of the fluid portion of the digesta in grey duiker was studied.[40] The mean fluid retention time was found to be 12 hr with an estimated 18 hr for solid particles.

It is obvious that the diet given to the animal profoundly influences the mean retention time, and physical properties seem particulary important. The *Perissodactyla* have a slightly lower retention time than the *Artiodactyla* of similar size. The faster flow rate compensates for the incomplete hindgut digestion and improves energy supply by a larger intake and turnover.

The species of plant eaten and particle size taken (either chewed or not) will also influence the rate of passage. The larger bulk and roughage feeders tend to fill the rumen and ruminate for periods to decrease the size of digesta particles. This causes the digesta to remain in the rumen until the particles are small enough to pass on to the rest of the gut, and leads to high mean retention times. The concentrate feeders and smaller antelopes consume highly nutritious berries, leaves, flowers, and tend to have limited rumen fermentation. The more fibrous part of the food is subjected to hindgut fermentation and the overall rate of passage is faster. This strategy however could lead to lower digestibility and a lack of sufficient energy if unsuitable diets are fed under experimental conditions.

The components of the diet given during the experiment will affect the rate of passage. Different sized particles move at different rates and influenced by the other particles present in the digesta. The quantity of feed given during the experiment also has an effect on the rate of passage. An increase in the amount of food given will require either an increase in gut volume or a faster rate of passage. The usual result is lower retention time. This decrease seems to apply to the entire gut but tends to be the greatest in the hindgut. In animals where the gut is more or less empty on occasion, the results of increased rations vary.

The age of the animal also affects retention time. Animals showing the greatest differences are those being weaned. Pregnant animals also show differences in rate of passage, most of the loss in retention time occurring in the rumen, but this may be regained once more in the caecum and proximal colon.

The frequency of feeding and time of dosing with marker may have an effect on the mean retention time. Other influences are digestibility, rumen fill, microbial activity, and water content of the digesta. These factors interact to control the rate of passage, which in turn influences many events in the digestive tract, and have been used to predict digestive function.[80,81] But to date, no such studies have been carried out with wild herbivores on natural diets.

V. SUMMARY

Mammalian herbivores can be classified either as foregut or hindgut fermenters. Enlarged compartments in the gut developed as a strategy to deal with increased fiber in the diet. Ruminating foregut fermenters can be divided into concentrate selecters, intermediate feeders, and bulk feeders, depending on the stomach structure and the type of diet utilized.

To hydrolyze β-glycosidic structural polysaccharides, herbivores acquired a range of digestive symbionts (bacteria and protozoa) which established themselves in the enlarged compartments. Fermentation provides a variety of substrates for the host. Volatile fatty acids usually satisfy a large part of the hosts' energy requirements. In the elephant and rhinoceros, the major contribution is from acetic acid. The ratio of volatile fatty acids varies with the type of food consumed and the time after feeding.

Digestibility of feedstuffs depends on the retention times and the microbial population present. The hippopotamus is capable of forestomach digestion equal to that of grazing ruminants even though there is no rumination. Retention times of foodstuffs vary widely and have not been well studied using natural foodstuffs.

Studies on the digestive processes in wild herbivores are increasing, but to date, the variety in species seems to have precluded in-depth work. Almost all aspects of digestion in wild herbivores need attention due to the diversity of species and variety of habitats in which they live.

REFERENCES

1. **Moir, R. J.,** *Proc. Symp. Herbivore Nutrition in the Tropics and Subtropics*, Gilchrist, F. M. C. and Mackie, R. I., Eds., Ad Donker, Johannesburg, 1984.
2. **Eloff, A. K.,** The Digestion in the Hyrax (*Procavia capensis* (Pallas 1766)), M.Sc. thesis, University of Pretoria, 1981.
3. **Clemens, E. T.,** Sites of organic acid production and patterns of digesta movement in the gastrointestinal tract of the rock hyrax, *J. Nutr.*, 107, 1954.
4. **Van Hoven, W., Prins, R. A., and Lankhorst, A.,** Fermentative digestion in the African elephant, *S. Afr. J. Wildl. Res.*, 11, 78, 1981.
5. **Bell, R. H. V.,** A grazing ecosystem in the Serengeti, *Sci. Am.*, 225, 86, 1971.
6. **Hofmann, R. R.,** *The Ruminant Stomach*, East African Monographs in Biology, East African Literature Bureau, Nairobi, 1973.
7. **Maloiy, G. M. O. and Clemens, E. T.,** Gastrointestinal osmolarity, electrolyte and organic acid composition in five species of East African herbivorous mammals, *J. Anim. Sci.*, 51, 917, 1980.
8. **Moir, R. J.,** Ruminant digestion and evolution, *Handb. Physiol.*, 6, 2673, 1968.
9. **Langer, P.,** Stomach, evolution in the *Artiodactyla, Mammalia*, 38, 295, 1974.
10. **Hoppe, P. P.,** Rumen fermentation and body weight in African ruminants, *Proc. Int. Congr. Game Biol.*, 13, 141, 1977.
11. **Hoppe, P. P., Quortrup, S. A., and Woodford, M. H.,** Rumen fermentation and food selection in East African sheep, goats, Thomson's gazelle and impalas, *J. Agric. Sci. Camb.*, 89, 129, 1977.
12. **Hoppe, P. P., Quortrup, S. A., and Woodford, M. H.,** Rumen fermentation and food selection in East African zebu cattle, wildebeest, Coke's hartebeest and topi, *J. Zool. London*, 181, 1, 1977.
13. **Maloiy, G. M. O., Clemens, E. T., and Kamau, J. M. Z.,** Aspects of digestion in *in vitro* rumen fermentation rate in six species of East African wild ruminants, *J. Zool. London*, 197, 345, 1982.
14. **Van Hoven, W.,** Comparative rumen functioning in Kalahari antelope, *S. Afr. J. Anim. Sci.*, 13, 209, 1983.
15. **Hungate, R. E.,** Microbial fermentation in certain mammals, *Science*, 130, 1192, 1959.

16. **Norden, M.,** Nutritional physiology and behaviour of the lesser mousedeer, Symp. on Animal Populations: Wildlife Management, Bogor, Indonesia, 1978.
17. **Langer, P.,** Functional anatomy of the stomach of *Hippopotamus amphibius* L. 1758, *S. Afr. J. Sci.*, 72, 12, 1976.
18. **Langer, P.,** Macroscopic anatomy of the stomach of the *Hippopotamidae* Gray, 1821, *Zbl. Vet. Med.*, 4, 334, 1975.
19. **Langer, P.,** Adaptational significance of the forestomach of the collared peccary, *Dicotyles tajacu* (L. 1758), (Mammalia:Artiodactyla), *Mammalia*, 43, 235, 1979.
20. **Yohoe, Y. and Yasumasu, I.,** Cellulose in invertebrates, *Comp. Biochem. Physiol.*, 13, 323, 1964.
21. **Sherman, V. B. D., McGowan, V., and Sneath, D. H. A.,** Approved lists of bacterial names, *Int. J. Syst. Bact.*, 30, 225, 1980.
22. **Hungate, R. E.,** *The Rumen and its Microbes*, Academic Press, New York, 1966.
23. **Oxford, A. E.,** A guide to rumen microbiology, *N.Z. Dep. Sci. Ind. Res. Bull.*, 160, 1964.
24. **Clarke, R. T. Z.,** The gut and its micro-organisms, in *Microbial Ecology of the Gut*, Clarke, R. T. J., and Cauchop, T., Eds., Academic Press, New York, 1977.
25. **Ogimoto, K. and Imai, S.,** *Atlas of Rumen Microbiology*, Japan Scientific Societies Press, Tokyo, 1981.
26. **Van Soest, P. J.,** *Nutritional Ecology of the Ruminant*, O & B Books, Corvallis, Oregon, 1982.
27. **Bryant, M. P.,** Bacterial species of the rumen, *Bact. Rev.*, 23, 125, 1959.
28. **Buchanan, R. E. and Gibbons, N. E., Eds.,** *Bergey's Manual of Determinative Bacteriology*, 8th ed., Williams & Wilkins, Baltimore, 1974.
29. **Hobson, P. N. E.,** Rumen bacteria, in *Methods in Microbiology*, Norris, J. R. and Ribbons, D. W., Eds., Academic Press, New York, 1969, 3B.
30. **Prins, R. A., Lankhorst, A., and van Hoven, W.,** Gastrointestinal fermentation in herbivores and the extent of plant cell wall digestion, *Proc. Symp. Herbivore Nutrition in the Tropics and Subtropics*, Gilchrist, F. M. C. and Mackie, R. E., Eds., Ad Donker, Johannesburg, 1984.
31. **Dogiel, V. A.,** Monographie der Familie *Ophryoscoleicidae*, *Arch. Prostistenk.*, 59, 1, 1927.
32. **Coleman, G. S.** Rumen ciliate protozoa, *Adv. Parasitol.*, 18, 121, 1980.
33. **Van Hoven, W. and Prins, R. A.,** Carbohydrate fermentation by the rumen ciliate *Dasytricha ruminantium*, *Protistologica*, 23, 599, 1977.
34. **Prins, R. A. and van Hoven, W.,** Carbohydrate fermentation by the rumen ciliate *Isotricha prostoma*, *Protistologica*, 13, 549, 1977.
35. **Van Hoven, W.,** Ciliate protozoa and aspects of the nutrition of the hippopotamus in the Kruger National Park, *S. Afr. J. Sci.*, 70, 107, 1974.
36. **Van Hoven, W.,** Digestion physiology in the stomach complex and hindgut of the hippopotamus *(Hippopotamus amphibius)*, *S. Afr. J. Wildl. Res.*, 8, 59, 1978.
37. **Van Hoven, W. and Boomker, E. A.,** Unpublished data, 1984.
38. **Hoppe, P. P., van Hoven, W., von Engelhardt, W., Prins, R. A., Lankhorst, W., and Gwynne, M. D.,** Pregastric and caecal fermentation in dik-dik (*Madoqua kirki*) and suni *(Nesotragus moschatus)*, *Comp. Biochem. Physiol.*, 75A, 517, 1983.
39. **Van Hoven, W. and Boomker, E. A.,** Feed utilization and digestion in the black wildebeest (*Connochaetes gnou*, Zimmerman, 1780) in the Golden Gate Highlands National Park, *S. Afr. J. Wildl. Res.*, 11, 35, 1981.
40. **Boomker, E. A.,** A Study of the Digestive Processes of the Common Duiker, *Sylvicapra grimmia* (Linn), M.Sc., thesis, University of Pretoria, 1981.
41. **Giesecke, D. and van Gylswyk, N. O.,** A study of feeding types and certain rumen functions in six species of S.A. wild ruminants, *J. Agric. Sci. (Cambridge)*, 85, 75, 1975.
42. **Parra, R.,** Comparison of foregut and hindgut fermentation in herbivores, in *Ecology of Arboreal Folivores*, Montgomery, G. G., Ed., Smithsonian Inst. Press, Washington, 1978, 205.
43. **Arman, P. and Field, C. R.,** Digestion in the hippopotamus, *East Afr. Wildl. J.*, 11, 9, 1973.
44. **Clemens, E. T. and Maloiy, G. M. O.,** The digestive physiology of the East African herbivores, the elephant, rhinoceros and hippopotamus, *J. Zool. London*, 198, 141, 1982.
45. **Hoppe, P. P., Gwynne, M. D., and van Hoven, W.,** Nutrients, protozoa and volatile fatty acids in the rumen of Harvey's red duiker *Cephalophus harveyi*, *S. Afr. J. Wildl. Res.*, 11, 110, 1981.
46. **Boomker, E. A.,** Volatile fatty acid production in the grey duiker *Sylvicapra grimmia*, *S. Afr. J. Anim. Sci.*, 13, 33, 1983.
47. **Maloiy, G. M. O., Clemens, E. T., and Kamau, J. M. Z.,** Aspects of digestion and *in vitro* rumen fermentation rate in six species of East African wild ruminants, *J. Zool. London*, 197, 345, 1982.
48. **Gasaway, W. C. and Coady, J. W.,** Review of energy requirements and rumen fermentation in moose and other ruminants, *Nat. Can.*, 101, 227, 1974.
49. **Short, H. L., Medin, D. E., and Anderson, A. E.,** Seasonal variation in volatile fatty acids in rumen of mule deer, *J. Wildl. Manage.*, 30, 466, 1966.

50. **Prins, R. A. and Geelen, M. J. H.,** Rumen characteristics of red deer, fallow deer and roe deer, *J. Wildl. Manage.*, 35, 673, 1971.
51. **Allo, A. A., Oh, J. H., Longhurst, W. H., and Connolly, G. E.,** Volatile fatty acid production in the digestive system of deer and sheep, *J. Wildl. Manage.*, 37, 202, 1973.
52. **Bruggeman, J., Drescher-Kaden, N., Schubert, R., Erbersdobler, H., and Giesecke, D.,** Comparative study in reindeer and white-tailed deer of Finland on rumen metabolism and fatty acids of adipose tissues, *Biol. Pap. Univ. Alaska*, 1, 290, 1976.
53. **Kay, R. N. B., Engelhardt, W. V., and White, R. G.,** The digestive physiology of wild ruminants, in *Digestive Physiology and Metabolism in Ruminants*, Ruckebusch, Y. and Thivend, P., Eds., MTP Press, Lancaster, England, 1979.
54. **Foose, T. J.,** Trophic Strategies of Ruminant versus Non-ruminant Ungulates, Ph.D. thesis, University of Chicago, 1982.
55. **Gasaway, W. C. and Coady, J. W.,** Review of energy requirements and rumen fermentation in moose and other ruminants, *Nat. Can.*, 101, 227, 1974.
56. **Short, H. L.,** Rumen fermentation and energy relationship in white-tailed deer, *J. Wildl. Manage.*, 27, 184, 1963.
57. **Wesley, D. E., Knox, K. L., and Nagy, J. G.,** Energy metabolism of pronghorn antelope, *J. Wildl. Manage.*, 37, 563, 1973.
58. **Rogerson, A.,** Energy utilization by the eland and wildebeest, in *Comparative Nutrition of Wild Animals*, Crawford, M. A., Ed., Academic Press, N.Y., 1968.
59. **Tilley, J. M. and Terry, R. A.,** A two stage technique for *in vitro* digestion of forage crops, *J. Br. Grassl. Soc.*, 18, 104, 1963.
60. **Ullrey, D. E., Robinson, P. T., and Whetler, P. A.,** Comparative digestibility studies with zoo herbivores, *AAZV Annual Proc.*, 1979.
61. **Schaefer, A. L., Young, B. A., and Chimwano, A. H.,** Ration digestion and retention times of digesta in domestic cattle (*Bos taurus*), American bison (*Bison bison*) and Tibetan yak *(Bos grunniens), Can. J. Zool.*, 56, 2355, 1978.
62. **Richmond, R. J., Hudson, R. J., and Christopherson, R. J.,** Comparison of forage intake and digestibility by American bison, yak and cattle, *Acta Theriol.*, 22, 225, 1977.
63. **Urness, P. J., Smith, A. D., and Watkins, R. K.,** Comparison of *in vitro* dry matter digestibility of mule deer forages, *J. Range Manage.*, 30, 119, 1977.
64. **Thomas, D. and Kroeger, P.,** Digestibility of plants in ruminal fluids of Barren-ground caribou, *Arctic*, 34, 321, 1981.
65. **McNeely, S. C. and Schemnitz, S. S.,** Chemical composition and *in vitro* digestibility of deer browse three years after a wildfire, *Southwest. Nat.*, 26, 365, 1981.
66. **Palmer, W. L. and Cowan, R. L.,** Comparison of deer and sheep digestive capacities, *J. Wildl. Manage.*, 43, 798, 1979.
67. **Mehra, U. P., Czerkawski, J. W., and Breckenridge, G.,** Incubation of hay with micro-organisms; two distinct phases of digestion, *Proc. Nutr. Soc.*, 40, 23A, 1981.
68. **Van Hoven, W. and Boomker, E. A.,** The influence of inoculum source on *in vitro* digestibility, *S. Afr. J. Anim. Sci.*, 13, 207, 1983.
69. **Arman, P. and Hopcraft, D.,** Nutritional studies on East African herbivores, *Br. J. Nutr.*, 33, 255, 1975.
70. **Boomker, E. A. and van Hoven, W.,** *In vitro* digestibility of plants normally consumed by the kudu, *Tragelaphus strepsiceros, S. Afr. J. Anim. Sci.*, 13, 206, 1983.
71. **Warner, A. C. I.,** Rate of passage of digesta through the gut of mammals and birds, *Nutr. Abstr. Rev. Series B*, 51, 789, 1981.
72. **Gill, J.,** Szybkose przechodzenia tresci przec praewod pokarmowy solma indyjskiego (*Elephus maximus* L.) w warunnkach ogrodu zoologicznego, *Acta Physiol. Pol.*, 11, 277, 1960.
73. **Wolter, R., Dutix, A., and Letourneau, J. C.,** Influence du mode de presentation du fourrage sur la vitesse du transit digestif chez le poney, *Ann. Zootech.*, 23, 293, 1974.
74. **Wolter, R., Durix, A., and Letourneau, J. C.,** Influence du mode de presentation dur aliment complet sur en vitesse du transit et la digestibilite chez le poney, *Ann. Zootech.*, 25, 181, 1976.
75. **Uden, P., Rounsaville, T. R., Wiggons, G. R., and Van Soest, P. J.,** The measurement of liquid and solid digesta retention in ruminants, equines and rabbits given timothy (*Phleum pratense*) hay, *Br. J. Nutr.*, 48, 329, 1982.
76. **Dean, R. E., Thorne, E. T., and Moore, T. D.,** Passage rate of alfalfa through the digestive tract of elk, *J. Wildl. Manage.*, 44, 272, 1980.
77. **Maloiy, G. M. O.,** Comparative studies of digestion and fermentation rate in the fore stomach of the one humped camel and the Zebu steer, *Res. Vet. Sci.*, 13, 476, 1972.
78. **Mautz, M. W. and Petrides, G. A.,** Food passage rate in the white-tailed deer, *J. Wildl. Manage.*, 35, 723, 1971.

79. **Hoppe, P. P. and Gwynne, M. D.,** Food retention times in the digestive tract of the suni antelope *(Nesotragus machatus), Saugetierkundliche vir Mitt.*, 26, 236, 1978.
80. **Mertens, D. R. and Ely, L. O.,** A dynamic model of fibre digestion and passage in the ruminant for evaluating forage quality, *J. Anim. Sci.*, 49, 1085, 1979.
81. **Orskov, E. R. and McDonald, I.,** The estimation of protein degradability in the rumen from incubation measurements weighted according to rate of passage, *J. Agric. Sci.*, 92, 499, 1979.

Chapter 6

MAINTENANCE METABOLISM

R. J. Hudson and R. J. Christopherson

TABLE OF CONTENTS

Animals are required to expend a large proportion of ingested energy for physiological homeostasis. In natural environments, additional energy must be allocated for activity and thermoregulation. Collectively, these comprise requirements for "ecological maintenance". This chapter examines physiological maintenance, the base to which costs of activity and thermoregulation are added.

I. MEASUREMENT

A. Techniques

The most direct way to determine maintenance requirements is energy balance trials using several feeding levels, preferably including those which cause energy gain and energy loss. Energy intake of individual animals is expressed on the basis of digestible or metabolizable energy and net energy retention ideally is assessed either by comparative slaughter techniques or less heroic measures of body energy content. Interpolation to zero energy retention provides an estimate of maintenance requirements (Figure 1). Alternatively, feeding levels can be progressively adjusted to maintain energy balance.

Weight stasis is often used as a proxy for energy equilibrium but body fat may be replaced by protein and water, contributing to maintenance energy about 25 kJ/g of fat converted. Particularly for immature animals, energy equilibrium usually implies an increase in body weight.[1] Feeding levels which maintain weight stasis depend on protein levels; diets low in protein must be supplied at levels which are several-fold higher.

If the efficiency with which energy is used for maintenance is known, energy requirements can be calculated from energy expenditures. These can be measured directly as heat loss using gradient-layer calorimetry[2] or, more often, indirectly as respiratory gas exchange which is based on the stoichiometric relationship between energy released during oxidation of organic substrates and O_2 consumption. The energetic equivalent (kJ) of a liter (ℓ) of oxygen depends upon the ratio CO_2 production/O_2 consumption (respiratory quotient, RQ), ranging from 19.606 kJ/ℓ at an RQ of 0.7 to 21.117 kJ/ℓ at an RQ of 1.0:[3]

$$\text{kJ}/\ell = 16.026 + 5.084\ \text{RQ} \tag{1}$$

Respiratory quotients of 0.7 indicate oxidation of fat while values of 1 accompany oxidation of carbohydrates. Depending on metabolic pathways, the RQ associated with protein oxidation is about 0.82. Occasionally, measured RQs are greater than 1. In nonfasted ruminants, CO_2 may be released from rumen fermentation as plant organic matter is converted to volatile fatty acids. High RQs also can occur when fat is being synthesized from carbohydrate substrates.

Ideally, energy expenditures calculated from the exchange of O_2 and CO_2 should be corrected for losses of combustible gases (CH_4) and small amounts of urinary nitrogen according to:[4]

$$H(\text{kJ}) = 16.18\ O_2 + 5.02\ CO_2 - 2.17\ CH_4 - 5.99\ N \tag{2}$$

where O_2, CO_2, and CH_4 are measured in liters at STP and N is measured in grams. However, there is very little error simply measuring O_2 consumption and applying an energetic equivalent of 20.45 kJ/ℓ which corresponds to a RQ of approximately 0.87.[5]

A variety of technical procedures have been used to measure gas exchange.[6] *Closed-circuit calorimetry* measures the change in concentrations of a fixed volume of air in a sealed chamber. *Open-circuit calorimetry* measures changes of oxygen concentrations in air drawn past the subject at a measured rate, or alternatively, the difference between ambient and expired air collected with a mask or tracheal fistula. Because it is less expensive and allows

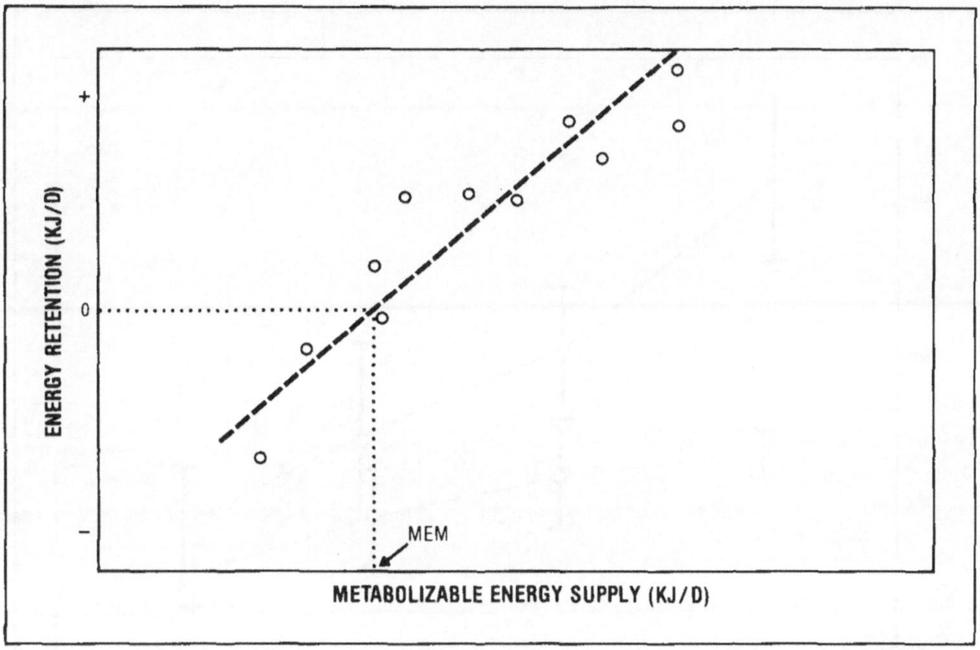

FIGURE 1. Estimation of maintenance energy requirements (MME) using balance trials. With a wide range in energy supply, there is a tendency for the relationship to be curvilinear reflecting diminishing returns at high levels of feeding.

animals to be exposed to natural environments, the latter method has been used widely in studies with captive wild herbivores. Nevertheless, techniques based on gas exchange impose constraints on the normal activities of animals and attempts have been made to develop alternatives.

The most direct is to use continuous infusions of radioactive bicarbonate to label the carbon dioxide pool.[7,8] However, CO_2 is removed from the body pool by means other than gas exchange so the technique must be carefully calibrated. A potentially better marker is doubly labeled water.[9] The hydrogen label allows estimation of oxygen lost as water, while the remaining oxygen is attributed to oxidation of substrates. The limitation has been the expense of the technique for large animals but the falling cost of the isotope and ability to measure smaller quantities may offer new opportunities.

Several cardiovascular respiratory parameters have been calibrated as indexes of energy expenditure. The most popular has been heart rate in spite of some initial rather disappointing results with both performance of equipment and strength of the relationship. However, the technology of biotelemetry has improved providing transmitters with long life and clear signals.[10,11] Although in some studies the correlation of heart rate and energy expenditure has been weak, in others it has been quite good in seasonally calibrated adult individuals.[12-19] Better results seem to be obtained with large confident species than smaller nervous ones. Our experience with wapiti and moose indicates much more stable relationships in adults than in calves. There appears to be little hope of deriving suitable general regressions that could be applied to wild populations but the technique may be quite useful in studies with individually calibrated research animals.

Heart rate is expected to be a satisfactory correlate only if stroke volume is either constant or changes with heart rate in a consistent manner. This generally is the case but under certain conditions, the relationship may break down. Stroke volume was shown to decrease during the act of eating in meal-fed sheep but arterio-venous differences in blood oxygen content

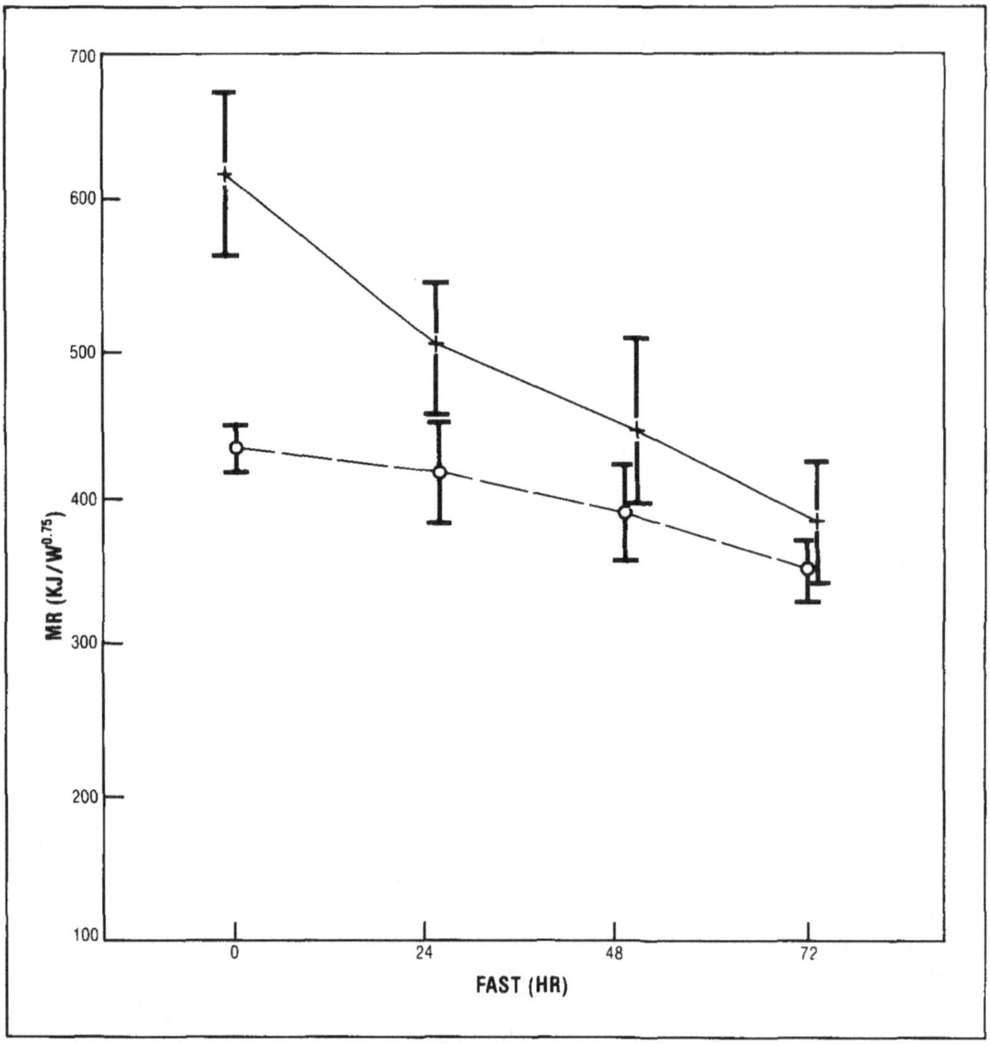

FIGURE 2. Decline in energy expenditures during fasting of wapiti previously maintained on high (+) or low (○) planes of nutrition. Data provided by Watkins, W. G., *Wildlife Productivity and Management Programme Annual Report*, Hudson, R. J., Ed., University of Alberta, Edmonton, 1983, 42.

increased, a change which would tend to compensate for lower stroke volumes.[20] Graham et al.[21] reported an increase in resting heat production of sheep during acclimation to intermittent cold in the absence of a change in heart rate. The relationship also may alter during stress since infusions of epinephrine have produced proportionally greater increases in heart rate than in metabolic rate.[22]

B. Standardization

A variety of factors influence heat production of animals at rest in a thermoneutral environment. Particularly important is the increment associated with digestion and metabolism of nutrients. To permit comparison, conditions of measurement must be carefully standardized.

Metabolic rate declines during fasting, ideally reaching a lower asymptote (Figure 2).[23,24] The length of the fast to achieve this baseline varies with body size, digestive anatomy, and prior feeding level. To minimize the effect of prior nutrition, an attempt is made to obtain a postabsorptive state, a rather difficult task in large animals with complex digestive systems. In studies of farm livestock, the convention is to measure heat production and urinary loss of energy on the third and fourth days of fasting, by which time the RQ has fallen to about 0.7 and methane production is at a minimum.[25]

In animals with copious energy reserves, there is generally little difference in heat production between days 3 and 4. However, in young ungulates, heat production may decline progressively until animals become comatose. Furthermore, the absolute value depends upon the level of feeding prior to the fast.[24] To obtain stable measurement, it is necessary to feed maintenance rations for at least three weeks prior to fasting.[25,56] Turner and Taylor[27] suggest periods which are considerably longer.

Basal metabolic rate (BMR) represents the minimal energy expenditure compatable with life. It is measured under standardized conditions, in which the subject is at rest, postabsorptive, in a thermoneutral environment, in the absence of emotional stress, disease or infection, and following a prolonged period of feeding at maintenance levels. Since such conditions are difficult, if not impossible, to achieve in ruminants, it is common to substitute the less pretentious term, *fasting metabolic rate* (FMR). Although many studies with wild herbivores attempt to standardize the fast, few maintain animals on maintenance planes of nutrition for the suggested 3-week period before fasting. For large herbivores, it is perhaps useful to use the term *standard metabolic rate* (SMR) for animals conditioned at maintenance before fasting and FMR where such conditions are not met.

The difference between SMR and FMR can be interpreted in various ways. It can be viewed simply as a spillover from metabolism of nutrients assimilated prior to the fast. However, Webster[28,29] proposed that it is a distinct entity associated with the priming of tissues to higher levels of metabolic activity. The difference also has been considered a form of "adaptive thermogenesis"[30] which spares energy in times of starvation and dissipates energy when it greatly exceeds requirements to obviate the penalty of extreme obesity.

Energy expenditure of nonfasted animals is termed *resting metabolic rate* (RMR). It requires only that animals be at rest in a defined thermal environment. It is greater than FMR by the heat increment of feeding which extends into postprandial periods. Estimates of this increment vary widely from 13% in bighorn sheep,[31] 20% in reindeer,[32] 25 to 30% in roe deer,[33] to over 50% in pronghorn antelope.[34] These differences are less likely to represent species-specific traits than the results of feeding levels prior to the fast and precise time since the last meal.

Webster[35] has outlined several factors that may contribute to the increased heat production associated with food intake. These include energy expenditures associated with eating and rumination, heat of fermentation arising from rumen microbial activity, energy expended by gut tissues in association with digestive processes, and increased metabolism of tissues other than the digestive tract. For sheep receiving a chopped hay diet, the sum of the estimated costs of eating, rumination, digestive tract metabolism, and heat of fermentation accounted for 141 out of a total heat increment of 340 kJ/MJ of metabolizable energy.[35] The residual 191 kJ/MJ must have occurred in tissues other than the gastrointestinal tract.

II. BIOLOGICAL BASIS

In addition to varying among genotypes, metabolic rates respond to a variety of environmental conditions. Causes for these patterns may be traced to the proportions of metabolically active tissues, neuroendocrine regulation, and ultimately to biochemical processes within cells.

A. Relative Size of Metabolically Active Tissues

Energy expenditures of animals represent the sum of expenditures of individual tissues, each of which may have different metabolic rates. A number of studies have traced differences in whole animal metabolic rate to body composition and the relative sizes of metabolically active organs. Generally, lean body mass is considered to be more metabolically active than fat because of the relatively high cost of protein turnover. This notion is supported by observations by Graham et al.[36] and Toutain et al.[37] who consider it useful to express MR as a function of fat-free body weight. Webster[38,39] noted that although the metabolic rate of fat Zucker rats was only 65% that of lean rats in relation to metabolic weight (kJ/$W^{0.75}$), the maintenance requirement was identical on the basis of body protein (kJ/$protein^{0.75}$). However, recent studies by McNiven[40] with domestic sheep suggest that the relationship may not be that general and may depend on the way in which differences in fatness are achieved. Turner and Taylor[27] conclude that the weight of critical evidence is that maintenance requirements are independent of body composition at least within genotypes.

Seemingly, a less debatable factor is the increased size of metabolically active organs such as the liver, heart, and tissues of the gastrointestinal tract which accompanies higher feeding levels. Their greater proportions may contribute to the higher energy expenditures and apparent maintenance costs of productive animals.[41,42]

B. Neuroendocrine Factors

Thermogenesis induced by cold exposure and diet have many common features and may be subject to similar neuroendocrinological controls.[30] Both factors have immediate and long term effects on energy metabolism and seem to be controlled proximally by the sympathetic nervous system.[43,44]

The sympathetic system stimulates metabolic heat production, cardiac output, and substrate mobilization. In small mammals, the principal mediator is norepinephrine, the adrenergic neurotransmitter, and the most important target tissue is brown fat. Epinephrine from the adrenal medulla has similar but less marked effects. But catecholamines also exert moderate calorigenic effects in adult ruminants[22] and may induce at least part of the heat increment of feeding, as well as the upward adjustment of metabolic rate after chronic exposure to cold environments.

A variety of hormones influence metabolic rate.[41] Thyroid hormones have well-documented effects on energy metabolism, exerting control on the mass and properties of mitochondria and the activity of the sodium pump.[46] However, their effect appears largely due to potentiating sympathetic stimulation. In northern ungulates, thyroid hormone levels tend to be high in summer and low in winter.[47-49] Treatment of reindeer with thyroxine (T4) and triiodothyronine (T3) stimulates feed intake but decreases weight gain suggesting an influence on energy expenditures.[50]

Other hormones, including growth hormone, adrenocorticotropin, and prolactin, appear to increase metabolic rate either by interacting with sympathetic or thyroid controls or by stimulating feed intake. The rate of heat production in cattle is increased for several hours following administration of growth hormone[51] and hydrocortisone.[52] Administration of prolactin to reindeer stimulates both feed intake and weight gain which implies a more moderate effect on energy expenditure than that of thyroid hormones.[50]

Since metabolic cycles of northern herbivores respond to photoperiod, considerable effort has been devoted to describing the complex endocrine changes that occur throughout the seasonal cycle.[53] Gonadal factors may play a role, but much of the cyclic pattern persists after castration in red deer,[54] reindeer,[55] and moose,[47] and is present in females as well as males. There is growing interest in prolactin because it is known to be controlled by photoperiod and appears correlated seasonally with feed intake.[56] However, correlations of circulating hormones and metabolic activity are unlikely to disclose controls since plasma concentrations may not always reflect the turnover rate of a hormone. The physiological effectiveness of hormones is a function of secretion rate, plasma binding proteins, tissue receptors, postreceptor cellular mechanisms, and metabolic clearance mechanisms.[57-79]

C. Biochemical Processes

Hormones and neurotransmitters exert their control ultimately on the metabolic activities of cells. Cellular biochemical processes requiring the expenditure of energy include the costs of ion transport, turnover of cellular constituents (particularly protein and lipid), and transport of precursors.[60-62]

Ion transport accounts for about 20 to 50% of the energy expenditure of many types of isolated cells.[42] The induction of resting metabolic rate during chronic cold exposure in sheep is closely associated with increased ion transport in isolated skeletal muscle fibers.[63] Energy expenditure associated with the sodium potassium pump in isolated intestinal epithelium and isolated hepatocytes is 10 to 42% higher in fed compared to fasted sheep.[63] Ion transport also may contribute to elevated metabolic rates of growing and lactating animals.[62]

Ion movement along electrochemical gradients followed by increased operation of active transport in the opposite direction is a true maintenance expenditure since no net work is performed in maintaining the gradient. The same can be said about other energy-requiring activities such as protein and fatty acid synthesis when this balances the rate of degradation. Approximately 20 to 25% of maintenance heat production appears associated with protein and fatty acid synthesis.[64] This contribution changes with physiological state which influences rates of protein[63] and lipid turnover.[64]

Hence, on biochemical grounds, it is valid to consider the elevated heat production associated with growth, pregnancy, and lactation as a cost of maintaining these productive states as well as supporting net biosynthesis of new tissue or secretions.

III. ALLOMETRIC SCALING

A. Among Species

Many size-dependent phenomena conform to the power law (Chapter 1). Basal metabolic rate is usually expressed on the basis of metabolically active mass ($W^{0.75}$), an empirical base derived from comparison of many species. Basal metabolism can be approximated from the following relationship:[67]

$$\text{BMR(kJ/d)} = 293\ \text{W(kg)}^{0.75} \tag{3}$$

Accordingly, a doubling of mass increases the total basal rate of mammals by 68% ($2^{0.75} = 1.68$) and decreases the metabolic rate of each unit of tissue by 16% ($2^{-0.25} = 0.84$).

The biological explanation for the slope of this relationship has been long disputed. Rubner's Surface Area Law proposed that metabolic rate should scale to heat loss, hence surface area and $W^{0.66}$. But the empirical slope is significantly different from this value, which led Kleiber[67] to believe that maintenance requirements compromised this thermoregulatory imperative with essential antigravity work (e.g., postural maintenance).[68] The problem is that heterothermic and aquatic vertebrates conform approximately to this rule.[69]

Other explanations trace to the proportionally greater amount of metabolically less active tissues which must be allocated to structural support in large animals. However, structural tissues such as skeleton and skin scale to much less than $W^{1.25}$ required to explain the phenomenon. Since protein turnover contributes significantly to metabolic rate, the proportions of lean and fatty tissue in animals of different sizes may be a contributing factor. But relative fatness does not appear to scale allometrically.

Perhaps one of the most probing biomechanical analyses has been offered by McMahon[70] who considered the importance of maintaining elastic similarity with increasing size. Elastic similarity requires a predictable relationship between volume, length, and cross-sectional area. The power output (hence energy requirements) of muscles and presumably all metabolic

Table 1
ALLOMETRIC SCALING OF METABOLIC RATE (MR) WITHIN SPECIES

Species	Sex	Age/Weight	Measurement	MR = aw^b		Ref.
				a	b	
Black-tailed deer	F	Fawn	RMR	623	0.75	72
	M	Fawn	RMR	335	0.94	72
Blesbok		10—45 kg	FMR	835	0.482	73
Cattle		100—600 kg	FMR	530	0.67	25
Pigs		Weaning—mature	FMR	719	0.63	74
Reindeer	M	17—56 kg	RMR	314	1.00	75
		61—79 kg	RMR	110—117	−0.47	75
		76—131 kg	RMR	98	1.20	75
Roe deer		Adult (winter)	RMR	810	0.51	33
		Fawn	RMR	332	0.93	33
Sheep (domestic)		1—6 weeks	FMR	706	0.58	76
		2—27 months	FMR	450	0.61	76

volumes depends solely upon cross-sectional area which scales identically to the metabolic exponent (i.e., α diameter^2 = $(W^{0.375})^2$ = $W^{0.75}$).

B. Within Species

Kleiber's equation (Equation 3) fits metabolic rates of mature animals from a wide range of taxonomic groups. Within a species, the exponent ranges quite widely[71] both below and above 0.75 (Table 1). Part of the explanation lies in the deviation of FMR from true BMR since diets and voluntary intakes of animals prior to fasting measurements are seldom carefully standardized.

When the range of weights is quite wide and is based largely on differences in age, the exponent is typically lower than the interspecies exponent. Since metabolic rates of young rapidly growing animals are high relative to mature ones, the slope is flattened. When nutritional history (i.e., body condition) rather than age contributes most to body weight differences, larger animals tend to have relatively high metabolic rates because of their greater feeding level. This contrives to increase the exponent to values exceeding 1.0. A rather unusual case of negative exponents can occur in animals (particularly males) which exhibit seasonal inappetence in response to photoperiod. In such circumstances, smaller animals may have high metabolic rates because their voluntary feed intake is not suppressed as much as that of larger individuals.

Turner and Taylor[27] review evidence that if the confounding effects of feeding level are removed, energy expenditures within a genotype scale to $W^{1.0}$. They argue that metabolic weight is a correlate of genetic maximum size rather than a consequence of phenotypic size. If metabolic intensity is encoded in the genome, expenditures per unit mass should be similar throughout the maturing process.

IV. TAXONOMIC PATTERNS

Although metabolic rates of a large number of ungulates can be approximated by a simple allometric relationship, some species are consistently over- or underestimated. The most exhaustive studies, of course, have been conducted with domestic animals. ARC[25] cites mean fasting metabolic rates of 319 kJ/kg$^{0.75}$·d for cattle and 210 kJ/kg$^{0.75}$·d for female sheep (15% higher for rams). Only a few wild ungulates have been studied and it is uncertain whether measured contrasts reflect relative alertness, the tendency to use young animals, or deviations from standard procedures rather than true physiological differences. It is not

Table 2
FASTING ENERGY EXPENDITURES OF UNGULATES

Species	Age	Sex	Season/status	Weight (kg)	Expenditure (kJ/W$^{0.75}$)	Ref.
Donkey	Adult	F		178	293	77
Peccary	Adult	M, F	Summer	20	290	78
			Winter	20	348	78
Pig					285	79
Mule deer	Yearling	M	Spring	31—42	372	80
Roe deer	Adult		Winter	18.5	396	81
	8—10 months		Winter	13.7	493	81
			Spring (moult)		490	33
			Summer		416	33
			Winter		382	33
	6—8 months	M			460	33
	Adult				435	33
	Adult	F	Winter (pregnant)	24	434	33
	6—9 months		Winter	14	460	33
White-tailed deer	Adult	B	Summer	59	601	82
			Winter	65	406	82
	Fawn	M	Summer	36	547	82
			Winter	31	377	82
	Adult		Summer	49	585	83
			Winter	68	339	83
Red deer	Adult			58	377	84
	Calf	M	Winter	40	400	85
Moose	Adult	M, F	Winter		460	86
			Winter (lying)		356	86
			Summer		787	86
			Summer (lying)		556	86
Caribou	9 months		Winter	5	405	32
	Calf			75	427	75
			Summer	50	536	87
Reindeer	9 months		Winter		481	32
	Adult		Summer	99	461	88
Pronghorn	2 months			9	490	34
	7.5 months			26	305	34
	12 months			27	313	34
	18 months			38	334	34
Dik-dik		F		4.5	240	89
		M		4.5	219	89
Bighorn sheep	Adult	M	Winter		431	90
		F	Winter		360	90
		M	Winter	68	298	31
		F	Winter	68	256	31
Domestic sheep					303	79
	Lamb	F	Preweaned		350	25
		M	Preweaned		400	25
	Adult	F			210	25
		M			240	25
Domestic goat	Adult				227	79
Eland					465	91
Hartebeest					216	92
					251	93
Mountain Goat	18 months	M	Winter	32	319	94
Wildebeest					436	91
Cattle (Beef)	Growing				332	79
	Adult				319	25
Cattle (Dairy)					490	79
Bison	Yearling	M	Spring	114—152	305	95

Table 3
RESTING ENERGY EXPENDITURES OF UNGULATES

Species	Age	Sex	Season/status	Weight (kg)	Expenditure (KJ/W$^{0.75}$)	Ref.
Donkey	Adult	F		178	353	77
			Hydrated	149	403	96
			Dehydrated		343	96
			Warm (27°C)	106	568	97
			Hot (40°C)	106	364	97
Llama	Adult	M	Warm (27°C)	115	467	97
			Hot (40°C)	115	264	97
Dromedary			Hydrated	407	215	98
			Dehydrated		166	98
Black-tailed deer	Fawn	M	Summer	5	479	73
		F	Summer	5	623	73
		M	Summer	10	655	73
		F	Summer	10	623	73
		M	Summer	15	648	73
		F	Summer	15	591	73
		M	Summer	20	646	73
		F	Summer	20	623	73
		M	Summer	25	644	73
		F	Summer	25	623	73
Roe deer	Adult		Winter	22	476	33
			Summer	23	543	33
			Lact (11 d)	24	948	33
		F	Lact (30 d)	23	637	33
	Fawn		Winter	13	559	33
Red deer	Adult	M	Fall	68	512	99
				58	500	84
Wapiti	Calf		Winter (lying)	100—150	510	100
		F	Winter (lying)	161	368	101
		F	Winter (standing)	157	456	101
					573—673	102

Moose	Calf	M	Winter	88	597	103
	Adult	M, F	Winter (lying)		435	86
			Summer (lying)		619	84
			Winter (lying)		552	84
			Summer (lying)		904	84
Reindeer	Adult	M	Winter	100	556	104
			Summer		523	88
Pronghorn	2 months			9	636	34
	7.5 months			26	477	34
	12 months			27	586	34
	18 months			38	469	34
Chevrotain	Adult	M, F		1.5	294	105
Dik-dik	Adult	M, F		4	354	106
Gazelle (Mountain)	Adult			24	529	107
Blesbok	15 d			7	485	73
	32 d			9	497	73
	95 d			15	378	73
	184 d			24	337	73
	337 d			32	348	73
	504 d			47	301	73
Hartebeest	8 months		16-hr fast	95—100	439	108
Bighorn sheep	Adult	M	Winter		303	109
		F	Winter		291	109
Bison	Calf		Winter	118	441	110
	Calf—yrlg		Winter	110	715	111
			Spring	160	738	111
			Summer	210	721	111
			Fall	298	635	111
Yak			Winter	111	467	110

Table 4
METABOLIZABLE ENERGY REQUIREMENTS FOR MAINTENANCE

Species	Age	Sex	Season/status	Weight (kg)	ME intake (kJ/$W^{0.75}$)	Ref.
Pig	Juvenile	M, F			475	112
Mule deer	Fawn	M, F	Winter		661	113
Red deer	Calf	M	Summer	35—40	500	114
			Winter	35—40	450	114
	Adult	F			529	84
		M	Indoors	50—110	570	115
			Outdoor	80—105	850	115
		F	Indoors		570	116
White-tailed deer	Yearling	F	Winter		548	117
	Fawn		Winter		524	118
		M, F	Oct		695	118
			Jan		456	118
			May		682	118
	Yearling	M, F	Aug		803	118
	Fawn	M, F	Dec—Apr		640	119
	Yearling	M, F	May—Oct		677	119
Hartebeest	Adult				293	92
Cattle (Beef)	Adult				466	25
	Calf		Preweaned		459	25
Cattle (Dairy)	Adult				470	120

always clear whether experimental subjects were standing or lying during measurements and details regarding energetic equivalents of O_2 are not always given. Subject to these uncertainties, several generalizations are offered on the basis of fasting and resting metabolic rates and metabolizable energy requirements for maintenance (Tables 2 to 4).

McNab[121] wondered whether metabolic rate may be subject to adaptive modification and how this may be related to ecological strategies. Comparisons with a wider range of mammals may give some anticipation of possible patterns. For reasons outlined below, species with adaptive reduction in metabolic rate are expected to be those in hot dry environments, those subsisting primarily on plants protected by secondary compounds, and possibly ruminants of particularly small size. High metabolic rates are expected in species exploiting strongly pulsed forage resources.

Water turnover is closely linked to metabolic rate.[93] Many desert-adapted species have relatively low metabolic rates and/or allow body temperature to drift to reduce water required for cooling (Chapter 8). Except for camels and donkeys, few desert-adapted wild ungulates have been studied. Camels have very low metabolic rates particularly when dehydrated.[98] Donkeys are close to the interspecies mean but are low in comparison with most wild ruminants.[96,97]

In some mammals, food habits appear important. Frugivorous, nectarivorous, and most carnivorous bats have basal rates greater than those expected from Kleiber's equation whereas insectivorous and sanguinivorous species are characterized by low metabolic rates.[122] A reduction in basal rate of arboreal folivores is correlated with the fraction of the diet consisting of leaves.[123] This appears associated with both the presence of toxic compounds or low nutritional value of leaves, and with the reduction in muscle mass associated with the sedentary life of folivores.

Large herbivores exploit a rather limited spectrum of resources. Low metabolic rates might be similarly expected where diets contain plants protected with secondary compounds and where coarse diets are consumed by a relatively small herbivore. But within such a restricted niche, it is not surprising that strong contrasts are not found. Although McNab[123] cites several examples of browsing ungulates (presumably exposed to higher levels of secondary compounds) with low metabolic rates, this pattern is not obvious from inspection of currently available information; it may even be contrary. Perhaps rumen fermentation serves as a sufficient detoxifying mechanism. Although foregut digestion offers a number of nutritional advantages, it has been suggested that it first evolved to enable exploitation of succulent but potentially toxic forage items.[124]

There is some evidence of low metabolic rates in small roughage grazers. The best example is domestic sheep and apparently bighorn sheep,[31,109] but medium-sized grazing antelope such as kongoni[92,93] and blesbok[73] also express low values. A particularly interesting species which has not yet been studied is the reedbuck which is small but feeds on coarse graminoid forages.

Even on highly digestible diets, there is a lower limit to which rumen fermentation can supply end-products for host metabolism. Intrinsic limits to microbial fermentation theoretically constrain the lowest possible size of ruminants, a limit which is approached by a number of species. This may be possible because of only partial dependence on ruminant digestion since many soluble carbohydrates are absorbed without fermentation. However, another possible solution is to adaptively lower metabolic rate. In this light, the unusually low metabolic rates of two microherbivores, the dik-dik and chevrotain, are of interest. Although the 72-hr fast imposed in one study on the dik-dik may have been rather severe for such a small-bodied animal,[89] low RMRs measured in another study indicate generally low metabolic intensity.[106] Part of the adaptiveness of low metabolism may be related to the arid environment in which the dik-dik lives. However, this cannot apply to the chevrotain which has a RMR close to the interspecies mean for FMR.[105]

Lowered metabolic rate might be expected to increase energetic efficiency. But there may be a penalty for such frugal use of resources. Metabolic rates are directly related to rate of production so that animals adapted for survival are often quite unproductive. The highest SMRs are expressed by improved breeds of cattle with a high potential for growth and milk production.[39] Among wild ungulates, the highest metabolic rates are found in northern *Cervidae*. These animals use a forage supply which is strongly pulsed and must grow and fatten rapidly during a short growing season.

There have been few inquiries into the significance of energy expenditure in population ecology. McNab[121] noted that logistic growth constants of mammals are much lower than those of passerine birds of equal mass, a difference he attributed to the lower metabolic rate of mammals. He offered the interesting comparison of two high-latitude lagomorphs. The snowshoe hare (*Lepus americanus*) has a high rate of population increase and is well known for its spectacular population fluctuations. This hare has a very high rate of metabolism. The arctic hare (*Lepus arcticus*) has a lower rate of increase and is the only lagomorph known to have a low basal metabolic rate. It may be possible to invoke this explanation for various ungulate taxa but any differences could have a variety of explanations. It certainly doesn't apply to sheep which have low metabolic rates but short gestations and generally high prolificacy.

V. DETERMINANTS

Metabolic rates, even when measured after nominal fasts, respond to a variety of factors which impose additional energy demands and stimulate feed intake. Attempts to integrate these factors in predictive equations have been only marginally successful because of complex interactions.[31,76]

A. Sex and Age

While in man and domestic animals the metabolic rate of males is generally higher than that of females, this apparently is not always the case with wild ruminants, perhaps because of the additional complexity of metabolic cycles. Silver et al.[82] did not find significant differences between white-tailed does and bucks in summer coats. Wesley et al.[125] found no obvious sex differences in pronghorn antelope. However, higher metabolic rates of males have been recorded for bighorn sheep[31,90] and black-tailed deer fawns.[72]

Neonates are born with metabolic rates similar to that of their dams but this rises in the first few hours to the typically high metabolic rates of young animals.[126,127] For calves on milk or milk replacers, ARC[25] suggests a value of 342 to 385 $kJ/kg^{0.75}$·day. For lambs, the preferred value is 350 $kJ/kg^{0.75}$·day. Resting metabolic rates decline rapidly with age in most species that have been studied.[34,73,128] The high metabolic rates of young animals and perhaps of males may be related to their relatively rapid tissue synthesis.

B. Reproductive Status

Metabolic rates of gestating and lactating females are typically high. For medium to large ungulates, the cost of maintaining the gravid uterus is about 85% of the total cost of pregnancy (i.e., maintenance plus energy accumulated in the conceptus).[129] Resting energy expenditures of pregnant females increase about 40% by term.[130]

High metabolic rates also occur in lactating ungulates[131] but this is confounded by often marked increases in forage intake which may have residual effects even after standardized fasts. Although RMR has been measured in lactating wild ungulates,[33] researchers are reluctant to expose their valuable subjects to fasts when they are rearing the next generation of research animals.

C. Thermal Environment

Prolonged exposure to low ambient temperatures induces an upward adjustment of thermoneutral metabolic rate. Webster et al.[132] found a gradual increase in domestic sheep exposed to winter weather. Similar adjustments occur in cattle[133] which are not simply due to higher feed intakes.

In temperate ungulates, metabolic rates cycle seasonally with lowest levels in winter. This has lead to the discounting of thermal factors in the induction of metabolic rate. However, Zervanos[78] documented higher thermoneutral metabolic rates of peccaries which seemed to have this basis. Chappel and Hudson[31] found a significant increase in bighorn sheep after prolonged exposure to low ambient temperatures.

VI. METABOLIC RHYTHMS

Seasonal cycles of energy metabolism are quite noticeable in northern ruminants. Silver et al.[82,83] observed values in white-tailed deer which were over 50% higher in summer than in winter. Somewhat narrower contrasts were observed by McEwan and Whitehead[75] in caribou. In bighorn sheep, the amplitude of this oscillation was about 40%.[108] Weiner[33] found seasonal variations in roe deer which amounted to less than 10%.

It would be wrong to conclude on the basis of such limited data that these represent stable species-specific differences because the ages of animals and nutritional regimes have not been standardized. Metabolic cycles are better developed in mature animals living in good nutritional environments.

Since voluntary food intake varies seasonally (Chapter 4) and since feeding levels influence metabolic rate, an important question is whether metabolic cycles are simply a shadow of energy supply. Results from a study of red deer maintained on rationed rather than *ad libitum* feed intake suggest that metabolic cycles respond largely to energy supply.[84] However, evidence that there is an independent rhythm of metabolic rate comes from domestic sheep, a species usually not recognized to be cyclic. Blaxter and Boyne[134] recorded a sinusoidal cycle of metabolic rate in sheep held on maintenance rations which had an amplitude of about 14% of the mean with a minimum in winter and maximum in summer.

VII. MAINTENANCE REQUIREMENTS

Metabolizable energy requirements for maintenance (MEM) for pen-fed herbivores can be measured most directly by feeding experimental groups at several energy levels. Such results are empirical but apply specifically to the circumstances of the trial. Since these estimates include a variable amount of activity associated with living in confinement, they are not a particularly good base for adding incremental costs of free existence. Maintenance requirements determined in this manner may differ from RMR measured at maintenance feeding levels since feeding trails average a number of activity states (e.g., feeding and sleep) which bracket expenditures associated with (usually) standing quietly for calorimetric measurements.

Because energy retention is difficult to determine in short-term trials and should be based on comparative slaughter techniques, the NRC net energy system is less suitable for work with wild herbivores than the ARC metabolizable energy system which is based on energy expenditures. MEMs are related to estimates of SMR by:

$$\text{MEM} = \text{SMR}/k_m$$

where k_m is the efficiency of utilization of metabolizable energy for maintenance. This efficiency ranges from 0.64 to 0.75 and varies consistently with diet.

One problem is that SMR is influenced by prior nutrition. Short periods of maintenance feeding do not appear to tune metabolism to a stable minimum. Although this variation is often treated as an artifact which masks true basal expenditures, maintenance requirements do vary with energy supplies. Ledger and Sayers[135] restricted cattle at several liveweights with feeding levels sufficient to maintain weight stasis. For large cattle, the amount of food required to maintain weight fell by 18% between the 3rd and 24th week. For small cattle, the apparent requirement fell by over 50% although this was confounded by changes in gut fill and body composition. Thus, maintenance requirements determined in the short-term trials used for feed evaluation often do not give good estimates of requirements under chronic feeding regimes.

Even greater difficulty is encountered when SMR and k_m are applied to the maintenance component to estimate total requirements of growing, gestating, or lactating animals since production invokes a cost which is reflected in increased fasting metabolism. This increment could be charged to maintenance or alternatively to production and incorporated in the efficiency term for growth (k_f).

ARC[25] attempts to attribute increased energy expenditures to production and uses the following convention to determine efficiencies for maintenance and growth. Mature animals (wethers) are fed the same diet at three different levels for periods of 4 weeks. Measurements of ME and RMR in the last week of each period are used to calculate the efficiency of retention of ME as the slope of the lines above (k_f) and below (k_m) maintenance (Figure 3). These efficiency terms are defined rigorously as the average slopes of the exponential relationship between heat production at fasting and maintenance (k_m) and that between maintenance and twice maintenance (k_f).

To estimate energy requirements for growing or fattening animals, the maintenance requirement is calculated conventionally from SMR and k_m. Additional requirements for production are estimated from the amount/energetic composition of gain and the efficiency term, k_f. Blaxter and Boyne[136] offer a more precise calculation which accounts for the fact that efficiency declines continuously with feeding level.

Given the paucity of bioenergetic information on wild herbivores, SMR usually is estimated from Kleiber's equation (Equation 3) which scales metabolic rate to $W^{0.75}$ This may be incremented to account for the generally high metabolic rate of wild herbivores and ad-

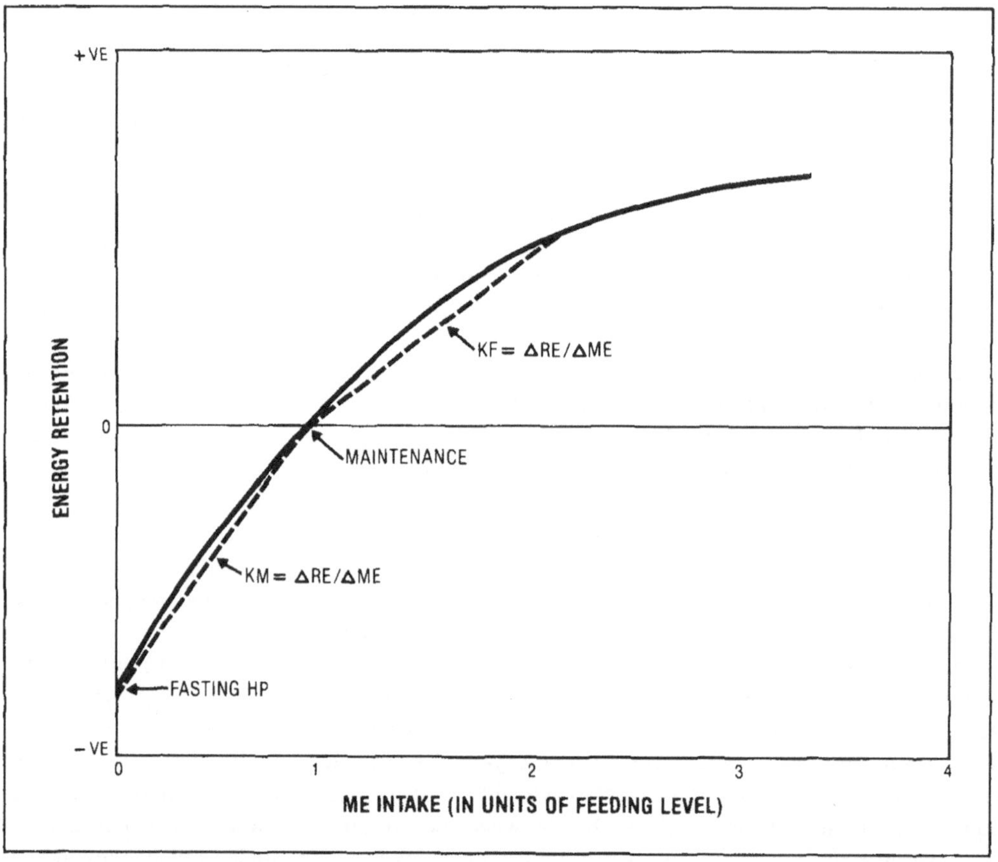

FIGURE 3. Relationship between ME intake and energy retention used to calculate the efficiencies of utilization of energy for maintenance (km) and growth (kf).

justments may be made for the effect of age. For domestic livestock, estimates from growing animals representing a relevant range of maturities are used to estimate the intercept and slope of the allometric equation. Because of the higher metabolism of young and rapidly growing animals, the relationship scales to values considerably lower than $W^{0.75}$. Thus, part of the cost of production is cryptically assigned to the maintenance component.

Webster et al.[137] claim that measured metabolic rates and hence this equation are irrelevant to the calculation of energy requirements of growing animals and recommends use of predicted fasting metabolism calculated from measurement of energy retention (RE):

$$FMR' = k_m (ME - RE/k_f)$$

where FMR′ is simply a parameter which correctly predicts RE at any level of ME intake. This allows the extensive body of information on various feeds compiled by ARC to be applied to animals in a variety of productive states. Predicted metabolic rates are generally higher than SMR by an amount which appears proportional to the intensity of production.

Predicted metabolic rate is simply a best-fit coefficient suitable for use in empirical feeding systems. Its application charges part of the cost of production to maintenance. Graham et al.[76] and Chappel and Hudson[31] have attempted to measure this directly by measuring metabolic rates of animals fasted from prevailing (and widely varying) feeding levels. However, this does not obviate the problem that fasting does not neatly separate energy expenditures associated with digestion from that related to tissue metabolism.

Turner and Taylor[27] advocate use of equilibrium maintenance requirements determined from long-term periods (e.g., 100 days) at constant feeding levels. They argue that this provides a stable base which is unconfounded by the work of production and has the convenient property of being proportional to $W^{1.0}$ within a genotype. When this base is used, differences in efficiencies of protein and fat deposition are exposed. Using conventional protocols, this difference is discounted on the grounds that estimates of kf are similar in young and fattening animals. But this may be an artifact of the short-term trials used to estimate energetic efficiencies, which give particularly biased estimates for young animals.

Precise determination of maintenance requirements is difficult particularly when the objective is to determine the additional requirements of production in a factorial calculation. Several conventions on how to partition maintenance and production components have been developed. Unfortunately, few studies on energy metabolism of wild herbivores have complied with any of these standards. This sometimes leaves us in the untenable position of applying efficiencies determined by one convention to maintenance expenditures from another.

VIII. CONCLUSIONS

Energy costs of physiological homeostasis can be crudely equated with energy expenditures of animals which are at rest in a thermoneutral environment and postabsorptive state. Metabolic rates measured under these circumstances scale strongly with body size in interspecific comparisons of adult animals. Specific environmental adaptations are evident in systemmatic deviations from the allometric relationship. The lowest metabolic rates are found in desert-adapted species and small herbivores subsisting on low quality forages.

Low metabolic rates are often equated with high efficiency since low energy requirements should mean more energy spared for production. Although species with low metabolic rates probably can survive on more limited resources, they seem less able to take advantage of nutritional opportunities. High metabolic intensity is required to support tissue synthesis. Metabolic cycles of northern ungulates may confer both advantages in strongly seasonal environments. Although cycles of feed intake, neuroendocrine status, and metabolic rate are synchronized by photoperiod, it is too early to conclude whether energy supply or demand drives the cycle.

Energy expenditures measured under standardized conditions provide an estimate of maintenance requirements where the efficiency of utilization of feed energy is known. But the estimate holds only under conditions similar to those of the measurement since apparent maintenance requirements adjust to prevailing energy supplies. It is difficult to estimate the maintenance component for animals which are growing, gestating, or lactating. Production invokes an increase in the size of metabolically active tissues and may prime the metabolic intensity of individual cells.

REFERENCES

1. **Holter, J. B. and Hayes, H. H.,** Growth in white-tailed deer fawns fed varying energy and constant protein, *J. Wildl. Manage.*, 41, 506, 1977.
2. **Pullar, J. D.,** Methods of calorimetry; direct, in *International Encyclopaedia of Food and Nutrition*, Vol. 17, Part I, Cuthbertson, D. P., Ed., Pergamon, Oxford, 1969, 417.
3. **Brody, S.,** *Bioenergetics and Growth*, Hafner Press, New York, 1945.
4. **Brouwer, E.,** Report of subcommittee on constant and factors, *Energy Metabolism, E.A.A.P.*, Academic Press, London, 1965.
5. **McLean, J. A.,** On the calculation of heat production from open circuit calorimetric measurements, *Br. J. Nutr.*, 27, 597, 1972.

6. **Young, B. A., Kerrigan, B., and Christopherson, R. J.,** A versatile respiratory pattern analyzer for studies of energy metabolism of livestock, *Can. J. Anim. Sci.*, 55, 17, 1975.
7. **Corbett, J. L., Farrell, D. J., Leng, R. A., McClymont, G. L., and Young, B. A.,** Determination of the energy expenditure of penned and grazing sheep from estimates of carbon dioxide entry rate, *Br. J. Nutr.*, 26, 277, 1971.
8. **Young, B. A. and McEwan, E. H.,** A method for measurement of energy expenditure in unrestrained reindeer and caribou, *Proc. Int. Reindeer Caribou Symp.*, Luick, J. R., Lent, P. C., Klein, D. R., and White, R. G., Eds., Biol. Pap. Univ. Alaska, Spec. Rep. 1, 1972 (1975), 355.
9. **Mullen, R. K.,** The D_2O^{18} method of measuring the energy metabolism of free-ranging animals, in *Ecological Energetics of Homeotherms*, Monogr. Ser. Vol 20, Gessaman, J. A., Ed., Utah State University Press, Logan, 1973, 32.
10. **Renecker, L. A., Hudson, R. J., and Freddy, D. J.,** Heart rate as an index of energy expenditure in moose, using two telemetry systems, *Int. Symp. Biotelem.*, 7, 1982.
11. **Johnston, R. H., MacArthur, R. A., and Geist, V.,** A biotelemetry system for monitoring heart rates in unrestrained ungulates, *Biotelem. Patient Monitg.*, 7, 188, 1980.
12. **Brockway, J. M. and McEwan, E. H.,** Oxygen uptake and cardiac performance in the sheep, *J. Physiol. (London)*, 20, 661, 1969.
13. **Webster, A. J. F.,** Continuous measurement of heart rate as an indicator of energy expenditure in sheep, *Br. J. Nutr.*, 21, 709, 1967.
14. **Holter, J. B., Urban, W. E., Hayes, H. H., and Silver, H.,** Predicting metabolic rate from telemetered heart rate in white-tailed deer, *J. Wildl. Manage.*, 40, 1976.
15. **Kautz, M. A.,** Energy Expenditure and Heart Rate of Active Mule Deer Fawns, M.S. thesis, Colorado State University, Fort Collins, 1978, 76p.
16. **Kautz, M. A., Mautz, W. W., and Carpenter, L. H.,** Heart rate as a predictor of energy expenditure of mule deer, *J. Wildl. Manage.*, 45, 715, 1981.
17. **Robbins, C. T., Cohen, Y., and Davitt, B. B.,** Energy expenditure by elk calves, *J. Wildl. Manage.*, 43, 455, 1979.
18. **Mautz, W. W. and Fair, J.,** Energy expenditure and heart rate for activities of white-tailed deer, *J. Wildl. Manage.*, 44, 333, 1980.
19. **Pauls, R. W., Hudson, R. J., and Sylven, S.,** Energy expenditure of free-ranging wapiti, *Univ. Alberta Feeders' Day Rep.*, 60, 87, 1981.
20. **Christopherson, R. J. and Webster, A. J. F.,** Changes during eating in oxygen consumption, cardiac function and body fluids of sheep, *J. Physiol. (London)*, 221, 441, 1972.
21. **Graham, A., Christopherson, R. J., and Thompson, J. R.,** Endocrine and metabolic changes in sheep associated with acclimation to constant or intermittent cold exposures, *Can J. Anim. Sci.*, 61, 81, 1981.
22. **Graham, A. and Christopherson, R. J.,** Effects of adrenaline and noradrenaline on the heat production of warm- and cold-acclimated sheep, *Can. J. Physiol. Pharmacol.*, 59, 985, 1981.
23. **Watkins, W. G.,** Fasting metabolism of wapiti on two planes of nutrition, in *Wildlife Productivity and Management Programme Annual Report*, Hudson, R. J., Ed., University of Alberta, Edmonton, 1983, 42.
24. **Marston, H. R.,** Energy transactions in the sheep. I. The basal heat production and heat increment, *Aust. J. Sci. Res.*, B1, 93, 1948.
25. **Agricultural Research Council,** *The Nutrient Requirements of Ruminant Livestock*, Commonwealth Agricultural Bureaux, Farnham Royal, England, 1980, 351.
26. **Blaxter, K. L.,** *The Energy Metabolism of Ruminants*, Hutchinson, London, 1967.
27. **Turner, H. G. and Taylor, C. S.,** Dynamic factors in models of energy utilization with particular reference to maintenance requirements of cattle, *World Rev. Nutr. Diet.*, 42, 135, 1983.
28. **Webster, A. J. F.,** Prediction of the energy requirements for growth in beef cattle, *World Rev. Nutr. Diet.*, 30, 189, 1978.
29. **Webster, A. J. F.,** Act of eating and its relation to the heat increment of feed in ruminants, *Environ. Physiol.*, 42, 1972.
30. **Rothwell, N. J. and Stock, M. J.,** Diet-induced thermogenesis, in *Mammalian Thermogenesis*, Girardier, L. and Stock, M. J., Eds., Chapman and Hall, London, 1983, 208.
31. **Chappel, R. W. and Hudson, R. J.,** Prediction of energy expenditures by Rocky Mountain bighorn sheep, *Can. J. Zool.*, 58, 1908, 1980.
32. **McEwan, E. H.,** Energy metabolism of barren ground caribou, *Can. J. Zool.*, 48, 391, 1970.
33. **Weiner, J.,** Energy metabolism of the roe deer, *Acta Theriol.*, 22, 3, 1977.
34. **Wesley, D. E., Knox, K. L., and Nagy, J. G.,** Energy metabolism of pronghorn antelopes, *J. Wildl. Manage.*, 37, 563, 1973.
35. **Webster, A. J. F.,** Energy costs of digestion and metabolism in the gut, in *Digestive Physiology and Metabolism in Ruminants*, Ruckebusch, Y. and Thivend, P., Eds., AVI Publ. Co., Westport, Conn., 1980, 469.

36. **Graham, N. McC.,** The metabolic rate of fasting sheep in relation to total and lean body weight, and the estimation of maintenance requirements, *Aust. J. Agric. Res.*, 18, 127, 1967.
37. **Toutain, P. L., Toutain, C., Webster, A. J. F., and McDonald, J. D.,** Sleep and activity, age and fattness, and the energy expenditure of confined sheep, *Br. J. Nutr.*, 38, 445, 1977.
38. **Webster, A. J. F.,** The energetic efficiency of metabolism, *Proc. Nutr. Soc.*, 40, 121, 1981.
39. **Webster, A. J. F.,** Energetics of maintenance and growth, in *Mammalian Thermogenesis*, Girardier, L. and Stock, M. J., Eds., Chapman and Hall, London, 1983, 178.
40. **McNiven, M. A.,** The effect of body fatness in energetic efficiency and fasting heat production in adult sheep, *Br. J. Nutr.*, 297, 1984.
41. **Koong, L. J., Nienaber, J. A., Pekas, J. C., and Yen, J. C.,** Effect of plane of nutrition on organ size and fasting heat production in pigs, *J. Nutr.*, 112, 1638, 1982.
42. **Smith, N. E. and Baldwin, R. L.,** Effects of breed, pregnancy and lactation on weight of organs and tissues in dairy cattle, *J. Dairy Sci.*, 57, 1055, 1974.
43. **Landsberg, L. and Young, J. B.,** Autonomic regulation of thermogenesis, in *Mammalian Thermogenesis*, Girardier, L. and Stock, M. J., Eds., Chapman and Hall, London, 1983, 99.
44. **Landsberg, L. and Young, J. B.,** Diet-induced changes in sympathoadrenal activity: implications for thermogenesis, *Life Sci.*, 28, 1801, 1981.
45. **Trenkle, A.,** Endocrine regulation of energy metabolism in ruminants, *Fed. Proc. Fed. Am. Soc. Exp. Biol.*, 2536, 1981.
46. **Himms-Hagen, J.,** Thyroid hormones and thermogenesis, in *Mammalian Thermogenesis*, Girardier, L. and Stock, M. J., Eds., Chapman and Hall, London, 1983, 141.
47. **Ryg, M.** , Seasonal changes in weight gain, growth hormone and thyroid hormones in intact and castrated male moose, *(Alces alces alces), Can. J. Zool.*, 60, 2941, 1982.
48. **Ryg, M. and Jacobsen, E.,** Seasonal changes in growth rate, feed intake, growth hormone and thyroid hormones in young male reindeer *(Rangifer tarandus tarandus), Can. J. Zool.*, 60, 15, 1982.
49. **Ryg, M. and Langvatn, R.,** Seasonal changes in weight gain, growth hormone and thyroid hormones in male red deer *(Cervus elaphus atlanticus), Can. J. Zool.*, 60, 2577, 1982.
50. **Ryg, M. and Jacobsen, E.,** Effects of thyroid hormones and prolactin on food intake and weight changes in young male reindeer *(Rangifer tarandus tarandus), Can. J. Zool.*, 60, 1562, 1982.
51. **Yousef, M. K. and Johnson, H. D.,** Calorigenesis of cattle as influenced by growth hormone and environmental temperature, *J. Anim. Sci.*, 25, 1076, 1966.
52. **Yousef, M. K. and Johnson, H. D.,** Calorigenesis of cattle as influenced by hydrocortisone and environmental temperatures, *J. Anim. Sci.*, 26, 1087, 1967.
53. **McMillin, J. M., Seal, U. S., and Karns, P. D.,** Hormonal correlates of hypophagia in white-tailed deer, *Fed. Proc. Fed. Am. Soc. Exp. Biol.*, 39, 2964, 1980.
54. **Brown, W. B., Forbes, J. M., Goodall, E. D., Kay, R. N. B., and Simpson, A. M.,** Effects of photoperiod on food intake, sexual condition and hormone concentrations in stags and rams, *J. Physiol. (London)*, 296, 58P, 1979.
55. **Ryg, M. and Jacobsen, E.,** Effects of castration on growth and food intake cycles in young male reindeer, *(Rangifer tarandus), Can. J. Zool.*, 60, 942, 1982.
56. **Suttie, J. M. and Simpson, A. M.,** Photoperiodic control of appetite, growth, antlers and endocrine status of red deer, *Biology of Deer Production*, Drew, K. R. and Fennessy, P. F., Eds., MAFF, New Zealand, in press.
57. **Roth, J. and Grunfeld, C.,** Endocrine systems: mechanisms of disease, target cells and receptors, in *Textbook of Endocrinology*, 6th ed., Williams, R. H., Ed., W.B. Saunders, Philadelphia, 1981, 15.
58. **Yousef, M. K. and Luick, J. R.,** Estimation of thyroxine secretion rate in reindeer (*Rangifer tarandus*): effects of sex, age and season, *Comp. Biochem. Physiol.*, 40, 789, 1971.
59. **Watkins, B. E., Nachreiner, R. F., Schmitt, S. M., Cooley, T. M., and Ullrey, D. E.,** Thyroid function in fed and fasted white-tailed deer fawns, *Can. J. Zool.*, 60, 1331, 1982.
60. **Milligan, L. P.,** Energetic efficiency and metabolic transformations, *Fed. Proc. Fed. Am. Soc. Exp. Biol.*, 30, 1454, 1971.
61. **Newsholme, E. A. and Crabtree, B.,** Substrate cycles in metabolic regulation and in heat generation, *Biochem. Soc. Symp.*, 41, 61, 1976.
62. **Milligan, L. P. and McBride, B. W.,** Energy costs of ion pumping by animal tissues, *Fed. Proc. Fed. Am Soc. Exp. Biol.*, 43, in press.
63. **Gregg, V. and Milligan, L. P.,** Role of Na^+, K^+-ATPase in muscular energy expenditure of warm- and cold-exposed sheep, *Can. J. Anim. Sci.*, 62, 123, 1982.
64. **Reeds, P. J., Wahle, K. W. J., and Haggarty, P.,** Energy costs of protein and fatty acid synthesis, *Proc. Nutr. Soc.*, 41, 155, 1982.
65. **Miller, B. G., Otto, W. R., Grimble, R. F., York, D. A. and Taylor, T. G.,** The relationship between protein turnover and energy balance in lean and genetically obese (ob/ob) mice, *Br. J. Nutr.*, 42, 185, 1979.

66. **Emery, R. S.,** Mobilization, turnover and deposition of adipose tissue lipids, in *Digestive Physiology and Metabolism in Ruminants*, Ruckebush, Y. and Thivend, P., Eds., A.V.I. Publ. Co., Westport, Connecticut, 1980, 541.
67. **Kleiber, M.,** *The Fire of Life*, Wiley-Interscience, New York, 1961.
68. **Gunther, B. and Morgado, E.,** Theory of biological similarity revisited, *J. Theor. Biol.*, 96, 543, 1981.
69. **Hughes, G. M.,** Dimensions and the respiration of lower vertebrates, in *Scale Effects in Animal Locomotion*, Pedley, T. J., Ed., Academic Press, London, 1977, 57.
70. **McMahon, T.,** Size and shape in biology, *Science*, 179, 1201, 1973.
71. **Thonney, M. L., Touchberry, R. W., Goodrich, R. D., and Meiske, J.,** Intraspecies relationships between fasting heat production and body weight: a re-evaluation of $W^{0.75}$, *J. Anim. Sci.*, 43, 692, 1976.
72. **Nordan, H. C., Cowan, I. McT., and Wood, A. J.,** The feed intake and heat production of young black-tailed deer, *Can. J. Zool.*, 48, 275, 1970.
73. **Meissner, H. H.,** A note on the energy requirements of young, growing blesbok, *S. Afr. J. Wildl. Res.*, 6, 51, 1976.
74. **Whittemore, C. T.,** Development of recommended energy and protein allowance for growing pigs, *Agric. Syst.*, 11, 159, 1983.
75. **McEwan, E. H. and Whitehead, P. E.,** Seasonal changes in the energy and nitrogen intake in reindeer and caribou, *Can. J. Zool.*, 48, 905, 1970.
76. **Graham, N. McC., Searle, T. W., and Griffiths, D. A.,** Basal metabolic rate in lambs and young sheep, *Aust. J. Agric. Res.*, 25, 957, 1974.
77. **Yousef, M. K. and Dill, D. B.,** Resting energy metabolism and cardiorespiratory activity in the burro *(Equus asinus)*, *J. Appl. Physiol.*, 27, 229, 1969.
78. **Zervanos, S. M.,** Seasonal effects of temperature on the respiratory metabolism of the collared peccary, *(Tayassu tajacu)*, *Comp. Biochem. Physiol.*, 50A, 365, 1975.
79. **National Research Council,** *Effect of Environment on Nutrient Requirements of Domestic Animals*, Part 1, National Academy Press, Washington, D.C., 1981.
80. **Knox, K. L., Nagy, J. G., and Brown, R. D.,** Water turnover in mule deer, *J. Wildl. Manage.*, 33, 389, 1969.
81. **Drozdz, A., Wiener, J., Gebczynska, J., and Krazinska, M.,** Some bioenergetic parameters of wild ruminants, *Pol. Ecol. Stud.*, 1, 85, 1975.
82. **Silver, H., Colovos, N. F., Holter, J. B., and Hayes, H. H.,** Fasting metabolism of white-tailed deer, *J. Wildl. Manage.*, 33, 490, 1969.
83. **Silver, H. H., Holter, J. B., Colovos, N. F., and Hayes, H. H.,** Effect of falling temperature on heat production in fasting white-tailed deer, *J. Wildl. Manage.*, 35, 37, 1971.
84. **Brockway, J. H. and Maloiy, G. M. O.,** Energy metabolism of red deer, *J. Physiol. (London)*, 194, 22P, 1968.
85. **Simpson, A. M., Webster, A. J. F., Smith, J. S., and Simpson, C. A.,** Energy and nitrogen metabolism of red deer in cold environments. Comparison with cattle and sheep, *Comp. Biochem. Physiol.*, 60, 251, 1978.
86. **Regelin, W. L., Schwartz, C. C., and Franzmann, A. W.,** Seasonal energy metabolism of adult moose, *J. Wildl. Manage.*, in press.
87. **Luick, B. R. and White, R. G.,** Indirect calorimetric measurements of the caribou calf, *Acta Zool. Fenn.*, 175, 89, 1983.
88. **White, R. G. and Yousef, M. K.,** Energy expenditure in reindeer walking on roads and on tundra, *Can. J. Zool.*, 56, 215, 1977.
89. **Kamau, J. M. Z. and Maloiy, G. M. O.,** The fasting metabolism of a small East African antelope, the dik-dik, *J. Physiol. (London)*, 319, 50P, 1981.
90. **Chappel, R. W. and Hudson, R. J.,** Energy cost of feeding in Rocky Mountain bighorn sheep, *Acta Theriol.*, 23, 359, 1978.
91. **Rogerson, A.,** Energy utilization by the eland and wildebeest, *Symp. Zool. Soc. London*, 21, 153, 1968.
92. **Stanley-Price, M. R.,** The nutritional ecology of Coke's hartebeest in Kenya, *J. Appl. Ecol.*, 15, 33, 1978.
93. **MacFarlane, W. V. and Howard, B.,** Comparative water and energy economy of wild and domestic mammals, *Symp. Zool. Soc. London*, 31, 261, 1972.
94. **Krog, H. and Monson, M.,** Notes on the metabolism of a mountain goat, *Am. J. Physiol.*, 178, 515, 1954.
95. **Knox, K. L., Wesley, D., Crownova, J. C., and Nagy, J.,** Comparative energy metabolism of American ungulates, *Proc. Symp. Energy Metab. Farm Animals*, Vol. 14, Hohenheim, H., Ed., European Association of Animal Production, Wageningen, 1974, 229.
96. **Maloiy, G. M. O.,** Temperature regulation in the Somali donkey *(Equus asinus)*, *Comp. Biochem. Physiol.*, 39A, 403, 1971.

97. **El-Nouty, F. D., Yousef, M. K., Magdub, A. B., and Johnson, H. D.,** Thyroid hormones and metabolic rate in burros (*Equus asinus*) and llamas (*Lama glama*): effects of environmental temperature, *Comp. Biochem. Physiol.*, 60A, 235, 1978.
98. **Schmidt-Nielsen, K., Crawford, E. C., Jr., Newsome, A. E., Rawson, K. S., and Hammel, H. T.,** Metabolic rate of camels: effect of body temperature and dehydration, *Am. J. Physiol.*, 212, 341, 1967.
99. **Brockway, J. M. and Gessaman, J. A.,** The energy cost of locomotion on the level and on gradients for the red deer, *Q. J. Exp. Physiol.*, 62, 333, 1977.
100. **Gates, C. C. and Hudson, R. J.,** Effects of posture and activity on metabolic responses of wapiti to cold, *J. Wildl. Manage.*, 43, 564, 1979.
101. **Gates, C. C. and Hudson, R. J.,** Energy costs of locomotion in wapiti, *Acta Theriol.*, 23, 365, 1978.
102. **Robbins, C. T., Cohen, Y., and Davitt, B. B.,** Energy expenditure by elk calves, *J. Wildl. Manage.*, 43, 445, 1979.
103. **Renecker, L. A., Hudson, R. J., Christopherson, M. K., and Arelis, C.,** Effect of posture, feeding low temperature and wind on energy expenditures of moose calves, *Proc. North Am. Moose Conf. Workshop*, Vol. 14, Lakehead University, Thunder Bay, Ontario, 1978, 126.
104. **Hammell, H. T., Houpt, T. R., Andersen, K. L., and Skjenneberg, S.,** Thermal and metabolic measurements on a reindeer at rest and in exercise, *Arctic Aeromed. Lab. (U.S.) Tech. Doc. Rep.* Aaltdr-61-54, 1962.
105. **Whittow, G. C., Seammel, C. A., Leong, M., and Rand, D.,** Temperature regulation in the smallest ungulate, the lesser mouse deer *(Tragulus javanicus), Comp. Biochem. Physiol.*, 56A, 23, 1977.
106. **Maskrey, M. and Hoppe, P. O.,** Thermoregulation and oxygen consumption in Kirk's dik dik (*Madoqua kirkii*) at ambient temperatures of 10-45°C, *Comp. Biochem. Physiol.*, 62A, 827, 1979.
107. **Taylor, C. R., Dmi'el, R., Shkolnik, A., Baharav, D., and Borut, A.,** Heat balance of running gazelles: strategies for conserving water in the desert, *Am. J. Physiol.*, 226, 439, 1974.
108. **Finch, V. A.,** Thermoregulation and heat balance of the east African eland and hartebeest, *Am. J. Physiol.*, 222, 1374, 1972.
109. **Chappel, R. W. and Hudson, R. J.,** Winter bioenergetics of Rocky Mountain bighorn sheep, *Can. J. Zool.*, 56, 2388, 1978.
110. **Christopherson, R. J., Hudson, R. J., and Richmond, R. J.,** Comparative winter bioenergetics of American bison, yak, Scottish Highland and Herford calves, *Acta Theriol.*, 23, 49, 1978.
111. **Christopherson, R. J., Hudson, R. J., and Christopherson, M. K.,** Seasonal energy expenditures and thermoregulatory responses of bison and cattle, *Can. J. Anim. Sci.*, 59, 611, 1979.
112. **Verstegen, M. W. A., Close, W. H., Stuart, I. B., and Mount, L. E.,** The effects of environmental temperature and plane of nutrition on heat loss, energy retention and deposition of protein and fat in groups of growing pigs, *Br. J. Nutr.*, 30, 21, 1973.
113. **Baker, D. L., Johnson, D. E., Carpenter, L. H., Wallmo, O. C., and Gill, R. B.,** Energy requirements of mule deer fawns in winter, *J. Wildl. Manage.*, 43, 162, 1979.
114. **Simpson, A. M., Webster, A. J. F., Smith, J. S., and Simpson, C. A.,** The efficiency of utilization of dietary energy for growth in sheep (*Ovis ovis*) and red deer *(Cervus elaphus), Comp. Biochem. Physiol.*, 59A, 99, 1978.
115. **Fennessy, P. E., Moore, G. H., and Carson, I. D.,** Energy requirements of red deer, *Proc. N. Z. Soc. Anim. Prod.*, 41, 167, 1981.
116. **Kay, R. N. B. and Staines, B. W.,** The nutrition of the red deer *(Cervus elaphus), Nutr. Abstr. Rev.*, 51, 601, 1981.
117. **Ullrey, D. E., Youat, W. G., Johnson, H. E., Fay, L. D., Schoepke, B. L., and Magee, W. T.,** Digestible and metabolizable energy requirements for winter maintenance of Michigan white-tailed does, *J. Wildl. Manage.*, 34, 863, 1970.
118. **Thompson, C. B., Holter, J. B., Hayes, H. H., Silver, H., and Urban, W. E.,** Nutrition of white-tailed deer. I. Energy requirements of fawns, *J. Wildl. Manage.*, 37, 301, 1973.
119. **Holter, J. B., Urban, W. E., and Hayes, H. H.,** Nutrition of northern white-tailed deer throughout the year, *J. Anim. Sci.*, 45, 365, 1977.
120. **van Es, A. H. J.,** Between animal variation in maintenance energy of cows, Centrum voor Landboupublikaties en Landbouwdoumentatie, Wagenin, 1961.
121. **McNab, B. K.,** Food habits, energetics and the population biology of mammals, *Am. Nat.*, 116, 106, 1980.
122. **McNab, B. K.,** The economics of temperature regulation in neotropical bats, *Comp. Biochem. Physiol.*, 31, 227, 1969.
123. **McNab, B. K.,** Energetics of arboreal folivores: physiological problems and ecological consequences of feeding on an ubiquitous food supply, *The Ecology of Arboreal Folivores*, Montgomery, G. G., Ed., Smithsonian Inst. Press, Washington, D.C., 1978, 153.
124. **Janis, C.,** The evolutionary status of the *Equidae* and the origins of rumen and cecal digestion, *Evolution*, 30, 757, 1976.

125. **Wesley, D. E., Knox, K. L., and Nagy, J. G.,** Energy flux and water kinetics in young pronghorn antelope, *J. Wildl. Manage.*, 34, 908, 1970.
126. **Wilkie, D. R.,** Metabolism and body size, in *Scale Effects in Animal Locomotion*, Pedley, T. J., Ed., Academic Press, London, 1977, chap. 2.
127. **Davies, G. S. and Mott, J. C.,** The increase in oxygen consumption of the lamb after birth, *J. Physiol. (London)*, 146, 295, 1959.
128. **Myrcha, A. and Jezvrtuki, W.,** Metabolic rate during postnatal development of wild boars, *Acta Theriol.*, 17, 443, 1972.
129. **Robinson, J. J., McDonald, J., Fraser, C., and Gordon, J. G.,** Studies on reproduction in prolific ewes. VI. The efficiency of energy utilization for conceptus growth, *J. Agric. Sci.*, 94, 331, 1980.
130. **Robbins, C. T.,** *Wildlife Feeding and Nutrition*, Academic Press, New York, 1983.
131. **Holter, J. B.,** Fasting heat production in lactating versus dry dairy cows, *J. Dairy Sci.*, 59, 755, 1976.
132. **Webster, A. J. F., Hicks, A. M., and Hays, F. L.,** Cold climate and cold temperature induced changes in the heat production and thermal insulation of sheep, *Can. J. Physiol. Pharmacol.*, 47, 553, 1969.
133. **Young, B. A.,** Effects of winter acclimatization on resting metabolism of beef cows, *Can J. Anim. Sci.*, 55, 619, 1975.
134. **Blaxter, K. L. and Boyne, A. W.,** Fasting and maintenance metabolism of sheep, *J. Agric. Sci.* (Cambridge), 99, 611, 1982.
135. **Ledger, H. P. and Sayers, A. R.,** The utilization of dietary energy by steers during periods of restricted food intake and subsequent realimentation. I. The effect of time on the maintenance requirements of steers held at constant live weights, *J. Agric. Sci.* (Cambridge), 88, 11, 1977.
136. **Blaxter, K. L. and Boyne, A. W.,** The estimation of the nutritive value of feeds as energy sources for ruminants and the derivation of feeding systems, *J. Agric. Sci.* (Cambridge), 90, 47, 1978.
137. **Webster, A. J. F., Brockway, J. M., and Smith, J. S.,** Prediction of the energy requirements for growth in beef cattle. I. The irrelevance of fasting metabolism, *Anim. Prod.*, 19, 127, 1974.

Chapter 7

INCREMENTAL COST OF ACTIVITY

S. G. Fancy and R. G. White

TABLE OF CONTENTS

Precise estimates of energy costs of behavioral activities are needed to fully interpret short- and long-term consequences of altered activity patterns within the context of daily energy budgets, as the behavioral implications may not always be apparent. Energy budgets themselves are used for comparing energetic aspects of life history strategies, behavior, and productivity of animal species (Chapter 1). The way various activities, e.g., locomotion, scale with respect to body weight and to other bioenergetic components gives insight into the constraints, or degree of plasticity, that a species may have in optimizing the allocation of energy to various functions on a daily, seasonal, and life-time basis.

Mathematical models of daily energy expenditure are based on the assumption that the cost of each activity over and above standard fasting metabolism (SMR, see Chapter 6) plus thermogenesis can be summed arithmetically. Evidence that these components are additive and give a sound estimate of total energy expenditure stems from laws of energy conservation and from the earliest studies of Rubner[1] to the more recent studies of Brody,[2] Blaxter,[3] and Kleiber.[4] Work with domestic grazers has been reviewed by Osuji,[5] who discusses this approach for estimating the maintenance energy requirements of grazing animals. Additive models are also used extensively for describing energy budgets of species, guilds, and trophic levels in ecosystem studies,[6-9] but because estimates of activity costs are lacking for some or many species in the array, the accuracy of the budget is frequently not known. Often, the results cannot be used to provide more than a rough estimate of daily energy flow.[9] Clearly, there is a need to generalize the arithmetic model across species and body sizes. In this respect, the role of scaling[10-12] and identification of species adaptations[13] has special importance.

The summation of several individual costs to give a mathematical expression of a complex behavioral trait, e.g., migration or food searching, can also be scaled if the allometry of individual costs are known.[14] This approach is important for examining predator-prey relations and concepts such as "foraging radius", optimum foraging, and resource partitioning.[15,16]

Until recently, techniques for checking the accuracy of energy budget calculations under natural conditions have not been available. The development of a mobile indirect calorimeter,[17] the carbon dioxide entry rate technique,[18] and judicious application of the heart-rate technique to calibrated animals[9,19-22] has allowed some empirical daily budgets of free-ranging animals to be determined and budgets based on estimated energy costs to be checked. Estimation of carbon dioxide output by the doubly-labeled water technique of Lifson and McClintock[23] (reviewed by Mullen[24]) is potentially the most useful of these methods. However, except for small species,[24,25] the technique has rarely been evaluated under field conditions and can be extremely expensive for use in large animals.[24] Studies using the mobile indirect calorimeter and carbon dioxide entry rate techniques with domestic sheep indicate that the energy expenditure of free existence, i.e., while grazing normally at a range of animal densities, is 1.86 to 2.56 times SMR for sheep maintaining body weight.[26] These studies do not give a breakdown of the contribution of activity costs to this added expenditure.

Using modern analytical equipment and well-trained experimental animals, the incremental costs of most activities can be measured with a high degree of accuracy. Activities can be divided into identifiable periods that may be differentiated at any level of resolution. The simplest differentiation includes active and inactive periods. Active states are usually further divided into feeding, standing, moving, specialized or social activities, and various types of predator avoidance. Subdivisions of the rest period may include lying (asleep, alert, or ruminating), standing, and predator avoidance for species which hide from their predators. Clearly, each of these behavioral activity periods is composed of a variable makeup of measurable components.

The objective of this chapter is to outline generalities that can currently be made about the activity increment. Several studies[19,20] have provided useful information on energy costs

Table 1
INCREMENTAL COST OF STANDING OVER LYING

Species	Increment (%)	Method	Ref.
Roe deer	22	Chamber	27
Pronghorn	21	Chamber	28
Caribou calves	12	Chamber	29
White-tailed deer	20	Chamber	30
Bighorn sheep	18	Chamber	31
Reindeer	10	Mask	32
Wapiti	19	Mask	33
	15	Mask	22
	25	HR telemetry	22
Moose calves	35	Chamber	34
Moose	25	Chamber	35
	21	HR telemetry	21

associated with general levels of activity, where an activity level may include several individual activities. To allow comparison between species, we have included only values for specific activity states (i.e., standing, eating and ruminating, locomotion and cratering). This chapter emphasizes locomotion, the most costly activity, dealing with its physiology, allometric scaling, adaptive mechanisms, and contribution to the daily energy budget.

I. STANDING

The incremental cost of standing over lying has been measured using respiration chambers, gas collection masks, and heart-rate telemetry (Table 1). In all cases, the cost of standing for wild herbivores has been higher than the 9% reported for humans, cattle, and sheep,[2] and appears to be higher for species with long legs such as moose and wapiti. A component of within- and between-species variation probably can be attributed to slight body movements. Therefore, the higher standing increments measured for wild herbivores may be attributed in part to their greater level of alertness. For example, the lowest increment of standing over lying measured by Weiner[27] for roe deer was 22% but slow movements during the standing position increased the increment to over 45%. Similarly, using a metabolism chamber, Regelin et al.[35] measured standing increments as high as 69% for alert, moving moose, compared to 25% for moose standing with minimal movement. Movement during the lying period also affects the incremental change, as heart-rate telemetry studies have predicted a 12% increase in metabolic rate for wapiti lying with head up compared to those lying with head down,[22] and the metabolic rates of bedded but alert moose were 7% higher than those for dozing moose.[21]

II. EATING AND RUMINATING

Foraging time is a large component of the daily activity budget of the animal. Measurement of the energy cost of eating is technically difficult because the energy cost of food prehension, mastication, and propulsion in the alimentary tract is usually confounded by heat arising from microbial fermentation, metabolism of absorbed end-products, and the marked increase in the rate of redistribution of body fluids between the lumen of the gut and the extravascular and vascular compartments.[5,36] Osuji[5] concludes from a review of available data that the additional costs of biting, chewing, salivating, and swallowing are minor in comparison to these sources of thermogenesis. Since the metabolic rate of the animal remains elevated

Table 2
ENERGY COST OF EATING ABOVE STANDING

	Energy cost		
Species	**(kJ/h·kg)**	**(kJ/h·kg$^{0.75}$)**	**Ref.**
Domestic cattle	1.84		37
Domestic sheep	2.26		38
	2.18		39
	3.47		40
	1.84		5
Bighorn sheep	1.84		41
Moose calves	2.18		34
Moose (below brisket ht.)		0.70	21
Moose (brisket-topline)		0.90	21
Moose (above topline)		1.50	21
Wapiti (grazing)		3.30	22
Wapiti (browsing)		4.60	22

throughout the period of the feeding bout, the energy cost of eating is related more to the amount of time spent feeding than to the type or amount of feed consumed.[5,36]

Values for the additional cost of eating above standing for wild herbivores are within the range of those reported for sheep and cattle (Table 2). Heart-rate telemetry studies have predicted slightly higher costs for browsing relative to grazing,[22] and for feeding above the topline of the animal relative to low feeding.[21] These studies found that foraging increased energy expenditure by 33% over lying with the head up for wapiti,[22] and by 28% for moose.[21]

Rumination increases the metabolic rate of both moose[21] and wapiti[22] by approximately 2% over lying with the head up. Osuji[5] reported a 1% increase in metabolic rate due to rumination for domestic sheep. If plant biomass is not limiting, ruminants are able to lower the energy cost of eating by taking larger bites during the feeding bout, and reducing particle size while ruminating in the lying position. This shortens the length of the feeding bout which serves to reduce total daily energy expenditure and the risk of predation.

III. LOCOMOTION

It is now well established that the relative cost of locomotion for large animals is less than that for small animals.[11,42] The physiological explanation for this apparent change in efficiency is complex, as several processes are involved. For example, if two animals of different body weight are moving at the same velocity, the corresponding muscles of the lighter animal will need to contract more often than those of the heavier animal, and thus the lighter animal will have a higher cost of transport per unit body weight. In addition, the muscles of large animals have a greater number of slow fibers (slower rates of actin-myosin cross bridging[43]), which operate more economically and efficiently than fast fibers;[44] this also represents an energy saving for larger animals. A. V. Hill[45] used dimensional analysis to predict that the mass-specific work per stride would be the same for large and small animals moving at "physiologically equivalent"[46] speeds. In agreement, it has been shown experimentally that the energy consumed by each gram of muscle per stride is indeed independent of body size for animals moving at the trot-gallop transition speed.[47] However, it was also found that the muscular efficiency (calculated as the ratio of total mechanical work produced to metabolic energy input) increased with body size. More recent studies show that the greater muscular efficiencies measured for large animals are due in part to the storage and recovery of elastic energy during a stride cycle.[47] During one part of each stride, tendons and other elastic elements are stretched, and the energy stored in them is then released in another part of the stride. Most of the energy of locomotion is spent in tensioning tendons that serve to store energy, and in doing other internal work, and comparatively little is used in accelerating and decelerating the limbs.[12,44,47] That the cost of swinging the limbs is a minor component of the energy cost of locomotion is exemplified by the finding that the increase in oxygen consumption with increasing speed is about the same for the cheetah, gazelle, and goat, in spite of considerable differences in the thickness and weight of their limbs.[48]

The energy cost of locomotion has been determined for a number of species by measuring their oxygen consumption (VO_2, mℓ O_2/g·hr) during exercise on a treadmill or various ground surfaces. For most animals studied, oxygen consumption has been described as a linear function of the velocity of travel (V, km/hr):

$$VO_2 = y + S \cdot V \tag{1}$$

where the slope s is the net cost of moving 1 g of body mass 1 km (mℓ O_2/g·km). The intercept y (mℓ O_2/g·hr) is the oxygen consumption while standing. If careful consecutive measurements of the same animal are made, the relationship between increasing oxygen cost and velocity of travel can be described as a series of curvilinear relationships, similar to those reported for horses[49] and elk calves[50] (Figure 1). Because the curvilinearity is very slight, and variability in data collection is not easily controlled in studies involving wild ungulates, the assumption of linearity is considered acceptable for most animals.[50]

Although the net cost of moving 1 g of body mass over 1 kilometer is independent of speed, the total cost per kilometer (i.e., net cost plus the overhead costs of metabolism while standing) declines curvilinearly with increasing speeds until a minimal asymptotic cost per unit of distance is achieved.[11]

A. Scaling to Body Size

Taylor et al.[11] published an allometric equation relating the oxygen cost of transport to body weight and velocity of travel for a wide variety of animals. This equation has recently been modified using data from additional research.[51] The published equation can be rearranged such that the oxygen cost of transport (s, mℓ O_2/g·km) for any biped or quadruped can be predicted by:

$$S = 0.533\ W^{-0.316} \tag{2}$$

where W is in kilograms. The oxygen cost for the whole animal (rather than on a per gram basis) is proportional to $W^{0.684}$. Ratios of the slopes predicted by Equation 2 to those measured for domestic and wild ungulates are shown in Table 3. Similar predictions can be made using the equation proposed by Cohen et al.[54] (Table 3). Both equations underestimate the values measured for very young animals, which appear to have higher oxygen costs than mature animals of equivalent body weight. For example, the net oxygen cost of horizontal movement (S, mℓ O_2/g·km) for young elk and mule deer can be better predicted by the equation:[50]

$$S = 0.615\ W^{-0.34} \tag{3}$$

A comparison of the net cost of walking for various species of ungulates is shown in Figure 2. With the exception of one value for domestic sheep, which was measured on sheep of low body weight taken directly from pasture,[52] adult domestic ungulates tend to have higher net costs of walking than do wild ungulates of similar body weight. The net

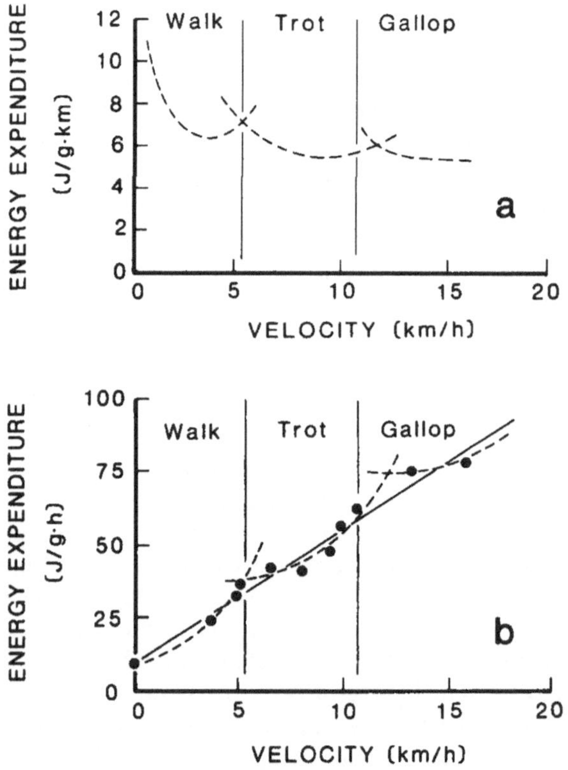

FIGURE 1. The energy cost of different gaits for a 42-kg elk calf plotted on a per hour and per kilometer basis. (Adapted from Parker, K. L., Ecological Energetics of Mule Deer and Elk: Locomotion and Thermoregulation, Ph.D. thesis, Washington State University, Pullman, 1983. With permission.)

cost of walking could be related to the actual daily distances moved by a species. For example, caribou from a migratory herd have a higher walking efficiency than do nonmigratory domestic reindeer and Svalbard reindeer (Table 3 and Figure 3). The latter subspecies evolved on small islands lacking predators and insect pests, and consequently, they travel shorter distances than either of the continental subspecies.

Allometric relationships have also been published to predict the y-intercept of Equation 1, which is the extrapolated oxygen cost at zero velocity, as a function of body weight.[11,51] However, the intercept values for wild ungulates depend largely on the time since the last meal and the type and amount of food consumed. Since the data for different animals and species were collected at noncomparable fasting or feeding levels, the prediction has a high error and may be of questionable value. Several authors[11] have reported that the measured oxygen cost of standing is less than that estimated by extrapolation to zero speed. However, the effect is not generalizable, as other researchers have measured standing values which correspond to the y-intercept determined by extrapolation.[32,54,56-57]

B. Vertical Movement

The cost of moving up and down slopes has been measured for elk and mule deer,[50] red deer,[56] and reindeer[32] (Tables 4 and 5). The cost of climbing is higher than that for horizontal movements because of the energy expended in working against gravity, whereas downslope movements are less expensive because gravitational potential energy is recovered as kinetic

Table 3
NET OXYGEN COST OF TRANSPORT FOR DOMESTIC AND WILD UNGULATES[a]

Species	Weight (kg)	y-intercept (mℓ O_2/g·h)	Slope (mℓ O_2/g·km)	Ratio:[b,f] observed slope ($0.533\ W^{-0.316}$)	Ratio:[c,g] observed slope ($6.35\ W^{-0.34}$)	Ref.
Suni	3.5	0.612	0.520	1.45	1.31	51
Dik-dik	4.35	0.230	0.400	1.19	1.09	51
Domestic sheep	23.0	0.360	0.230	1.16	1.10	51
Desert gazelle	23.2	0.540	0.160	0.81	0.77	48
Domestic sheep	30.0		0.140	0.77	0.73	52
WT deer fawns	39.0	0.700	0.320	1.91	1.83	53
Elk calves	42.0	0.493	0.228	1.39	1.34	50
	48.0	0.380	0.220	1.40	1.35	54
Caribou calves	54.0	0.569	0.090	0.60	0.58	55
Red deer	68.3	0.368	0.129	0.92	0.89	56
Caribou[d]	75.5	0.575	0.094	0.69	0.67	57
Svalbard reindeer	76.0	0.014	0.168	1.24	1.21	58
Reindeer	89.1	0.268	0.129	1.00	0.98	58
Wildebeest	92.0	0.248	0.092	0.72	0.71	59
Reindeer	93.0	0.403	0.099	0.78	0.76	32
Caribou[e]	95.2	0.398	0.082	0.65	0.64	57
Caribou	109	0.240	0.095	0.79	0.77	55
Waterbuck	114	0.166	0.130	1.09	1.07	51
Elk	125		0.119	1.03	1.01	50
	150		0.112	1.02	1.01	50
	157	0.066	0.076	0.70	0.70	33
Eland	213	0.259	0.083	0.85	0.85	51
Zebu cattle	254	0.094	0.096	1.04	1.04	51
Pony	270		0.086	0.95	0.95	2
Domestic cow	450		0.099	1.28	1.30	2
Horse	700		0.076	1.13	1.16	2

[a] Values were recalculated using the energy equivalent of 20.21 J/mℓ O_2, or the value stated in the original reference.
[b] Ratio of observed slopes to those predicted by $0.533\ W^{-0.316}$ (W in kg).[51]
[c] Ratio of observed slopes to those predicted by $6.35\ W^{-0.34}$ (W in g).[54]
[d] Yearling male during summer.
[e] 1 ½-year-old female during winter.
[f] Mean ± SD: 1.02 ± 0.30.
[g] Mean ± SD: 0.99 ± 0.29.

energy during descent.[53] However, the amount of potential energy recovered during descent decreases with body weight and increasing angle of descent.[32,50,54,60] When expressed as the oxygen cost per degree incline, i.e., mℓ O_2/(g·km·degree incline), the costs of ascent (0.025) and descent (0.005) tend to be independent of body weight.[54] Nevertheless, there is considerable variation in the efficiencies of these processes. For example, the downhill efficiency of elk calves ranged from 87% for a 25-kg calf to 25% for a 150-kg individual.[50] The efficiency of moving downslope is computed by subtracting the net energy cost of downhill movement (J/kg·m) from that measured for horizontal movement, and dividing by 9.79 (J of potential energy stored per kilogram per meter of vertical movement).

The energetic cost of lifting one kilogram of body weight one vertical meter averages 24.0 J/kg·m (Table 5), for an average efficiency of 39.6%. The net cost is independent of body weight and the angle of ascent with the more moderate slope angles examined thus far. An exception was found to differ for reindeer climbing slopes of 2.9 and 5.1 degrees.[32] However, this is probably explained by differences in the softness of the outdoor trails used

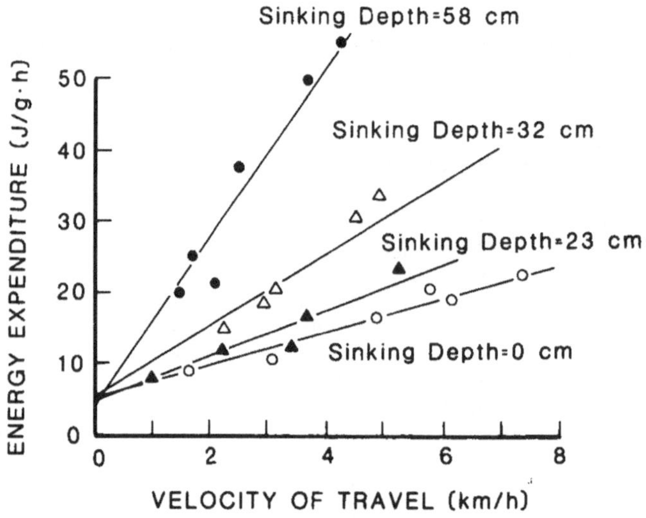

FIGURE 2. Effect of sinking depth on the energy cost of walking in snow for a 100-kg elk calf. (From Parker, K. L., Ecological Energetics of Mule Deer and Elk: Locomotion and Thermoregulation, Ph.D. thesis, Washington State University, Pullman. With permission.)

Table 4
ENERGY RECOVERED DURING DOWNHILL MOVEMENT

Species	Body weight (kg)	Incline (deg)	Energy recovered (J/kg·m)	Efficiency (%)	Ref.
Mule deer	19	14.3	7.36	75	59
Reindeer	96	2.9	6.95	71	32
	96	5.1	5.77	59	32
Elk calves	50	14.3	5.27	54	59
	100	14.3	3.56	35	59
	150	14.3	2.76	28	59
Burro	254	5.7	4.69	48	60
	254	9.6	0.96	10	60

Table 5
ENERGY COST OF LIFTING 1 KG OF BODY WEIGHT 1 VERTICAL METER

Species	Body weight (kg)	Incline (deg)	Energy cost[a] (J/kg·m)	Ref.
Mule deer	14—24	0—14.3	25.06	50
Domestic sheep	41	0—5.1	26.61	61
Elk calves	48	0—11.7	12.95	54
	28—145	0—14.3	23.97	50
Red deer	68	0—14	21.46	56
Reindeer	96.5	0—5.1	25.30	32
Domestic cattle	120—191	0—6.0	26.44	61
Burro	253.5	0—9.6	30.12	60

[a] Mean: 23.99.

during the experiments. Since ungulates expend the same amount of energy per unit of weight for uphill movement regardless of size,[63] a smaller animal will spend less energy climbing a given slope than will a larger animal. However, large animals may pick steeper ascents as the terrain becomes steeper, since the more vertical path may be less costly than traveling a much greater total distance when using a shallower trail angle.[64]

C. Effect of Soft Surfaces

Most of the measurements of locomotion costs of ungulates have been made using treadmills or on hard ground surfaces. The energy cost of transport is much higher when animals move across spongy or soft surfaces such as bogs, mires, or sand, and on snow. For example, reindeer expended 13% more energy when walking on dry tundra than on hard-packed roads, and 30% more when walking on wet tundra (sinking depth, 6 to 10 cm).[32] Measurements of the energy cost of moving through snow have recently been made for elk, mule deer, and caribou.[50,57] The depth to which an animal sinks in snow, and to a lesser extent the snow hardness and/or density, greatly influences the energy cost of walking (Figure 2). A 100-kg elk calf sinking to a depth of 23 cm in snow with a density of 0.2 g/cm^3 spends 36% more energy than that required for walking on a snow-free surface. The net cost of walking for an elk sinking to a depth of 58 cm (82% of brisket height) is 410% greater than that when snow is absent.[50]

The energy cost of moving on snow increases exponentially as the sinking depth approaches chest height (Figure 4), and thus mammals with the greatest chest heights and least foot loads (body weight divided by the area of the four feet contacting the ground or snow surface) are generally best adapted for walking in snow. The selective advantage of long legs in regions of deep snow is obvious, although long legs may also facilitate foraging and are not necessarily an adaptation only to deep snow.[65] Nasimovitch[66] reported that reindeer from regions of deep, soft snow had longer legs than reindeer from tundra areas, and moose in interior Alaska have longer legs than moose on the Kenai Peninsula where snow is shallower.[65] Within a species, females and younger animals are at a disadvantage in snow because of their shorter legs, but this is partially compensated for by lighter foot loads.

Dense and/or crusted snow increases the energy cost of locomotion both by increasing the drag on the legs or body and by causing the animal to lift its legs to a higher level. This influence of snow density on locomotion costs becomes more pronounced as the sinking depth approaches the brisket height of the animal (Figure 4). As shown in Figure 4 using our data for caribou, a snow crust which does not support the animal's weight can greatly exaggerate energy costs.[50,53,67] We measured a 568% increase in the net cost of locomotion on one snow crust which almost supported the caribou's weight before collapsing. Our data suggest that the cost of locomotion for caribou walking in light, uncrusted snow increased with relative sinking depth in a manner similar to other deer species. However, in certain snow types caribou will have a lower relative sinking depth than the other species because of their lighter foot loading.

Snow hardness determines the force which must be exerted by an animal to move on snow and the capacity of the snow to support the animal.[65] If snow hardness consistently exceeds foot loads, animals can walk on top of the snow or will sink into it for only a fraction of its total depth. With the exception of musk deer, caribou have the lightest foot loads measured for wild ungulates (Table 6). Species with the lowest foot loads have lateral metapodial digits which increase the surface area of the foot for supporting the weight of the body in soft snow.[69] In certain snow conditions, caribou will be supported by snow crusts or will sink in less than other species. The relatively large front hooves are also an adaptation for digging feeding craters in snow.[68,71] Within a species, females and younger animals may be at an advantage in crusted snows which support their weight but not that of males or adults with higher foot loads. The practical significance of these advantages as

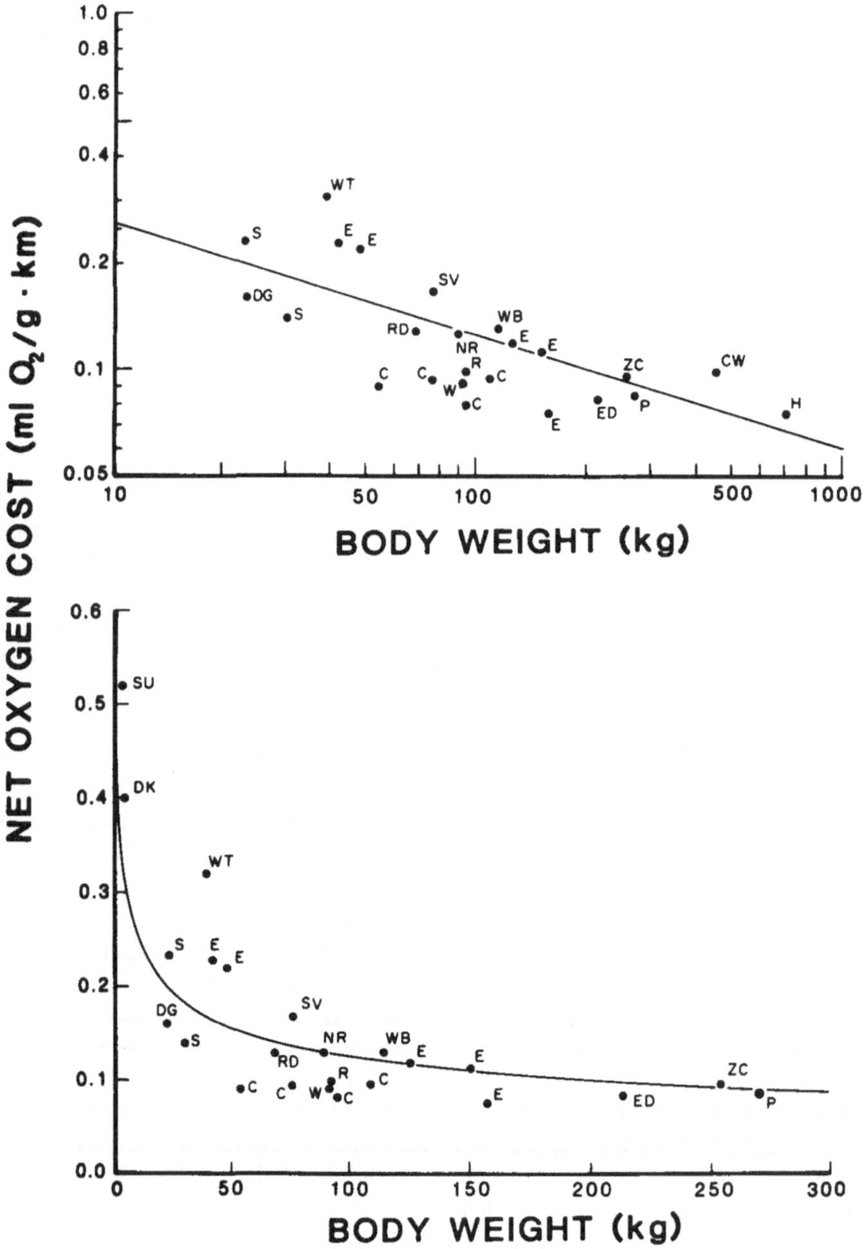

FIGURE 3. Relationship between the net oxygen cost of walking and body weight for different wild and domestic ungulates, plotted on logarithmic and linear scales. The interspecies line shown in each figure is calculated as $W^{-0.316}$ (see Reference 51). The species represented by the various symbols are as follows: SU = suni; DK = dik-dik; DG = desert gazelle; S = domestic sheep; WT = white-tailed deer fawns; E = wapiti; C = caribou; RD = red deer; SV = Svalbard reindeer; NR = Norwegian reindeer; R = reindeer; W = wildebeest; WB = waterbuck; ED = eland; ZC = zebu cattle; P = pony; CW = domestic cow; H = horse. Data are shown in Table 3.

aids to survival, however, depends on the relative frequency of occurrence and duration of snow having these certain characteristics.[69]

A number of behavioral adaptations also serve to reduce the costs of moving through snow. For example, in deep snow, caribou and moose will walk in single file along packed trails, thereby reducing the sinking depth of following animals to a few centimeters.[72] White-

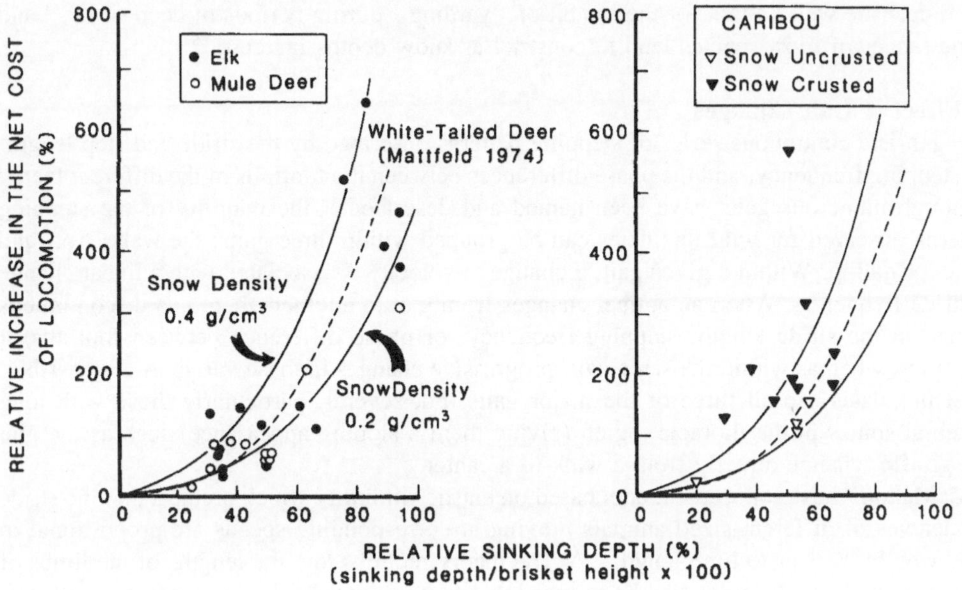

FIGURE 4. Relative increase in the net cost of locomotion with increasing sinking depth for four species of wild ungulates. The effect of increasing snow density on the cost of walking in uncrusted snow is shown in the left figure. (From Parker, K. L., Ecological Energetics of Mule Deer and Elk: Locomotion and Thermoregulation, Ph.D. thesis, Washington State University, Pullman, 1983. With permission.) The effect of a snow crust which does not support the animal's weight is shown in the right figure. (Fancy, S. G. and White, R. G., unpublished.)

Table 6
FOOT LOADINGS OF WILD UNGULATES

Species	Foot loading (g/cm^2)	Ref.
Musk deer	80—120	66
Caribou, calves	125	68
Caribou, yearlings	139—142	57
Caribou, adults	150	68
Chamois	200	66
Mule deer	259—350[a]	50
Elk, calves	305—427[b]	50
Moose, calves	317	65
Roe deer	300—600	66
Moose, adult cows	432—593	65
Moose, Asiatic	400—500	66
Moose, calf	512	66
White-tailed deer	520	69
Elk, adult cows	646—676	69
Moose, adult bulls	710	69
Moose, adults	701—789	70
Elk, adults	780	69
Bison, adults	920	69

[a] 34—63 kg body weight; foot loading predicted as 46 $kg^{0.49}$.
[b] 76—149 kg body weight; 35.0 $kg^{0.50}$.

tailed deer are well known for their habit of "yarding" during periods of deep snow,[53] and home ranges of most species tend to constrict as snow depths increase.[65,73]

D. Effect of Gait Changes

A gait is a continuous series of stepping patterns described by the stride and step length, the stepping frequency, and the phase differences between the footfalls of the different feet.[74] Although numerous gaits have been named and described,[75] the majority of the stepping patterns observed for wild ungulates can be grouped within three gaits: the walk, trot, and canter or gallop. Within a given gait, a change in velocity is associated with a linear change in stride frequency. When an animal changes from gait to another, there is a discontinuous change in the stride length, stepping frequency, or phase difference between footfalls of each foot, whereas within the same gait, progressive changes in these variables are possible. Most ungulates use all three of the major gaits, but several, particularly those with long vertebral spines in the thoracic region (giving them a sloping appearance) such as the gnu and giraffe, change directly from a walk to a canter.[74]

McMahon[76] has proposed a model based on elastic similarity which predicts that the stride frequencies of different-sized animals moving at corresponding speeds are proportional to body weight$^{-0.125}$ or to limb length$^{-0.5}$. The theory requires that the lengths of the limbs of similar animals of different size be proportional to body weight$^{0.25}$. In agreement with this theory, the stride frequencies of mice, rats, and horses at the trot-gallop transition speed are proportional to body weight$^{-0.14}$,[46] and the stride frequency for each gait of several African mammals is proportional to shoulder height^{-05}.[74]

Stride length does not show such simple interspecies scaling relations to body weight. However, simple relationships between stride or step length, body size (hip height), and speed can be formulated and applied to a wide range of animals by using Froude numbers.[77] The Froude number F for a running animal is $v^2/g{\cdot}h$ where v is the running velocity, g the acceleration due to gravity and h the hip height of the animal. The relative stride length (stride length/hip height) of African ungulates running at near maximum speeds is proportional to $F^{0.39}$.[77]

These scaling relationships become important to the effect of body size on more complex behavioral responses. An example is the concept of "foraging radius"[14] which is the maximum radius between foraging locations and a central location such as a den or watering site. The foraging radius for a species is calculated from its metabolic rate, locomotion costs, and the velocity used to move between the grazing and watering sites. Pennycuick[14] found that the foraging radius of several African mammals was proportional to body weight$^{0.4}$, which is quite a different porportionality to that for locomotion (body weight$^{0.68}$).

Hoyt and Taylor,[49] working with horses, found that there was a speed for each gait where the oxygen cost to move a given body mass over 1 kilometer reached a minimum value. The minimum values were about the same for a walk, trot, and a gallop. In addition, a horse moving freely over a marked grid selected speeds within each gait around the energetically optimal speed. Parker[50] similarly measured an energetically optimal speed for different gaits in an elk calf (Figure 1a). Although it has not been shown that wild ungulates select speeds which are energetically optimal, different species have been found to use a restricted range of speeds within each gait. For example, Thomson's gazelles walk at speeds very close to 3.1 km/hr, whereas for gnus this speed is 3.6 km/hr.[74] White-tailed deer were found to select walking speeds of 4.0 km/hr[53] which is the average speed for caribou during spring migration and summer movements.[57,72] On a theoretical basis, the cruising speed of animals for any particular gait should be proportional to body weight$^{0.13}$.[14]

E. Contribution to Daily Energy Expenditures

If transportation costs comprise the major part of an animal's total daily energy expenditure, then natural selection should strongly favor morphological, physiological, and behavioral adaptations which increase locomotory efficiency. If, on the other hand, transportation costs are only a minor component of the daily energy budget, then there would be less selection for increased efficiency. Using data from several sources, an estimate was made of the daily cost of locomotion in relation to SMR for two species of wild ungulates. The analysis included animals of different body weights with energy expenditure values which fall on the interspecies line.[51] These calculations have been simplified by assuming only horizontal movements on hard surfaces, as daily movements would vary greatly depending on snow and terrain features. In this theoretical exercise, the energy cost of tranport for a 92-kg wildebeest was determined to be 0.092 mℓ O_2/g·km,[51] which can be converted to 1.86 J/g·km using an energy equivalent of 20.21 J/mℓ O_2. The mean distance traveled by a migratory wildebeest population was 10 km/day, averaged over an entire year.[14] Thus, the net daily cost of locomotion was 1711 kJ/day. The additional cost of standing for this animal was 0.069 mℓ O_2/s·kg,[51] which converts to 5.02 J/g·hr. If we assume that the wildebeest walked at a mean speed of 3.53 km/hr,[14] then 2.83 hr were spent traveling. The total cost of locomotion, including both the cost of standing and the additional incremental cost of locomotion, was therefore 3020 kJ/day. If the SMR of this animal is estimated to be 8701 kJ/day using Kleiber's equation,[4] then the daily cost of locomotion for the wildebeest is 35% of SMR. Similar calculations for a 180-kg adult wildebeest give an estimate of 42% of SMR.

Female caribou of the Porcupine Herd have been observed to move an average distance of 16.8 km/day during spring migration.[72] We have developed a computer simulation model based on energy expenditure data from captive caribou. If the activity budgets for the Porcupine Herd caribou during 1983 spring migration are used as inputs, and assuming a body weight of 90 kg, the model predicts that the lead animals (breaking trail in slightly crusted snow, sinking depth approximately 15 cm[72]) will have locomotion costs of approximately 82% of SMR, 134% higher than those for the 92-kg wildebeest. Caribou following packed trails would have daily locomotion costs of approximately 33% of SMR, 6% lower than for the similar-sized wildebeest. Clearly, terrain surfaces are important components for assessing the importance of locomotion costs.

The total daily cost of locomotion in relation to SMR has been calculated for animals of different body weights and daily distance traveled (Table 7). These calculations assume that the energy values for these animals follow Equation 2, and further assume a mean walking speed of 3.6 km/hr. SMR is calculated from Kleiber's equation,[4] but since most wild ungulates studied to date have SMRs much higher than the values predicted by that equation, SMR (kJ/day) is also calculated as:

$$\text{SMR} = 377\ W^{0.75} \tag{4}$$

Using this equation, the total daily locomotion costs for animals walking 5 km/day are 16 to 18% of SMR (Table 7), whereas locomotion costs for animals walking 15 km/day are 48 to 53% of SMR. These costs would be much higher if upslope movements or movements on snow or other soft surfaces were involved.

These calculations reconfirm observations that locomotion costs rarely exceed the combined energy expenditures associated with SMR and the heat increment of feeding. However, locomotion makes up a substantial component of the extra energy required for free existence, which is 1.9 to 2.6 times SMR in domestic sheep at pasture.[26]

IV. CRATERING

Several species, particularly caribou, obtain a large portion of their winter food supply by digging beneath the snow cover. We have recently measured the cost of cratering by

Table 7
DAILY COST OF LOCOMOTION AS A PERCENT OF SMR[a]

Weight (kg)	Distance (km/d)	Energy cost (kJ/d)	% of SMR[b]	% of SMR[c]
50	5	1246	23	18
	10	2491	45	35
	15	3737	68	53
100	5	2008	22	17
	10	4016	43	34
	15	6023	65	51
200	5	3237	21	16
	10	6473	42	32
	15	9710	62	48

[a] Oxygen cost of walking is calculated from Equation 2 using the energy equivalent of 20.21 J/mℓ O_2. Calculations assume that no vertical movements occur, and that the animal moves at a mean speed of 3.6 km/hr.
[b] SMR calculated as 292.9 $W^{0.75}$, where W is in kg (Reference 4).
[c] SMR calculated as 376.6 $W^{0.75}$, which more closely predicts the values measured for wild ungulates.

caribou using heart-rate telemetry with calibrated animals. In light, uncrusted snow with a density of approximately 0.18 g/cm^3, the mean cost per pawing was 118 J. The mean cost per pawing in denser (0.36 g/cm^3) snow with a thin but hard (hardness = 1400 to 1600 g/cm^2) crust was 219 J. The cost of cratering in very dense and hard snow beneath a snow-machine trail was 481 J per pawing. Therefore, a fourfold variability in energy cost per pawing can occur, depending on snow conditions.

Cratering mechanics have also been determined using vector analysis of cratering events recorded on movie film.[68] Thing[68] estimated the cost per pawing based on the length and weight of the various limb segments on caribou carcasses, the displacement vectors of the limb segments, and the timing of limb movements. The estimated cost of the slow sweeping movement used by caribou in soft snow was 166 J per pawing. His estimated cost of one pawing in hard or crusted snow, in which caribou use a downward chisel-like movement of the hoof, was 334 J. These values do not take into account the efficiency of converting feed energy into mechanical energy. Using Thing's mean value of 19.6 pawings per minute for wild caribou, and our estimates of cratering costs, we can calculate the hourly cost of cratering above standing for a 95-kg caribou to be 1.46 kJ/kg·hr in light, uncrusted snow and 2.71 kJ/kg·hr in denser snow with a thin crust. By comparison, the net cost of walking on level, snow-free ground at 2 km/hr for these caribou is approximately 3.63 kJ/kg·hr. The cost per limb movement for cratering is greater than for walking, but because the frequency of limb movements is much lower while cratering, the rate of energy expenditure is comparable to that during a slow walk.

V. CONCLUSION

The total daily energy expenditure of an animal can be calculated by summing SMR, the heat increment of feeding, and the daily cost of thermoregulation and activity. The total daily cost of activity is usually less than the combined costs of SMR and the heat increment of feeding, but in certain situations, activity costs can be high. Most of the increase in energy expenditure of activity results from locomotion costs, which can be very high for animals walking long distances, breaking trail in deep snow, or climbing slopes. Wild herbivores tend to be more efficient at walking than are domestic species, and there is some evidence that walking efficiency is directly proportional to the average daily distance walked by an animal population.

Many useful scaling relationships have been found by comparing data between different species. Several relationships can sometimes be combined to form new relationships between body dimensions or weight and complex behavioral traits. Scaling relationships allow predictions to be made for species for which data are lacking, and give insight into the constraints imposed on animals in adapting to their environment.

Recent developments in isotope and heart-rate telemetry techniques have allowed estimates of the energy cost of free existence to be made. Data obtained using these techniques can be used to verify the predictions of computer simulation models, which can be powerful tools when accurate input data are available. When combined with field observations of animal behavior, data on energy costs greatly improve our understanding of the ecology of wild herbivores.

ACKNOWLEDGMENTS

We would like to thank Drs. R. D. Cameron, D. R. Klein, K. L. Parker, and C. T. Robbins for reviewing a draft of this chapter. We also thank Don Borchert for preparing the figures. Research and preparation of this chapter were supported by the NSF (Grant No. GB-29342), State of Alaska funds for applied reindeer research and for organized research through the Institute of Arctic Biology, and ACST Grant No. 46-82.

REFERENCES

1. **Rubner, M.,** Uber den Einfluss der Korpergrosse auf Stoffund Kraftwechsel, *Z. Biol. (Munich)*, 19, 535, 1883.
2. **Brody, S.,** *Bioenergetics and Growth*, Hafner, New York, 1945.
3. **Blaxter, K. L.,** *The Energy Metabolism of Ruminants*, Charles C Thomas, Springfield, Ill., 1962.
4. **Kleiber, M.,** *The Fire of Life: An Introduction to Animal Energetics*, R. E. Krieger, Huntington, N.Y., 1975.
5. **Osuji, P. O.,** The physiology of eating and the energy expenditure of the ruminant at pasture, *J. Range Manage.*, 27, 437, 1974.
6. **Golley, F. B.,** Energy dynamics of a food chain of an old-field community, *Ecol. Monogr.*, 30, 187, 1964.
7. **Kendeigh, S. C.,** Monthly variations in the energy budget of the house sparrow throughout the year, in Proc. Gen. Meet. Working Groups on Granivorous Birds, Kendeigh, S. C. and Pinowski, J., Eds., IBP, PT Section, Warszawa.
8. **Ricklefs, R. E.,** *Ecology*, Chiron Press, New York, 1979.
9. **Gessaman, J. A.,** Methods of estimating the energy cost of free existence, in *Ecological Energetics of Homeotherms*, Gessaman, J. A., Ed., Monogr. Ser. 20, Utah State University Press, Logan, 1973, 3.
10. **Schmidt-Nielsen, K.,** *How Animals Work*, Cambridge University Press, London, 1972.
11. **Taylor, C. R., Schmidt-Nielsen, K., and Raab, J. L.,** Scaling of energetic cost of running to body size in mammals, *Am. J. Physiol.*, 291, 1104, 1970.
12. **Alexander, R. McN.,** Terrestrial locomotion, in *Mechanics and Energetics of Animal Locomotion*, Alexander, R. McN. and Goldspink, G., Eds., Chapman and Hall, London, 1977, 168.
13. **Prosser, C. L., Ed.,** *Comparative Animal Physiology*, W.B. Saunders, Philadelphia, 1973.
14. **Pennycuick, C. J.,** Energy costs of locomotion and the concept of "foraging radius", in *Serengeti: Dynamics of an Ecosystem*, Sinclair, A. R. E. and Norton-Griffiths, M., Eds., University of Chicago Press, Chicago, 1979, 164.
15. **Taylor, C. R.,** Energy cost of locomotion, in *Comparative Physiology*, Bolis, L., Schmidt-Nielsen, K., and Maddrell, S. H. P., Eds., North Holland, Amsterdam, 1973.
16. **Sinclair, A. R. E. and Norton-Griffiths, M., Eds.,** *Serengeti: Dynamics of an Ecosystem*, University of Chicago Press, Chicago, 1979.

17. **Corbett, J. L., Leng, R. A., and Young, B. A.**, Measurement of energy expenditure by grazing sheep and the amount of energy supplied by volatile fatty acids produced in the rumen, in *Energy Metabolism of Farm Animals*, Blaxter, K. L., Kielanowski, J., and Thorbek, G., Eds., Oriel Press, Newcastle-Upon-Tyne, 1969.
18. **Young, B. A., Leng, R. A., White, R. G., McClymont, G. L., and Corbett, J. L.**, Estimation of energy expenditure from measurements of carbon dioxide entry rate, in *Energy Metabolism of Farm Animals*, Blaxter, K. L., Kielanowski, J., and Thorbek, G., Eds., Oriel Press, Newcastle-Upon-Tyne, 1969, 435.
19. **Mautz, W. W. and Fair, J.**, Energy expenditure and heart rate for activities of white-tailed deer, *J. Wildl. Manage.*, 44, 333, 1980.
20. **Kautz, M. A., Van Dyne, G. M., Carpenter, L. H., and Mautz, W. W.**, Energy cost for activities of mule deer fawns, *J. Wildl. Manage.*, 46, 704, 1982.
21. **Renecker, L. A. and Hudson, R. J.**, Winter activity budgets of free-ranging moose, using a calibrated heart rate index, in Proc. 4th Int. Conf. on Wildlife Biotelemetry, Halifax, Nova Scotia, 1983, 187.
22. **Pauls, R. W., Hudson, R. J., and Sylven, S.**, Energy expenditure of free-ranging wapiti, *Univ. Alberta Feeders Day Rep.*, 60, 87, 1981.
23. **Lifson, N. and McClintock, R.**, Theory and use of the turnover rates of body water for measuring energy and material balance, *J. Theor. Biol.*, 12, 46, 1966.
24. **Mullen, R. K.**, The $D_2{}^{18}O$ method of measuring the energy metabolism of free-living animals, in *Ecological Energetics of Homeotherms*, Gessaman, J. A., Ed., Monogr. Ser. 20, Utah State University, Logan, 1973, 32.
25. **Holleman, D. F., White, R. G., and Feist, D. D.**, Seasonal energy and water metabolism in free-living Alaskan voles, *J. Mammal.*, 63, 293, 1982.
26. **Young, B. A. and Corbett, J. L.**, Maintenance energy requirements of grazing sheep in relation to herbage availability. I. Calorimetric estimates, *Aust. J. Agric. Res.*, 23, 57, 1972.
27. **Weiner, J.**, Energy metabolism of the roe deer, *Acta Theriol.*, 22, 3, 1977.
28. **Wesley, D. E., Knox, K. L., and Nagy, J. G.**, Energy metabolism of pronghorn antelopes, *J. Wildl. Manage.*, 37, 563, 1973.
29. **Luick, B. R. and White, R. G.**, Indirect calorimetric measurements of the caribou calf, *Acta Zool. Fenn.*, 175, 89, 1982.
30. **Silver, H., Holter, J. B., Colovos, N. F., and Hayes, H. H.**, unpublished data, cited by **Weiner, J.**, Energy metabolism of the roe deer, *Acta Theriol.*, 22, 3, 1977.
31. **Chappel, R. W. and Hudson, R. J.**, Energy cost of standing in Rocky Mountain bighorn sheep, *J. Wildl. Manage.*, 43, 261, 1979.
32. **White, R. G. and Yousef, M. K.**, Energy expenditure in reindeer walking on roads and on tundra, *Can. J. Zool.*, 56, 215, 1977.
33. **Gates, C. C. and Hudson, R. J.**, Energy costs of locomotion in wapiti, *Acta Theriol.*, 23, 365, 1978.
34. **Renecker, L. A., Hudson, R. J., Christopherson, M. K., and Arelis, C.**, Effect of posture, feeding, low temperature and wind on energy expenditures of moose calves, *Proc. North Am. Moose Conf. Workshop*, 14, 126, 1978.
35. **Regelin, W. L., Schwartz, C. C., and Franzmann, A. W.**, Energy Expenditure of Moose on the Kenai National Wildlife Refuge, Annu. Prog. Rep., U.S. Fish and Wildlife Service and Alaska Dept. Fish and Game, 1981.
36. **Webster, A. J. F.**, Act of eating and its relation to the heat increment of feed in ruminants, in Bioenergetics International Symposium on Environmental Physiology, American Soc. Exp. Biol., Dublin, 1972, 42.
37. **Adam, I., Young, B. A., and Nicol, A. M.**, Energy cost to cattle of ingesting feed, *Univ. Alberta Feeders Day Rep.*, 58, 38, 1979.
38. **Graham, N. McC.**, Energy cost of feeding activities and energy expenditure of grazing sheep, *Aust. J. Agr. Res.*, 15, 969, 1964.
39. **Young, B. A.**, Energy expenditure and respiratory activity of sheep during feeding, *Aust. J. Agr. Res.*, 17, 355, 1966.
40. **Webster, A. J. F. and Hayes, F. L.**, Effects of beta-adrenergic blockade on the heart rate and energy expenditure of sheep during feeding and during acute cold exposure, *Can. J. Physiol. Pharmacol.*, 46, 577, 1968.
41. **Chappel, R. W. and Hudson, R. J.**, Energy cost of feeding in Rocky Mountain bighorn sheep, *Acta Theriol.*, 23, 359, 1978.
42. **Schmidt-Nielsen, K.**, Locomotion: energy cost of swimming, flying and running, *Science*, 177, 222, 1971.
43. **Goldspink, G.**, Design of muscles in relation to locomotion, in *Mechanics and Energetics of Animal Locomotion*, Alexander, R. McN. and Goldspink, G., Eds., Chapman and Hall, London, 1977, 1.
44. **Goldspink, G.**, Muscle energetics, in *Mechanics and Energetics of Animal Locomotion*, Alexander, R. McN. and Goldspink, G., Eds., Chapman and Hall, London, 1977, 57.
45. **Hill, A. V.**, The dimensions of animals and their muscular dynamics, *Science Prog.*, 38, 209, 1950.

46. **Heglund, N. C., Taylor, C. R., and McMahon, T. A.,** Scaling stride frequency and gait to animals size: mice to horses, *Science*, 186, 1112, 1974.
47. **Heglund, N. C., Fedak, M. A., Taylor, C. R., and Cavagna, G. A.,** Energetics and mechanics of terrestrial locomotion. IV. Total mechanical energy changes as a function of speed and body size in birds and mammals, *J. Exp. Biol.*, 97, 57, 1982.
48. **Taylor, C. R., Shkolnik, A., Dmiel, R., Baharav, D., and Borut, A.,** Running in cheetahs, gazelles and goats: energy cost and limb configuration, *Am. J. Physiol.*, 227, 848, 1974.
49. **Hoyt, D. F. and Taylor, C. R.,** Gait and energetics of locomotion in horses, *Nature (London)*, 292, 239, 1981.
50. **Parker, K. L.,** Ecological Energetics of Mule Deer and Elk: Locomotion and Thermoregulation, Ph.D. thesis, Washington State University, Pullman, 1983.
51. **Taylor, C. R., Heglund, N. C., and Maloiy, G. M. O.,** Energetics and mechanics of terrestrial locomotion, *J. Exp. Biol.*, 97, 1, 1982.
52. **Farrell, D. J., Leng, R. A., and Corbett, J. L.,** Under-nutrition in grazing sheep. II. Calorimetric measurements on sheep taken from pasture, *Aust. J. Agric. Res.*, 23, 466, 1972.
53. **Mattfeld, G. F.,** The Energetics of Winter Foraging by White-tailed Deer: Perspective on Winter Concentration, Ph.D. thesis, State Univ. New York, Syracuse, 1974.
54. **Cohen, Y., Robbins, C. T., and Davitt, B. B.,** Oxygen utilization by elk calves during horizontal and vertical locomotion compared to other species, *Comp. Biochem. Physiol.*, 61A, 43, 1978.
55. **Luick, B. R.,** unpublished data.
56. **Brockway, J. M. and Gessaman, J. A.,** The energy cost of locomotion on the level and on gradients for the red deer *(Cervus elaphus)*, *Q. J. Physiol.*, 62, 333, 1977.
57. **Fancy, S. G. and White, R. G.,** unpublished data.
58. **Nilssen, K. J., Johnsen, H. K., Rognmo, A., and Blix, A. S.,** Heart rate and energy expenditure in resting and running Svalbard and Norwegian reindeer, *Am. J. Physiol.*, 24, R963, 1984.
59. **Parker, K. L.,** unpublished data.
60. **Yousef, M. K., Dill, D. B., and Freeland, D. V.,** Energetic cost of grade walking in man and burro, *Equus asinus*: desert and mountain, *J. Appl. Physiol.*, 53, 337, 1972.
61. **Clapperton, J. L.,** The effect of walking upon the utilization of food by sheep, *Br. J. Nutr.*, 18, 39, 1964.
62. **Ribeiro, J. M. de C. R., Brockway, J. M., and Webster, A. J. F.,** A note on the energy cost of walking in cattle, *Anim. Prod.* 25, 107, 1977.
63. **Taylor, C. R., Caldwell, S. L., and Rowntree, V. J.,** Running up and down hills: some consequences of size, *Science*, 178, 1096, 1972.
64. **Reichman, O. J. and Aitchison, S.,** Mammal trails on mountain slopes: optimal paths in relation to slope angle and body weight, *Am. Nat.*, 117, 416, 1981.
65. **Coady, J. W.,** Influence of snow on behaviour of moose, *Nat. Can.*, 101, 417, 1974.
66. **Nasimovich, A. A.,** The Role of the Regime of Snow Cover in the Life of Ungulates in the USSR, Akad. Nauk SSSR, Moskva, 1955 (trans. from Russian), Canadian Wildlife Service, Ottawa.
67. **Hepburn, R. L.,** Effects of Snow Cover on Mobility and Local Distribution of Deer in Algonquin Park, M.A. thesis, University of Toronto, Toronto, 1959.
68. **Thing, H.,** Behavior, mechanics and energetics associated with winter cratering by caribou in northwestern Alaska, *Biol. Pap. Univ. Alaska*, 18, 1977.
69. **Kelsall, J. P. and Telfer, E. S.,** Studies of the physiological adaptation of big game for snow, in *Proc. Snow and Ice in Relation to Wildlife and Recreation Symposium*, Haugan, A. O., Ed., Iowa State University, Ames, 1971, 134.
70. **Kelsall, J. P.,** Structural adaptations of moose and deer for snow, *J. Mammal.*, 50, 302, 1969.
71. **Pruitt, W. O., Jr.,** Snow as a factor in the winter ecology of the barren ground caribou, *Arctic*, 12, 158, 1959.
72. **Duquette, L. S.,** Patterns of Activity and Their Implications to the Energy Budget of Migrating Caribou, M.S. thesis, University of Alaska, Fairbanks, 1984.
73. **Craighead, J. J., Craighead, F. C., Jr., Ruff, R. L., and O'Gara, B. W.,** Home ranges and activity patterns of nonmigratory elk of the Madison Drainage herd as determined by biotelemetry, *Wildl. Monogr.*, 33, 1973.
74. **Pennycuick, C. J.,** On the running of the gnu (*Connochaetes taurinus*) and other animals, *J. Exp. Biol.*, 63, 775, 1975.
75. **Hildebrand, M.,** Symmetrical gaits of horses, *Science*, 150, 701, 1965.
76. **McMahon, T. A.,** Using body size to understand the structural design of animals: quadrapedal locomotion, *J. Appl. Physiol.*, 39, 619, 1975.
77. **Alexander, R. McN., Langman, V. A., and Jayes, A. S.,** Fast locomotion of some African ungulates, *J. Zool. London*, 183, 291, 1977.

Chapter 8

THERMOREGULATION IN UNGULATES

K. L. Parker and C. T. Robbins

TABLE OF CONTENTS

Thermoregulation, or the ability to regulate body temperature within acceptable limits, is a major factor in ungulate ecology. Wild ungulates occupy tropical to arctic environments where daily and seasonal temperature variations may be minimal to in excess of 100°C. Because of such a wide distribution, numerous mechanisms are employed by ungulates to thermoregulate in such diverse environments. Behavioral, morphological, and physiological adaptations alter heat production or loss to regulate core temperature. Heat transfer processes of conduction, convection, radiation, and evaporation depend primarily on the temperature gradient between the animal and its environment,[1] its effective surface area,[2-4] and pelage resistance.[3,5] The concept of thermoneutrality, however, is not self-evident as there are no "neutral" conditions for the feed-back system of temperature regulation.[6] Hence, thermoneutral environments have been analyzed relative to metabolism, evaporative heat loss, thermal comfort, preferred range, and animal productivity.[7]

Traditionally and most frequently, thermoregulation has been studied using the Scholander approach of delineating the limits of the thermoneutral zone (TNZ).[8] This zone of minimum heat production refers to the range of air temperatures over which metabolic rate of the resting animal is minimal, constant, and independent of environmental temperature. A thermal balance is achieved through physical or vasomotor regulation of heat loss. The lower limit of the TNZ, or lower critical temperature (LCT), separates chemical from physical body temperature regulation and depends on the interplay between an animal's insulation and its heat production. Below this limit, heat losses exceed heat production in the TNZ. Thus, metabolic rate must increase to maintain acceptable deep body temperatures and prevent impairment of tissue function. The upper limit of the thermoneutral zone, or upper critical temperature (UCT), has classically been defined as that temperature above which metabolic rate increases and which is usually associated with hyperthermia.

Alternative thermoregulatory zones or environments are difficult to measure because of the subjectivity involved[9-11] and have only been minimally researched. Therefore, the conventionally accepted concept of thermoneutrality has been used in these analyses. Comfort zones and preferred environments, nonetheless, should also be measured to most effectively manage ungulate habitat.

In the majority of studies, air temperature (T_a) alone has been used to describe the thermal environment experienced by the animal. As a thermal index, it is useful in thermally simple laboratory conditions but inadequate when considering heat exchange in the natural environment. Operative temperature more appropriately describes the effective temperature (T_e) experienced by the animal in complex outdoor environments.[12] This environmental temperature is a biophysical expression which incorporates the thermal consequences of air temperature, low windspeeds, and thermal radiation for a particular animal[13] in the following equation:

$$T_e = T_a + (r_e/\rho c_p)R_{ni} \qquad (1)$$

where r_e is a parallel equivalent resistance to convective and radiative heat transfer, ρc_p is the volumetric specific heat of air, and R_{ni} is the net absorbed solar and thermal radiation.

T_e is not a fixed entity determined from a particular combination of T_a, wind, and incident radiation. Rather, it describes a thermal environment in which the same T_e can be attained from different environmental variables. Theoretically, T_e determined in the environment should evoke similar thermal responses in an animal as T_a measured in the metabolic chamber with no wind or solar radiation. Standard operative temperature (T_{es})[12,14] should be employed in those situations in which wind chill is an important input to the thermoregulatory response. Rarely, however, have all variables necessary for the calculation of operative temperature been monitored and, hence, many of the responses of ungulates presented here have been recorded relative to air temperature.

I. BODY TEMPERATURE

In most wild ungulates, core temperatures (T_b) range between 36 and 40°C with no apparent relationship to body weight (Table 1). Caution must be taken in any direct comparison of body temperatures between species as T_b can vary with operative temperature,[15] daily and seasonal metabolic rhythms, and animal age. Nychthemeral changes range from less than 1°C in the guanaco and hippopotamus[26,28] to 6 to 8°C in the oryx and eland.[17] Seasonal variations have been recorded for mule deer, elk, and collared peccary.[22,37] Newborn and young animals typically exhibit higher and more variable T_bs than do adults. For example, body temperatures of newborn white-tailed deer fawns range from 29.6 to 40°C daily.[46]

The extent to which thermolability serves as an advantage in stressful thermal environments is unknown. A rise in body temperature during the day which adds a considerable thermal buffer under changing environmental conditions has two useful consequences: heat storage as a water conservation mechanism and a reduced air to body temperature gradient which decreases heat absorption. Tolerance to large fluctuations in internal temperatures may provide greater freedom of movement in hot environments.[47] However, for those species that consistently regulate at relatively low body temperatures, such as the hippopotamus, high rates of evaporative water loss and low rates of metabolism may restrict exploitation of some terrestrial habitats.[48]

II. SKIN TEMPERATURE

While the core temperature of an animal is closely regulated, surface skin temperatures fluctuate greatly between body regions and in response to changes in operative temperatures. Skin temperatures play a major role in heat loss, cutaneous evaporation,[30,37,49,50] and cold perception for thermogenesis.[51] Knowledge of the range of surface skin temperatures exhibited by an animal is useful only to assess the feasible responses of that animal to a particular set of environmental conditions. Direct comparisons between species are invalid for animals not subjected to similar operative environments.

Both pelaged and almost hairless species, such as the elephant, rhinoceros, and hippopotamus, regulate skin temperatures within a similar range (Table 2), although marked variations (5.4°C) have been noted between pelaged and shaved areas of collared peccaries.[22] For those species inhabiting extremely cold environments, the 20°C difference between foot and trunk skin temperatures serves an important role in heat conservation. Such low temperatures of the extremities necessitate increased concentrations of unsaturated fatty acids with low melting points such that adipose tissues remain soft under cold conditions for effective operation of the legs.[55] Overall skin temperatures may be fairly accurately estimated from ratios developed between average and trunk temperatures.[56]

In general, surface skin temperatures vary with operative temperature, although the extent of coupling depends on insulation quality.[37] Effects of water,[54] solar radiation,[22,37,50] and wind[45] have seldom been addressed. Because of the high conductivity of water, immersion in water may reduce skin temperatures markedly from dry skin temperatures.[54] Skin temperatures of ungulates exposed to solar radiation average up to 6.4°C higher than those in the shade[22,37] while moderate windspeeds (4 to 6 m/sec) reduce skin temperatures by up to 8.5°C from still conditions.[45] Mule deer and elk standing in snow and exposed to solar radiation had skin temperatures in winter at a T_a of -30°C equal to those of $+30$°C in summer.[37] Skin temperatures may actually be higher for animals in winter than in summer because of the increased potential of the winter coat to prevent heat loss. Fur temperatures fluctuate to even greater degrees than skin temperatures and may be equivalent to or higher than air temperatures in cold and hot environments, respectively.[49,57,58]

Table 1
BODY TEMPERATURES (T_b) OF WILD UNGULATES AT REST

Species	T_b (°C)	T_a (°C)	Season	Condition	Location	Wt. (kg)	Ref.
Tropical							
Black rhinoceros	37.7—39.9	12—25.5	—	Field	Kenya	200	15
(Diceros bicornis)	37.4—37.8	19.5	—	Laboratory	NY (USA)	862	16
Buffalo	36.9—40.1	10—24.5	—	Field	Kenya	350	15
(Syncerus caffer)	36.6—40.0	22—40	—	Laboratory	Kenya	198	17
Burro	36.5—39.0	19—36.9	—	Field	NV (USA)	178	18
(Equus asinus)							
Bushpig	37.8—38.7	15.6—26.7	—	Laboratory	Rhodesia	—	19
(Potamochoerus porcus)							
Camel	34.8—41.2	25—38.3	—	Field	Australia	290	20
(Camelus dromedarius)	35.1—39.1	21—35.5	—	Field	Kenya	467	15
	35.5—38.5	0—18	Winter	Field	Algeria	—	21
Collared peccary	37.5—40.9	27—44	Summer	Field	AZ (USA)	21	22
(Tayassu tajacu)	36.4—40.3	4—18	Winter	Field	AZ (USA)	21	22
Eland	38.3—39.9	14.5—26.5	—	Field	Uganda	200	15
(Taurotragus oryx)	33.9—41.9	22—40	—	Laboratory	Kenya	130	23
	36.3—39.6	22—40	—	Laboratory	Kenya	133	17
Elephant	36.2—36.7	19.5	—	Laboratory	NY (USA)	2790	16
(Elephus indicus)	32.5—37.5	18.3—35	—	Field	Zambia and S. Africa	—	24
Giraffe	37.8—39.1	10.5—21.5	—	Field	Kenya	400	15
(Giraffa camelopardalis)	36.4—39.2	16—24	—	Field	Kenya	600	25
Grant's gazelle	38.2—40.8	22—40	—	Laboratory	Kenya	24	17
(Gazella granti)							
Guanaco	38.4—39.0	26—45	Winter	Laboratory	Chile	58	26
(Lama guanicoe)							
Hartebeest	38.5—40.4	22—50	—	Laboratory	Kenya	87.5	27
(Alcelaphus buselaphus)							
Hippopotamus	34.7—36.1	18.4—32.2	Yearly	Field	Uganda	—	28
(Hippopotamus amphibius)							

Impala	38.3—40.4	22—50	—	Laboratory	Kenya	38	27
(Aepyceros melampus)							
Kirk's dik-dik	38.5—42.0	10—45	—	Laboratory	Kenya	4.0	29
(Madoqua kirkii)	38.4—40.7	20—40	—	Laboratory	Kenya	4.5	30
Lesser mouse deer	37.5—40.8	10—40	—	Laboratory	Malasia	1.3	31
(Tragulus javanicus)							
Oryx	37.2—40.8	22—40	—	Laboratory	Kenya	100	17
(Oryx beisa)	35.7—42.1	22—40	—	Laboratory	Kenya	—	32
Somali donkey	35.7—40.1	20—50	—	Laboratory	Kenya	153	33
(Equus asinus)							
Suni	39.0—40.7	20—40	—	Laboratory	Kenya	2.5	30
(Nesotragus moschatus)							
Thomson's gazelle	37.6—41.6	22—40	—	Laboratory	Kenya	12.7	17
(Gazella thomsoni)							
Warthog	35.0—39.1	15.6—26.7	—	Laboratory	Rhodesia	—	19
(Phacocoerus aethiopicus)							
Waterbuck	38.3—40.3	25—45	—	Laboratory	Uganda	100	34
(Kobus defassa)							
White rhinoceros	33.6—37.5	5—25	—	Field	Uganda	—	35
(Ceratotherium simum)							
Wildebeest	37.8—40.0	22—40	—	Laboratory	Kenya	160	17
(Connochaetes taurinus)							
Temperate							
Bison	38.2—38.7	−30—10	Winter	Laboratory	ALTA (CAN)	91—290	36
(Bison bison)							
Elk	37.8—39.7	2—34	Summer	Field	WA (USA)	144	37
(Cervus elaphus)	37.5—40.0	−29—24	Winter	Field	WA (USA)	92	37
Moose	38.0—38.6	−20—35	Yearly	Field	ALTA (CAN)	322	38
(Alces alces)							
Mountain goat	37.8—39.0	10—20	Winter	Field	AK (USA)	40—140	39
(Oreamnos americanus)							
Mule deer	37.2—40.3	2—34	Summer	Field	WA (USA)	49	37
(Odocoileus hemionus)	37.0—39.3	−38—23	Winter	Field	WA (USA)	61	37
Red deer	38.4—39.1	4—16	Winter	Laboratory	Scotland	46	40
(Cervus elaphus)	39.3—40.9	20—45	Winter	Laboratory	England	95	41
White-tailed deer	37.7—38.3	−20—40	Winter	Laboratory	NH (USA)	—	42
(Odocoileus virginianus)	37.7—39.9	−19—36.9	Winter	Laboratory	NH (USA)	—	43

Table 1 (continued)
BODY TEMPERATURES (T_b) OF WILD UNGULATES AT REST

Species	T_b (°C)	T_a (°C)	Season	Condition	Location	Wt. (kg)	Ref.
Arctic							
Caribou	38.8—40.0	−3—20	Winter	Laboratory	NWT (CAN)	5.5	44
(Rangifer tarandus)	38.1—41.0	−1—6	Winter	Field	NWT (CAN)	5.9	44
	38.5—40.0	−26— −12	Winter	Field	AK (USA)	42—120	39
Muskox	40.0	5	Winter	Field	AK (USA)	40—140	39
(Ovibos moschatus)							
Reindeer	38.0—39.3	−45—15	Winter	—	USSR	—	45
(Rangifer tarandus)	38.4—40.2	0—30	Summer	—	USSR	—	45
	37.5—39.8	−46—15	Winter	Field	AK (USA)	35—110	39

Table 2
SKIN TEMPERATURES (T_s) OF TROPICAL, TEMPERATE, AND ARCTIC UNGULATES AT REST

Species	T_s (°C)	T_a (°C)	Season	Condition	Location	Wt. (kg)	Ref.
Tropical							
Black rhinoceros	24.1—33.4	19.5	—	Laboratory	NY (USA)	862	16
Collared peccary	34.8—40.9	27—44	Summer	Field	AZ (USA)	21	22
	27.9—42.3	4—18	Winter	Field	AZ (USA)	21	22
Elephant	22.7—32.5	16—19.5	—	Laboratory	NY (USA)	4191	52
	22.5—28.1 (foot)	16—19.5	—	Laboratory	NY (USA)	4191	52
Hippopotamus	21.2—30.9	19.5	—	Laboratory	NY (USA)	2086	16
	28.2—35.0	23.1—29.7	—	Field	Uganda	—	53
	14.5—29.5	9.8—14.8	—	Laboratory	Poland	2000	54
Kirk's dik-dik	25.4—39.9	20—40	—	Laboratory	Kenya	4.5	30
Suni	35.8—40.6	20—40	—	Laboratory	Kenya	2.5	30
Temperate							
Elk	27.5—42.9	2—34	Summer	Field	WA (USA)	144	37
	18.7—44.3	−29—24	Winter	Field	WA (USA)	92	37
Mule deer	25.0—42.3	2—34	Summer	Field	WA (USA)	49	37
	24.9—45.1	−38—23	Winter	Field	WA (USA)	61	37
Red deer	25.8—33.1	4—16	Winter	Laboratory	Scotland	46	40
Arctic							
Caribou	31.2—36.8	−3—20	Winter	Laboratory	NWT (CAN)	5.5	44
	26.5—35.2 (foot)	−3—20	Winter	Laboratory	NWT (CAN)	5.5	44
	26.6—31.7	−1—6	Winter	Field	NWT (CAN)	5.9	44
	10.0—26.0 (foot)	−1—6	Winter	Field	NWT (CAN)	5.9	44
	27.2—36.8	−45—26	Winter	Field	NWT (CAN)	31.8	44
	6.3—35.0 (foot)	−45—26	Winter	Field	NWT (CAN)	31.8	44
Reindeer	18.8—39.4	−40—27	Winter	—	USSR	—	45
	−8.2—26.8 (foot)	−40—27	Winter	—	USSR	—	45

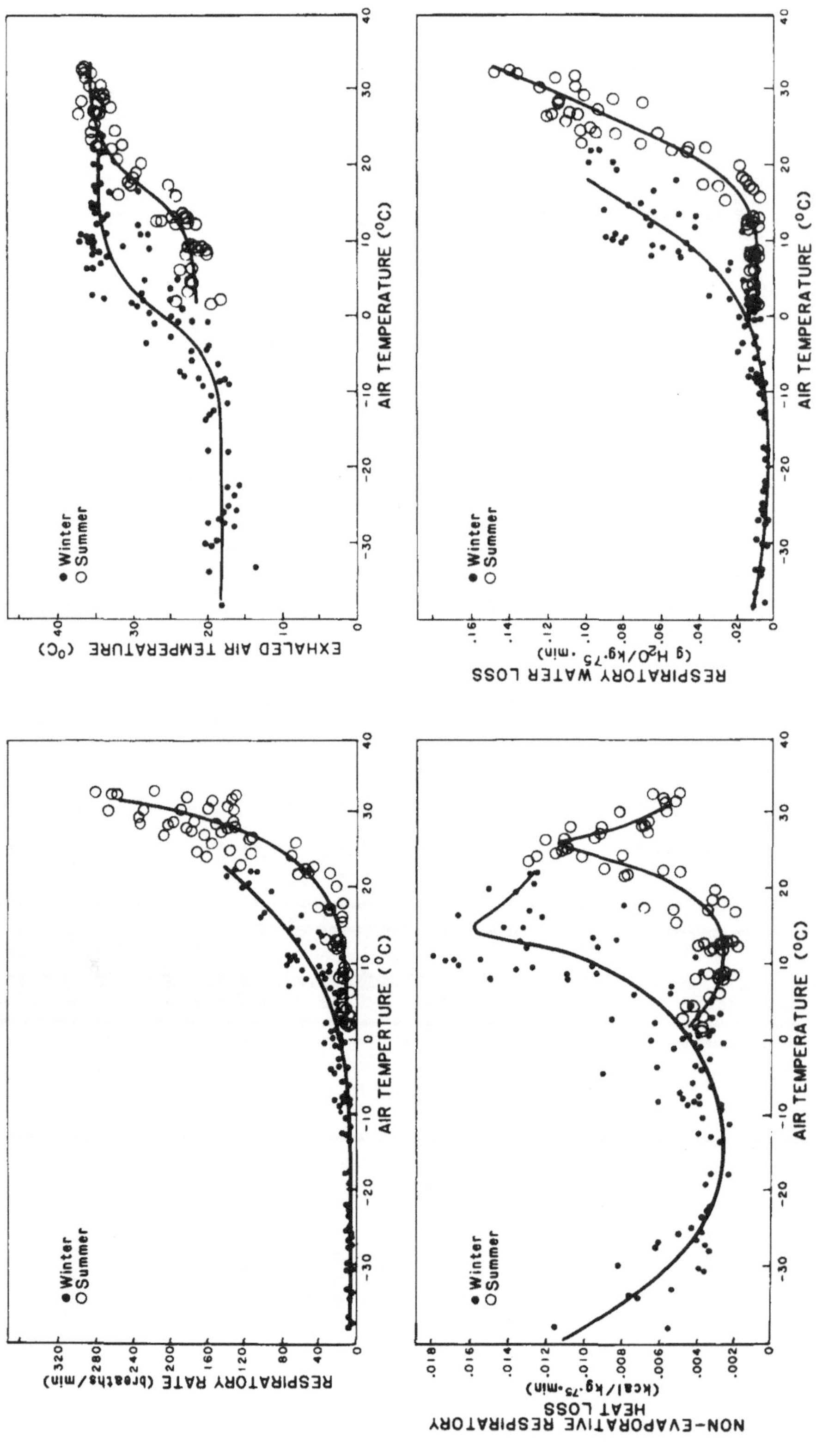

FIGURE 1. Effect of air temperature on respiratory rate, exhaled air temperature, non-evaporative respiratory heat loss, and respiratory water loss of mule deer in winter and summer. (From Parker, K. L. and Robbins, C. T., *Can. J. Zool.*, 62, 1409, 1985. With permission.)

III. RESPIRATORY HEAT AND WATER LOSS

Respiratory rates increase exponentially at high air and operative temperatures[29,31,33,37,59] (Figure 1). An increase in ventilation from a relatively constant frequency occurs before there is any increase in deep body temperature.[60] This increase in respiratory rate may be precipitated by warming of the skin alone.[61,62] Respiratory rates range as low as 4/min in camels (T_a = 25°C)[20] to 500/min in the lesser mouse deer (T_a = 36°C)[31] and Kirk's dik-dik (T_a = 40°C).[29] In general, there is a tendency for maximum respiratory frequency in panting species to be inversely related to body size.[57] Significant differences in respiratory rate also occur between animals in summer and winter pelage. The range of critical air temperatures within which respiratory cooling becomes necessary in winter is substantially lower than in summer[37] (Figure 1). In conjunction with frequency, expired air volumes also increase with higher operative temperatures.

Second phase breathing, or a decline in respiratory rate accompanied by increased tidal and minute volumes at very high temperatures has been documented in domestic ungulates[63,64] but not confirmed in wild species.[65] This deeper, slower type of respiration may represent the breakdown of thermal polypnea as a physiological mechanism which permits maximal respiratory evaporative cooling and minimal disturbance of blood gases.[66]

Despite the importance of panting as an evaporative cooling mechanism, few studies have dealt with its physiology and energetic cost. The cost of panting (200/min) (T_a = 40°C) in the ox has been estimated to be approximately 11% of the total energy expenditure.[64] For smaller species, however, in which the frequency of respiration may be higher than that of the ox, this energetic cost may be underestimated by the above value.

Exhaled air temperature (T_{ex}), as a component of respiratory heat loss, has been recorded for several ungulates, including the elephant, giraffe, waterbuck, wildebeest, and camel.[25,52,67] Although the relationship between T_{ex} and T_a has been assumed to be linear,[25,67] a sigmoid expression has been observed over a broad range of air temperatures in mule deer, elk, and reindeer[37,68] (Figure 1).

Under very cold ambient conditions, exhaled air temperatures as low as 20°C below core temperatures provide a mechanism for heat conservation.[37,68,69] Similarily, at high air temperatures, particularly in desert-dwelling species, the exhalation of air at temperatures below core temperatures plays a primary role in the recovery of respiratory water.[52,67] In winter, T_{ex} is often higher than in summer at the same high operative temperature, allowing more heat to be dissipated respiratorily when the increased pelage of temperate and arctic ungulates reduces cutaneous heat loss.

Non-evaporative respiratory heat loss due solely to the temperature difference between inspired and expired air varies with respiratory rate and exhaled volume. In mule deer and elk, this response is a parabolic function below an upper threshold temperature (Figure 1) and may represent 5 to 15% of the metabolic heat production.[37] Increasing respiratory heat loss at colder temperatures due to slightly greater exhaled air volumes[37] corresponds closely with lower critical temperatures.[69] At very high air temperatures, respiratory heat loss declines as the temperature gradient between inspired and expired air decreases. Therefore, respiratory evaporative loss, as an exponential function of air temperature (Figure 1), represents a substantially greater means of heat dissipation in hot environments than does non-evaporative loss.

IV. CUTANEOUS EVAPORATION

Sweat glands have been documented in most artiodactyls and perissodactyls,[70-72] but not in proboscids.[53,72] There are two distinct types of sweat glands: *epitrichial*, or associated with hair follicles and *atrichial*, opening directly onto the skin's surface.[73,74] Most ungulates

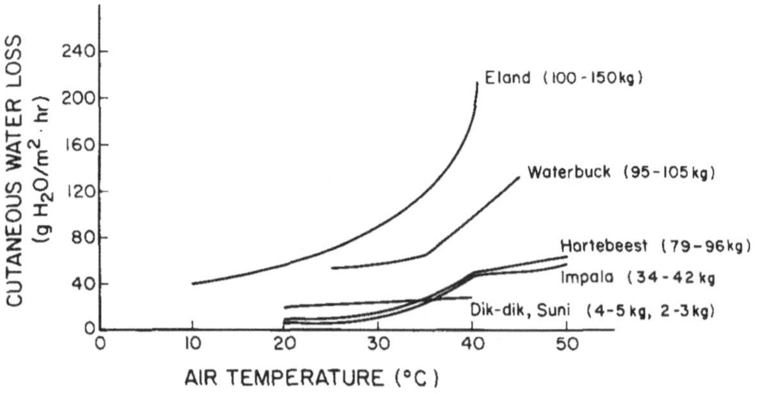

FIGURE 2. Cutaneous water loss measured in the metabolic chamber as a function of air temperature for eland,[23] waterbuck,[34] hartebeest, impala,[27] dik-dik, and suni.[30]

possess epitrichial glands[60] under adrenergic neuron control.[70] Because the total number of sweat glands is fixed at birth, sweat gland density decreases as the animal grows. This decline is counterbalanced by an increase in volume and rate of discharge in the individual sweat gland.[57]

The functional significance of sweat glands varies extensively. Some species, such as the rhinoceros, sweat only on exertion.[47] Others, such as the gazelle (*Gazella gazella*) which has sweat glands over most of its body surface, do not sweat even when running at high speeds.[75] In species which sweat in response to hot ambient conditions, the control and rates of evaporation are variable.[70,76] In addition, water loss measurements are affected by the site of measurement,[57,77,78] animal age, feed intake, and activity,[60] seasonal type and amount of pelage,[37,41] and animal size.[70]

Both core sensitivity[79] and surface skin temperature[30,37,49,50] provide stimuli for the regulation of sweating. There are typically three distinct sweating patterns which do not correlate with phylogenetic position of the species.[70] A cyclical periodic discharge due to synchronous contractions of myoepithelial structures surrounding the gland[74] occurs in the duiker (*Silvicapra grimmia*), Grant's and Thomson's gazelles, oryx, and wildebeest.[70] Evaporative cooling from the skin may be negligible in these species.[80] Waterbuck and eland exhibit a continuous but stepwise rise in discharge in contrast to the continuous but immediate rise in discharge recorded in buffalo, llama (*Lama glama*), and red deer.[41,70,76]

Cutaneous water loss per unit area at high air temperatures increases with increasing body size (Figures 2 and 3). Smaller animals dissipate most of their excess heat via panting whereas large animals depend upon sweating. Sweating in large ungulates which are characterized by relatively small surface to body weight ratios and small convective heat losses may have evolved to dissipate heat during exercise.[81] In general, as passive diffusion of water through the skin does not exceed 50 g $H_2O/m^2 \cdot hr$,[76] higher rates of cutaneous evaporation are attributed to true thermal sweating. The hippopotamus, however, while not possessing sweat glands comparable to other animals, exhibits an unusually high rate of evaporation from the skin (268 g $H_2O/m^2 \cdot hr$) ascribed to the wicklike action of a mucopolysaccharide which permits the transfer of water from deeper layers of the skin to the surface.[82]

Solar radiation is an important factor affecting cutaneous evaporation[83,84] and is, unfortunately, ignored in metabolic chamber studies. For example, cutaneous water loss measured in hartebeest outdoors (T_a = 30°C)[49] averaged 5.4 times higher than when monitored in a chamber (T_a = 30°C).[27] Hence, in the natural radiant environment, cutaneous evaporation for ungulates in summer pelage also increases with increasing body weight (Figure 3), but

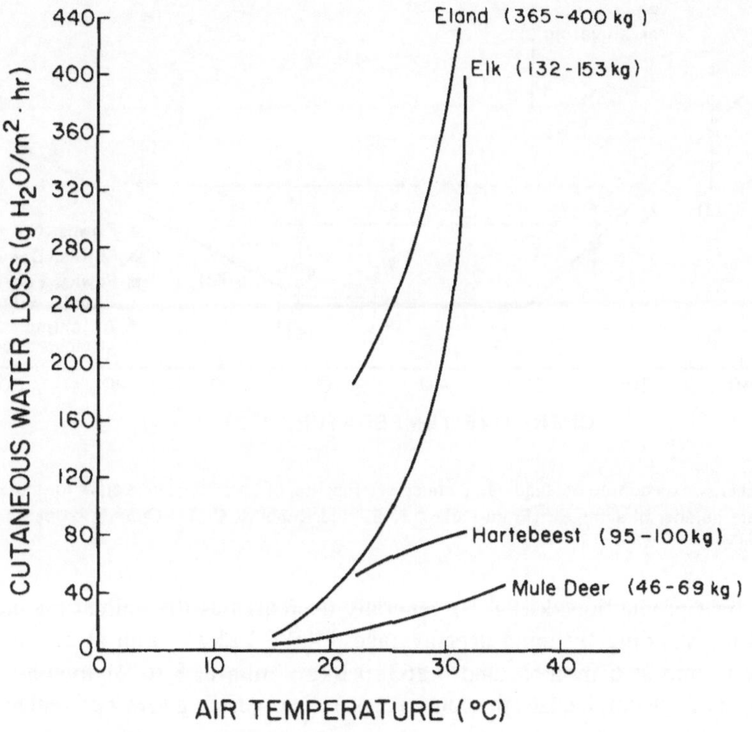

FIGURE 3. Cutaneous water loss measured outdoors as a function of air temperature for elk and mule deer in summer pelage[37] and eland and hartebeest.[49]

to substantially higher levels than that in the metabolic chamber. A thin layer of summer hair is probably the best physical compromise for allowing skin heat dissipation, providing solar load protection under hot ambient conditions, and preventing excessive heat loss by radiation and convection in cold conditions.[85]

The advantages and disadvantages of sweating and panting for evaporative cooling have been reviewed by Taylor.[81] A most obvious consideration is that cutaneous evaporation allows for simultaneous foraging and evaporative cooling. Whether an animal pants or sweats, however, does not necessarily dictate the environment in which it is found. The eland and buffalo which are capable of large amounts of skin moisture loss seek shade at midday in areas where water intake is restricted.[17,49] The hartebeest, using panting as a primary means of evaporative cooling, remains exposed to the hot sun throughout the day apparently due to an extremely high reflectance from the light-tipped hair coat.[49] In those species that neither pant nor sweat under hot ambient conditions, alternative strategies for thermoregulation exist. The Grant's gazelle, for example, reduces its environmental heat load by allowing core temperature to rise quickly above even very high air temperatures.[17]

V. METABOLIC HEAT PRODUCTION

An animal's metabolic heat production may be altered by its adaptational history, level of activity, body posture, plane of nutrition, and age. For example, acclimatization results in a 9% decrease in metabolic rate for cattle.[51] Lying postures potentially reduce heat loss through the extremities and may decrease LCTs for moose by more than 10°C.[86] The critical ambient temperature at which heat production increases for fasted pronghorn *(Antilocapra americana)* is 12 to 25°C higher than for those on *ad libitum* intake.[87]

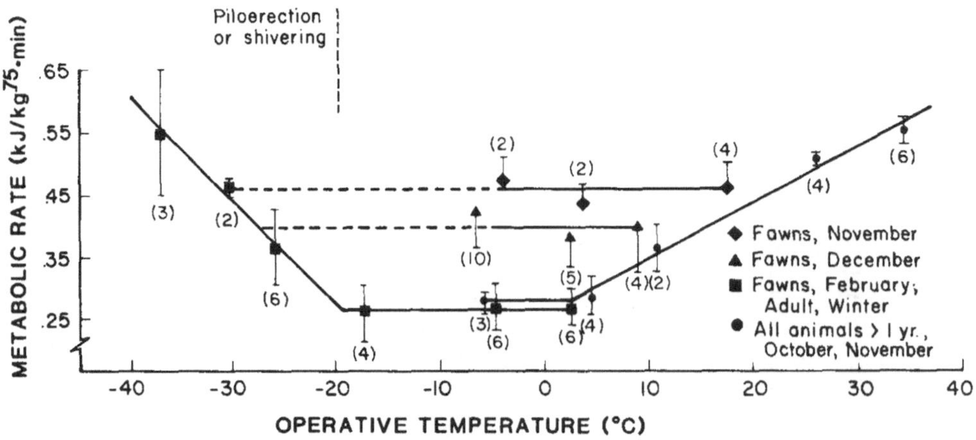

FIGURE 4. Energy expenditure of standing in winter as a function of operative tempeature for mule deer. Values in parentheses are number of samples. (From Parker, K. L. and Robbins, C. T., *Can. J. Zool.*, 62, 1409, 1985. With permission.)

Metabolic heat production (kJ/kg$^{0.75}$) generally decreases as the animal matures. Energy expenditures for standing by mule deer average 0.4585 kJ/kg$^{0.75}$•min at 5 months of age, 0.3971 kJ/kg$^{0.75}$•min at 6 months, and 0.2638 kJ/kg$^{0.75}$•min at 8 to 21 months (Figure 4). The importance of age emphasizes the care necessary in predicting lower critical temperatures and an ecological strategy employed in an environment with limited energy resources. Extrapolation of the average 5-month-old fawn's thermally neutral heat production yields a LCT of −30°C. This apparent benefit, however, is countered by the fawn's continual high metabolic cost. With a lower thermally neutral metabolism, the adult animal must increase metabolism only during periods when operative temperatures are less than −20°C. Thermogenesis and shivering occur below the LCT when animals are exposed to extremely cold nighttime winter temperatures and cease with warming of the pelage by direct beam solar radiation.[37,86] Hence, if limited energy resources have been an important selective factor, a lower LCT attained through increased insulation or a higher LCT combined with short-term increases in metabolism and active pursuit of favorable thermal environments would be more advantageous than the continuous high metabolism necessary to attain an extremely cold LCT if insulation is unchanged. However, an increase in insulation is potentially counterbalanced by a decrease in the upper critical temperature.

Seasonal changes in pelage and metabolism are exemplified in mule deer.[37] Adult mule deer decrease thermally neutral heat production in winter by 11.5% from summer. Lower critical temperatures are more than 30°C lower in winter than in summer, providing some indication of the capacity of this species to adapt to seasonal changes in the environment.

Nontraditional temperature-metabolism relationships have also been recorded in wild ungulates. Reductions in metabolic rate in winter down to a LCT without a classical thermoneutral zone have been observed in elk,[37] moose,[86] bison,[88] and to a lesser extent, bighorn sheep *(Ovis canadensis)*[89] and waterbuck.[34] Such reductions in metabolic response apparently involve primarily peripheral cooling since body temperatures do not fall consistently with decreasing metabolism.[37,88] Moen[90] notes that white-tailed deer in outdoor conditions become lethargic whereas those confined in chambers exhibit a more traditional response. A decrease in metabolic heat production at low air temperatures may reduce energy expenditures at a time when decreased forage availability coincides with higher thermoregulatory demands.[88] This may also allow the animal to reach much lower operative temperatures before metabolic rate is increased.

A reduction in metabolic heat production at temperatures above the UCT has been observed in elk,[37] moose,[91] Kirk's dik-dik,[29] and to a lesser degree, Merino sheep.[92] Decreased metabolic rate at high temperatures, which counteracts the whole body Q_{10} effect, effectively extends the TNZ, providing a wider range of thermal environments to which the animal is adapted.

VI. THERMAL RESISTANCE AND CONDUCTANCE

To minimize thermoregulatory costs and maintain thermal equilibrium, an animal must adjust its thermal resistance to heat transfer as operative temperature changes. Hair coat and peripheral tissue resistances are altered via counter current heat exchange, differential blood flow to the extremities, vasomotor control of skin and tissue temperatures, changes in evaporation, development of adipose tissue or undercoat hairs, and piloerection. The boundary layer of still air adhering to the body surface also provides added insulation. Inherent in the calculation of thermal resistance are the animal's heat increment of feeding, seasonal pelage effects, metabolic rate, adipose tissue depots based on age and weight, and cutaneous water loss.[13] Whole body thermal resistance, therefore, serves as a potential indicator of heat loss, thermal balance, and winter hardiness.[40,93]

The importance of using measurements made on live animals in the calculation of body resistance is contrasted to the potential error when using pelt conductivities. Because the physiological adjustments of the living animal are lacking, coat resistance values determined from measurements of heat flow through isolated pelt samples may actually be greater than total body resistance.[94] The irregular shape of live animals, shivering-induced disturbances in the boundary layer, and contributions of less well-insulated extremities all tend to decrease the effective insulation of the coat.[89]

Within the thermoneutral environment, an animal's thermal resistance declines as a function of increasing operative temperature.[13,37] The operative temperature at which resistance attains a maximum value corresponds to the animal's LCT. Maximum resistances in summer are, therefore, less than those in winter. Although resistances below a LCT have been assumed to be constant,[95] a decrease in thermal resistance may result from vasodilation and increased heat flow to prevent tissue damage in the extremities, the effects of shivering on convective heat transfer,[96] and/or increased respiratory heat loss at low air temperatures.[37]

Thermal conductance is a measure of the heat flow from an animal to its surroundings and depends largely on the thermal resistance afforded by the hair coat and entrapped air.[57] Conductance increases inversely with coat depth, density, and length.[95,97] Because larger animals already have reduced physiological demands due to their size, pelage insulation is of greatest importance to the small homeotherm.[48]

Although thermal conductance varies with body region,[2,95,97] average whole body dry conductance (C) in wild herbivores (Table 3) can be estimated by Newton's Law of Cooling:[101]

$$C = M/(T_b - T_a) \qquad (2)$$

where M is the rate of metabolic heat production and Ta is taken to be an animal's LCT. The heat transfer process is grossly oversimplified by this estimate since the boundary layer (r_e) and evaporative (λE) components of an animal's energy budget are not included.[13] However, r_e is minimal in comparison to tissue and coat resistances and λE is relatively unimportant at low air temperatures. As in the typical Scholander temperature-metabolism model,[8] dry conductance assumes that the metabolism curve below an animal's LCT extrapolates to body temperature at zero metabolism.

Thermal conductance may also be calculated from the slope of an animal's temperature-metabolism relationship below its LCT. These values are similar to those based on the dry conductance Equation 2 in all tropical species (Table 3). In several large North American

Table 3
METABOLIC HEAT PRODUCTION AND THERMAL CONDUCTANCE OF TROPICAL, TEMPERATE, AND ARCTIC UNGULATES

Species	Wt. (kg)	Activity	Feed level	Season	Location	T_b (°C)	UCT (°C)	LCT (°C)	Metabolism (kJ/kg$^{0.75}$·day)	Conductance (kJ/kg$^{0.75}$·day·°C)[a]	Ref.
Tropical											
Eland	150	—	*Ad libitum*	—	Kenya	36.3	30	20	405.5	24.9 (22.3)	23
Kirk's dik-dik	4	Standing	*Ad libitum*	—	Kenya	39.0	35	25	359.1	25.4 (19.9)	29
Lesser mouse deer	1.6	—	—	—	Malaysia	38.4	30	26.6	293.4	24.9 (28.9)	31
Waterbuck	100	—	*Ad libitum*	—	Kenya	38.3	37	13	408.6	16.1 (12.8)	34
Temperate											
Bighorn sheep	70.5	Resting	Fasted	Jan	ALTA(CAN)	38.0[b]	—	−10	462.3	9.5 (18.8)	89
Elk (wapiti)	125	Standing	*Ad libitum*	Jan—May	ALTA(CAN)	38.6[b]	—	−10	663.0	13.6	98
	125	Lying	*Ad libitum*	Jan—May	ALTA(CAN)	38.6[b]	—	−10	510.0	10.5	98
	91.7	Standing	*Ad libitum*	Jan	WY(USA)	38.6	15	−20	320.2	5.5 (14.5)	37
	143.5	Standing	*Ad libitum*	Summer	WA(USA)	38.2	—	9	504.7	17.3	37
Mule deer	61	Standing	*Ad libitum*	Winter	WA(USA)	38.0	5	−20	380.4	6.6 (23.5)	37
	48.7	Standing	*Ad libitum*	Summer	WA(USA)	38.3	25	12.5	546.2	21.2 (7.8)	37
Moose	87.5	Standing	Fasted	Oct—Jan	ALTA(CAN)	38.6[b]	—	−20	585.2	10.0	86
Pronghorn	37.8	—	*Ad libitum*	Dec—Jan	CO(USA)	38.0[b]	—	−16	484.9	9.0 (6.3)	87
	37.8	—	Fasted	Dec—Jan	CO(USA)	38.0[b]	—	0	317.7	8.4	87
Red deer	46.0	—	*Ad libitum*	Jan	Scotland	38.8	—	5	526.2	15.4	40
White-tailed deer	65.3[b]	—	Fasted	Feb	NH(USA)	37.9	—	−10	384.6	8.0 (5.9)	43
	65.3[b]	Lying	Fasted, fed	Winter	NH(USA)	37.9	25	5	248.7	7.6	42
Arctic											
Caribou	31.8	Standing	—	Mar	NWT(CAN)	38.0[b]	≈20	≈ −40	598.9	7.7	44
Reindeer	96.5[b]	—	—	Jan—Feb	USSR	38.4	15	< −40	496.6	5.9	45
	96.5[b]	—	—	Jan—Feb	USSR	38.4	—	5[b]	756.3	22.7	45

[a] Conductances calculated from Newton's Law of Cooling (see text, values not enclosed by parentheses) and from the slope of the temperature-metabolism relationship below LCT (values in parentheses).

[b] T_b estimated from mule deer and elk;[37] white-tailed deer and reindeer body weights estimated from Silver et al.[99] and White and Yousef,[100] respectively; LCT of reindeer from Johnsen and Blix.[69]

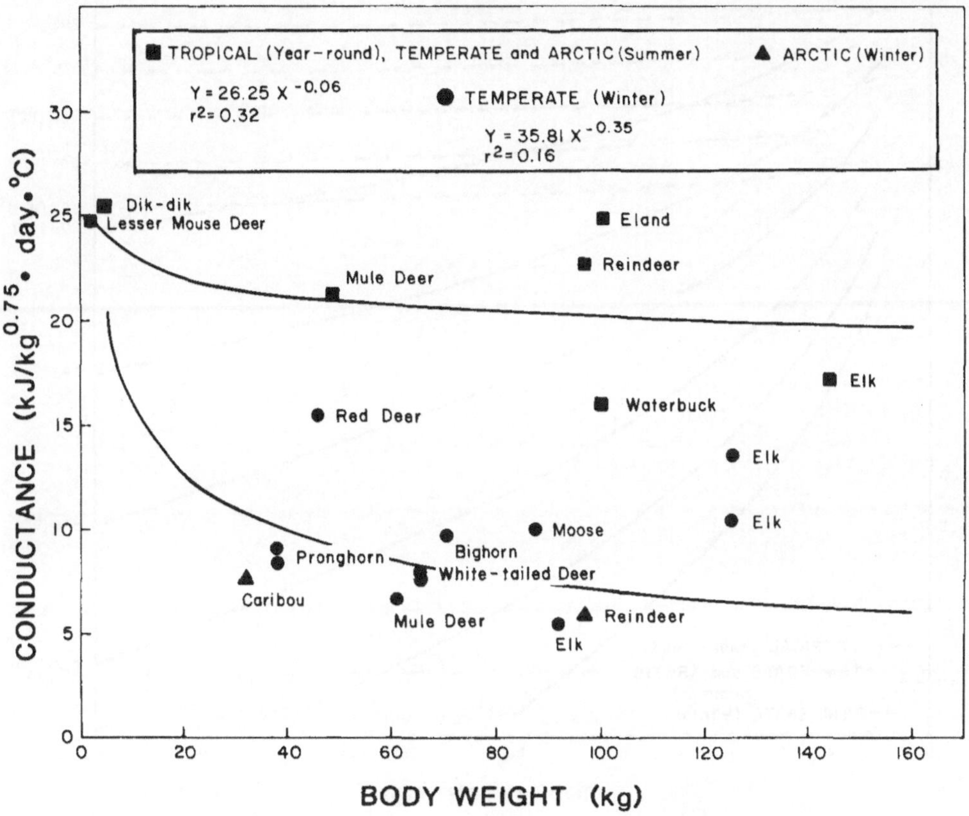

FIGURE 5. Effect of body weight on thermal conductance for tropical, temperate, and arctic artiodactyls (from Table 3). No distinction is made between seasonal conductances for tropical species; data for temperate and arctic ungulates in summer have been combined with tropical species.[85] Heat loss measurements for elk in winter[98] are not included in the regression for temperate species; values appear excessively high and may reflect the compounding effects of increasing age and seasonal pelage variation during the period of metabolic measurements. Postural effects, however, are noted: conductance for the lying animal declines 30% from standing. Caribou and reindeer values have also been omitted from the temperate regression as these values seem slightly peripheral to the relationship for temperate species.

ungulates, e.g., bighorn sheep, mule deer, and elk, the curve below thermoneutrality extrapolates to a temperature less than core temperature, contrary to the tendency for small mammals and birds.[101] This pattern results from various combinations of body temperature, metabolism, and whole body thermal resistance with decreasing operative temperatures. Conductances based on the temperature-metabolism slope are substantially greater than those predicted as dry conductance, reflecting the steeper slope. For those species in which this slope extends to temperatures greater than T_b, conductances are less than those based on Newton's Law of Cooling.

Thermal conductance is significantly greater for tropical ungulates than for their temperate counterparts in winter, which in turn is slightly greater than that of arctic species in winter (Figure 5). Consequently, the tropical animal must increase metabolism drastically for even a small decline in air temperature below the LCT.[102] For temperate and arctic ungulates in summer pelage, conductance is greater than two times that in winter and not significantly different from tropical animals[85] (Table 3, Figure 5).

Conductance appears to decline slightly with increasing body weight (Figure 5), in accordance with data for a variety of smaller mammals.[103,104] If insulation quality is similar between ungulates of varying size, differences in surface area to volume ratios support a

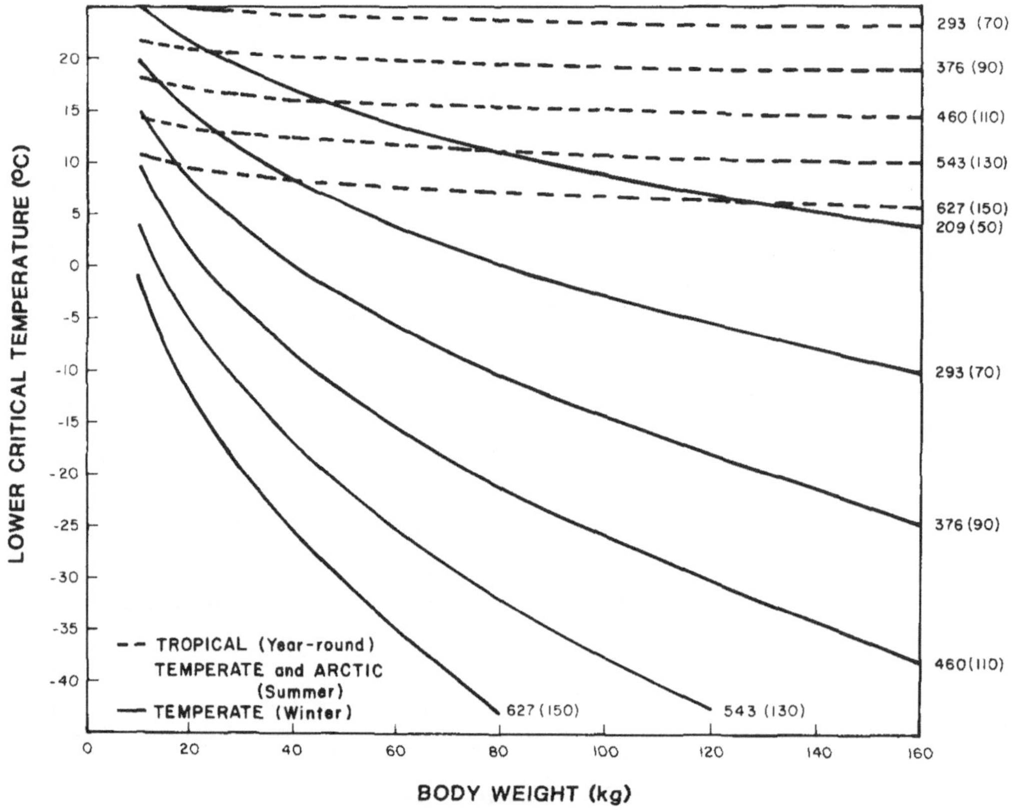

FIGURE 6. Prediction of lower critical temperatures as a function of body weight and metabolic heat production, assuming thermal conductance (kJ/kg$^{0.75}$·day·°C) decreases curvilinearily with weight (kg) in both tropical ($Y = 26.25X^{-0.06}$) and temperate ($Y = 35.81X^{-0.35}$) artiodactyls (Figure 5). Values on right are metabolic rates in kJ/kg$^{0.75}$·day and (kcal/kg$^{0.75}$·day).

decline in conductance. Such differences due to surface area-volume ratios become even larger if the smaller animal must have a relatively shallow coat for movement efficiency. Because of the large inherent variation and relatively small range of available body weights (Figure 5), thermal conductance may not, however, differ significantly between ungulate species. The relatively small differences in heat loss potential created by surface area to volume ratios in different-sized ungulates living in similar environments may be easily compensated for by alterations in insulation quality. Conductances in ungulates average 21.8 ± 3.78 kJ/kg$^{0.75}$·day·°C for tropical species year-round and temperate and arctic species in summer pelage, 8.9 ± 2.8 kJ/kg$^{0.75}$·day·°C for temperate species in winter, and 6.8 ± 1.3 kJ/kg$^{0.75}$·day·°C for arctic species in winter.

In all calculations of conductance, it should be recognized that the compounding effects of wind and precipitation result in additional increases in heat loss. For example, heat loss through sheep fleece and calf hair theoretically increases 41 to 96% when animals are exposed to simulated rain (T_a = 6 to 7°C).[105] Windspeeds up to 10 m/s decrease effective insulation of bighorn sheep and caribou by 41 to 87%.[44,89] These observations further suggest the need for understanding and measuring operative temperatures.

Estimates of thermal conductances, nevertheless, facilitate the prediction of lower critical temperatures for those animals in which energetic measurements are not available. If thermal conductance declines with increasing body size (Figure 5), LCTs also theoretically decline with increasing body weight in tropical and temperate artiodactyls (Figure 6). Values are

Table 4
EFFECT OF METABOLIC RATE ON LOWER CRITICAL TEMPERATURES (°C)

	Metabolism ($kJ/kg^{0.75}\cdot day$)					
	209	293	376	460	543	627
Tropical	28.6	24.8	20.9	17.1	13.3	9.4
Temperate	14.7	5.3	−4.1	−13.5	−23.9	−32.2
Arctic	7.5	−4.8	−17.1	−29.4	−41.7	−54.0

Note: This table assumes thermal conductance averages 21.8 $kJ/kg^{0.75}\cdot day\cdot °C$ for tropical species year-round and temperate and arctic species in summer pelage, 8.9 $kJ/kg^{0.75}\cdot day\cdot °C$ for temperate species in winter, and 6.8 $kJ/kg^{0.75}\cdot day\cdot °C$ for arctic species in winter (see text).

significantly higher for tropical species than for temperature ones, reflecting their reduced insulation. Similarly, based on average conductances without regard to body weight, LCTs are highest for tropical ungulates (Table 4). Metabolic rate varies inversely with LCT for all species as previously exemplified for mule deer (Figure 5). Hence, thermal conductance and/or resistance provide insight for the prediction of activity patterns, microhabitat use, and energy costs of thermoregulation.[106]

VII. THE TRANSIENT STATE

Further progress in understanding ungulate thermoregulation and applying field observations to answer management questions depends on detailed mathematical analyses of variables affecting heat flow between the animal and its environment. The energy budget of an animal under static conditions, for example, has been quantified by the following expression:[13]

$$M - \lambda E - h_r = [\rho c_p(T_b - T_e)/r_{Hb}] - r_e \qquad (3)$$

where λE is the combined respiratory and cutaneous latent heat loss and r_{Hb} is the thermal resistance of skin and pelage. Non-evaporative respiratory heat loss (h_r) should also be added to Campbell's original equation.[37] Despite limitations at high operative temperatures when T_e exceeds T_b or ($\lambda E + h_r$) is greater than M, Equation 3 may also be rearranged to determine the whole body thermal resistance of an animal. Such calculations, however, usually assume a constant body temperature and minimal heat storage and provide only an instantaneous assessment of an animal's energy state.

Animals are not static components of an environment under a particular set of operative conditions. Because of activity patterns and movement locations, a variety of nonuniform microsites are encountered and the nonsteady or transient energetic state becomes of great importance in free-ranging species. An estimate of heat storage (q) in the animal is presented by Campbell:[13]

$$q = V/A\left(\rho_b c_b \frac{dT_b}{dt}\right) \qquad (4)$$

where V and A are the body volume and area, respectively, ρ_b and c_b are the body density and specific heat, and t is the time interval. Caution should be exercised in using rectal temperature as a measure of deep body and arterial blood temperatures.[60] Equation 4 assumes that the thermal conductance and heat capacity of the animal's core are large when compared to that of the coat and peripheral tissues. The time that an animal can actually remain in a particular microclimate within acceptable body temperatures can be calculated:[13]

$$(T_{b2} - T_e - T_o)/(T_{b1} - T_e - T_o) = -t/\tau \qquad (5)$$

where T_{b2} is the new equilibrium core temperature; T_{b1} the initial body temperature; T_o an offset temperature based on the physiological responses of the animal (thermal resistance, metabolic heat production, and latent heat loss); and τ a time constant incorporating the effects of heat storage.[13,107]

Therefore, animals with large time constants which are maximized by a large volume, small surface area, and large thermal resistance would be expected to endure exposures to environments outside acceptable operative conditions for relatively longer periods of time. Small animals would respond much more quickly to environmental changes and be more restricted by behavioral thermoregulation. Taylor[81] estimated that a 4°C rise in body temperature would occur in approximately 5 hr in the larger eland as compared to 1 hr in a small gazelle. The relatively small-sized peccary (21 kg) in fact requires only 1 hr to increase core temperature by 3°C whereas the camel requires 10 hr.[22] Further studies are necessary to examine the ecological implications of the transient state relative to an animal's energy budget.

VIII. CONCLUSION

While we have discussed physiological and morphological characteristics that affect thermoregulation, many behavioral responses are important mechanisms of temperature regulation under natural conditions. The repertoire of thermoregulatory behaviors in ungulates is generally restricted by body size but, nevertheless, includes postural adjustments,[42] shelter-seeking relative to shade[45] and protection from cold,[108] burrowing,[19] wallowing,[82] and even gregarious huddling during hot[109] and cold[23] conditions. In some cases, foraging patterns may be restricted by high ambient temperatures.[23,47] Metabolic heat generated during activity may substitute for thermoregulatory costs at low air temperatures.[110]

The importance of thermal constraints in modifying or determining activity patterns and habitat selection has been recognized in the concept of thermal cover. Based on observations of free-ranging ungulates purposefully selecting cover, bedding sites, and postures for apparent thermal benefits,[111,112] thermal cover is now considered a manageable entity of ungulate habitat.[113] Further studies of the use of thermal cover are necessary to distinguish between energetic requirement and behavioral preference.

REFERENCES

1. **Christopherson, R. J. and Young, B. A.,** Heat flow between large terrestrial animals and the cold environment, *Can. J. Chem. Eng.*, 59, 181, 1981.
2. **Morrison, P.,** Insulative flexibility in the guanaco, *J. Mammal.*, 47, 18, 1966.
3. **Moen, A. N.,** *Wildlife Ecology: An Analytical Approach*, W. H. Freeman, San Francisco, 1973, chap. 6.
4. **Chappel, M. A.,** Insulation, radiation, and convection in small arctic mammals, *J. Mammal.*, 61, 268, 1980.
5. **Hammel, H. T.,** Thermal properties of fur, *Am. J. Physiol.*, 182, 369, 1955.
6. **Aschoff, J.,** personal communication in **Mount, L. E.,** The concept of thermoneutrality, in *Heat Loss from Animals and Man*, Monteith, J. L. and Mount, L. E., Eds., Butterworths, London, 1974, chap. 21.

7. **Hill, J. R.,** personal communication in **Mount, L. E.,** The concept of thermoneutrality, in *Heat Loss from Animals and Man,* Monteith, J. L. and Mount, L. E., Eds., Butterworths, London, 1974, chap. 21.
8. **Scholander, P. F., Hock, R., Walters, V., Johnson, F., and Irving, L.,** Heat regulation in some arctic and tropical mammals and birds, *Biol. Bull.,* 99, 237, 1950.
9. **Alexander,** personal communication in **Mount, L. E.,** The concept of thermoneutrality, in *Heat Loss from Animals and Man,* Monteith, J. L. and Mount, L. E., Eds., Butterworths, London, 1974, chap. 21.
10. **Bianca, W.,** personal communication in **Mount, L. E.,** The concept of thermoneutrality, in *Heat Loss from Animals and Man,* Monteith, J. L. and Mount, L. E., Eds., Butterworths, London, 1974, chap. 21.
11. **Webster, A. J. F.,** personal communication in **Mount, L.E.,** The concept of thermoneutrality, in *Heat Loss from Animals and Man,* Monteith, J. L. and Mount, L. E., Eds., Butterworths, London, 1974, chap. 21.
12. **Bakken, G. S.,** How many equivalent black-body temperatures are there?, *J. Therm. Biol.,* 6, 59, 1981.
13. **Campbell, G. S.,** *An Introduction to Environmental Biophysics,* Springer-Verlag, Berlin, 1977, chap. 7.
14. **Bakken, G. S.,** The use of standard temperature in the study of the thermal energetics of birds, *Physiol. Zool.,* 53, 108, 1980.
15. **Bligh, J. and Harthoorn, A. M.,** Continuous radiotelemetric records of the deep body temperature of some unrestrained African mammals under near-natural conditions, *J. Physiol. (London),* 176, 145, 1965.
16. **Benedict, F. G., Fox, E. L., and Baker, M. L.,** The surface temperature of the elephant, rhinoceros, and hippopotamus, *Am. J. Physiol.,* 56, 464, 1921.
17. **Taylor, C. R.,** Strategies of temperature regulation: effect on evaporation in East African ungulates, *Am. J. Physiol.,* 219, 1131, 1970.
18. **Yousef, M. K. and Dill, D. B.,** Resting energy metabolism and cardiorespiratory activity in the burro *(Equus asinus), J. Appl. Physiol.,* 27, 229, 1969.
19. **Sowls, L. K. and Phelps, R. J.,** Body temperature of juvenile warthogs and bushpigs, *J. Mammal.,* 47, 134, 1966.
20. **Schmidt-Nielsen, K., Crawford, E. C., and Hammel, H. T.,** Respiratory water loss in camels, *Proc. R. Soc. London,* 211, 291, 1981.
21. **Sergent, E. and Lheritier, A.,** Note sur la temperature rectale des dromedaires, *C. R. Soc. Biol.,* 82, 172, 1919.
22. **Zervanos, S. M. and Hadley, N. F.,** Adaptational biology and energy relationships of the collared peccary *(Tayassu tajacu), Ecology,* 54, 759, 1973.
23. **Taylor, C. R. and Lyman, C. P.,** A comparative study of the environmental physiology of an East African antelope, the eland, and the hereford steer, *Physiol. Zool.,* 40, 280, 1967.
24. **Elder, W. H. and Rodgers, D. H.,** Body temperature in the African elephant as related to ambient temperature, *Mammalia,* 39, 395, 1975.
25. **Langman, V. A., Maloiy, G. M. O., Schmidt-Nielsen, K., and Schroter, R. C.,** Nasal heat exchange in the giraffe and other large mammals, *Resp. Physiol.,* 37, 325, 1979.
26. **Rosenmann, M. and Morrison, P.,** Physiological response to heat and dehydration in the guanaco, *Physiol. Zool.,* 36, 45, 1963.
27. **Maloiy, G. M. O. and Hopcraft, D.,** Thermoregulation and water relations of two East African antelopes: the hartebeest and impala, *Comp. Biochem. Physiol.,* 38A, 525, 1971.
28. **Luck, C. P. and Wright, P. G.,** The body temperature of the hippopotamus, *J. Physiol. (London),* 147, 53P, 1959.
29. **Maskrey, M. and Hoppe, P. P.,** Thermoregulation and oxygen consumption in Kirk's dik-dik *(Madoqua kirkii)* at ambient temperatures of 10-45°C, *Comp. Biochem. Physiol.,* 62A, 827, 1979.
30. **Musewe, V. O., Maloiy, G. M. O., and Kanja, J. K.,** Evaporative water loss in two small African antelopes: the dik-dik and the suni, *Comp. Biochem. Physiol.,* 53C, 17, 1976.
31. **Whittow, G. C., Seammel, C. A., Leong, M., and Rand, D.,** Temperature regulation in the smallest ungulate, the lesser mouse deer *(Tragulus javanicus), Comp. Biochem. Physiol.,* 56A, 23, 1977.
32. **Taylor, C. R.,** The eland and the oryx, *Sci. Am.,* 220, 88, 1969.
33. **Maloiy, G. M. O.,** Temperature regulation in the Somali donkey *(Equus asinus), Comp. Biochem. Physiol.,* 39A, 403, 1971.
34. **Taylor, C. R., Spinage, C. A., and Lyman, C. P.,** Water relations of the waterbuck, an East African antelope, *Am. J. Physiol.,* 217, 630, 1969.
35. **Allbrook, D. B., Harthoorn, A. M., Luck, C. P., and Wright, P. G.,** Temperature regulation in the white rhinoceros, *J. Physiol. (London),* 143, 51P, 1958.
36. **Christopherson, R. J., Hudson, R. J., and Christophersen, M. K.,** Seasonal energy expenditures and thermoregulatory responses of bison and cattle, *Can. J. Anim. Sci.,* 59, 611, 1979.
37. **Parker, K. L. and Robbins, C. T.,** Thermoregulation in mule deer and elk, *Can. J. Zool.,* 62, 1409, 1985.

38. **Renecker, L. A.,** Diurnal changes in bioenergetic parameters of moose, in *Wildlife Productivity and Management Program*, Department of Animal Science Annual Report, University of Alberta, Edmonton, 1982, 74.
39. **Irving, L. and Krog, J.,** Body temperatures of arctic and subarctic birds and mammals, *J. Appl. Physiol.*, 6, 667, 1954.
40. **Simpson, A. M., Webster, A. J. F., Smith, J. S., and Simpson, C. A.,** Energy and nitrogen metabolism of red deer *(Cervus elaphus)* in cold environments; a comparison with cattle and sheep, *Comp. Biochem. Physiol.*, 60, 251, 1978.
41. **Johnson, K. G., Maloiy, G. M. O., and Bligh, J.,** Sweat gland function in the red deer *(Cervus elaphus)*, *Am. J. Physiol.*, 233, 604, 1972.
42. **Holter, J. B., Urban, W. E., Jr., Hayes, H. H., Silver, H., and Skutt, H. R.,** Ambient temperature effects on physiological traits of white-tailed deer, *Can. J. Zool.*, 53, 679, 1975.
43. **Silver, H.,** Effect of level of intake and environmental temperature on heat production, measurement of physiological parameters and/or activities of deer, New Hampshire Fish and Game Dept., Job Progr. Rep. #W-51-R-6, 1973.
44. **Hart, J. S., Heroux, O., Cottle, W. H., and Mills, C. A.,** The influence of climate on metabolic and thermal responses of infant caribou, *Can. J. Zool.*, 39, 845, 1961.
45. **Segal, A. N.,** Thermoregulation in reindeer *(Rangifer tarandus)* (in Russian), *Zool. Zh.*, 59, 1718, 1980.
46. **Jacobsen, N. K.,** Physiology, Behavior, and Thermal Transactions of White-tailed Deer, Ph.D. thesis, Cornell University, Ithaca, N.Y., 1973.
47. **Luck, C. P. and Wright, P. G.,** Aspects of the body temperature, and habitat of large animals, in *Biometerology*, Pergamon Press, Oxford, 1962, 334.
48. **Spotilla, J. R. and Gates, D. M.,** Body size, insulation, and optimum body temperature of homeotherms, in *Perspectives of Biophysical Ecology*, Gates, D. M. and Schmerl, R. B., Eds., Springer-Verlag, New York, 1975, chap. 17.
49. **Finch, V. A.,** Energy exchanges with the environment of two East African antelopes, the eland and the hartebeest, *Symp. Zool. Soc. London*, 31, 315, 1972.
50. **Isabirye, J. K. and Robertshaw, D.,** The sweat gland response to solar radiation in domestic animals, *Biometeorology*, 5, 67, 1972.
51. **Webster, A. J. F.,** Heat Loss from cattle with particular emphasis on the effects of cold, in *Heat Loss from Animals and Man*, Monteith, J. L. and Mount, L. E., Eds., Butterworths, London, 1974, chap. 10.
52. **Benedict, F. G.,** *The Physiology of the Elephant*, Carnegie Inst. Wash., Washington, D.C., 1936, 136.
53. **Wright, P. G.,** Wild animals in the tropics, in *The Biology of Survival*, Edholm, O. G., Ed., Academic Press, New York, 1964, 17.
54. **Cena, K.,** Thermoregulation in the hippopotamus, *Int. J. Biometeor.*, 8, 57, 1964.
55. **Irving, L., Schmidt-Nielsen, K., and Abrahamsen, N.,** On the melting points of animal fat in cold climates, *Physiol. Zool.*, 30, 93, 1956.
56. **Webster, A. J. F. and Blaxter, K. L.,** The thermal regulation of 2 breeds of sheep exposed to air temperatures below freezing point, *Res. Vet. Sci.*, 7, 466, 1966.
57. **Bianca, W.,** Thermoregulation, in *Adaptation of Domestic Animals*, Hafez, E. S. E., Ed., Lea & Febiger, Philadelphia, 1968, chap. 7.
58. **Grojean, R. E., Sousa, J. A., and Henry, M. C.,** Utilization of solar radiation by polar animals: an optical model for pelts, *Appl. Opt.*, 19, 339, 1980.
59. **Yousef, M. K.,** Thermoregulation in wild ungulates, in *Progress in Biometeorology, Division B; The Effects of Weather and Climate on Animals*, Vol. 1, Part 1, Period 1963-1973, Tromp, S. W. and Bouma, J. J., Eds., Swets and Zeitlinger, B. V., Amsterdam, 1976, chap. 1, sec. 4g.
60. **Whittow, G. C.,** Ungulates, in *Comparative Physiology of Thermoregulation*, Vol. 2, Whittow, G. C., Ed., Academic Press, New York, 1971, chap. 3.
61. **Bligh, J.,** Evaporative heat loss in hot arid environments, *Symp. Zool. Soc. London*, 31, 357, 1972.
62. **Findlay, J. D. and Ingram, D. L.,** Brain temperature as a factor in the control of thermal polypnoea in the ox, *J. Physiol. (London)*, 155, 72, 1961.
63. **Bianca, W.,** The relation between respiratory rate and heart rate in the calf subjected to severe heat stress, *J. Agric. Sci.* (Cambridge), 51, 321, 1958.
64. **Whittow, G. C. and Findlay, J. D.,** Oxygen cost of thermal panting, *Am. J. Physiol.*, 214, 94, 1968.
65. **Parker, K. L.,** Ecological Energetics of Mule Deer and Elk: Locomotion and Thermoregulation, Ph.D. thesis, Washington State University, Pullman, 1983.
66. **Whittow, G. C., Sturkie, P. D. S., and Stein, G., Jr.,** Cardiovascular changes associated with thermal polypnea in the chicken, *Am. J. Physiol.*, 207, 1349, 1964.
67. **Schmidt-Nielsen, K., Schroter, R. C., and Shkolnik, A.,** Desaturation of exhaled air in camels, *Proc. R. Soc. London Ser. B*, 211, 305, 1981.
68. **Blix, A. S. and Johnsen, H. K.,** Aspects of nasal heat exchange in resting reindeer, *J. Physiol. (London)*, 340, 445, 1983.

69. **Johnsen, H. K. and Blix, A. S.,** Nasal heat exchange in Norwegian reindeer *(Rangifer tarandus tarandus), Acta Physiol. Scand. Suppl.*, 508, 45, 1982.
70. **Robertshaw, D. and Taylor, C. R.,** A comparison of sweat gland activity in eight species of East African bovids, *J. Physiol. (London)*, 203, 135, 1969.
71. **Ghobrial, L. I.,** A comparative study of the integument of the camel, Dorcas gazelle, and jerboa in relation to desert life, *J. Zool. London*, 160, 509, 1970.
72. **Sokolov, V. E.,** *Mammal Skin*, University of California Press, Berkeley, 1982, chap. 2.
73. **Weiner, J. S. and Hellman, K.,** The sweat glands, *Biol. Rev.*, 35, 141, 1960.
74. **Bligh, J.,** A thesis concerning the process of secretion and discharge of sweat, *Environ. Res.*, 1, 28, 1967.
75. **Taylor, C. R., Dmi'el, R., Shkolnik, A., Baharav, D., and Borut, A.,** Heat balance of running gazelles: strategies for conserving water in the desert, *Am. J. Physiol.*, 226, 439, 1974.
76. **Allen, T. E. and Bligh, J.,** A comparative study of the temporal patterns of cutaneous water loss of domesticated animals with epitricial sweat glands, *Comp. Biochem. Physiol.*, 31, 347, 1969.
77. **Rai, A. K., Mehta, B. S., and Singh, M.,** Thermal sweating in different body regions of sheep, *J. Agric. Sci. (Cambridge)*, 92, 511, 1979.
78. **Gatenby, R. M.,** Evaporimeter measuring sweat rate of cattle outdoors, *J. Therm. Biol.*, 5, 21, 1980.
79. **Finch, V. A., Bennett, I. L., and Holmes, C. R.,** Sweating in cattle and its relation to rectal temperature, tolerance of sun and metabolic rate, *J. Agric. Sci. (Cambridge)*, 99, 479, 1982.
80. **Robertshaw, D.,** The pattern and control of sweating in the sheep and the goat, *J. Physiol. (London)*, 198, 531, 1968.
81. **Taylor, C. R.,** Exercise and environmental heat loads: different mechanisms for solving different problems?, in *MTP International Review of Science, Physiology Series Two, Environmental Physiology*, Robertshaw, D., Ed., University Park Press, Baltimore, 1977, chap. 4.
82. **Luck, C. P. and Wright, P. G.,** Aspects of the anatomy and physiology of the skin of the hippopotamus *(H. amphibius), Q. J. Exp. Physiol.*, 49, 1, 1963.
83. **Murray, D. M.,** A comparison of cutaneous evaporation rates in cattle exposed to heat in a climate laboratory and in the field, *J. Agric. Sci.* (Cambridge), 66, 175, 1966.
84. **Allen, T. E., Bennett, J. W., Donegan, S. M., and Hutchinson, J. C. D.,** Moisture, its accumulation and site of evaporation in the coats of sweating cattle, *J. Agric. Sci.* (Cambridge), 74, 247, 1970.
85. **MacFarlane, W. V.,** Terrestrial animals in dry heat: ungulates, in *Handbook of Physiology, Adaptation to the Environment*, Section 4, Dill, D. B., Adolph, E. F., and Wilber, C. G., Eds., American Physiology Society, Washington, D. C., 1964, chap. 33.
86. **Renecker, L. A., Hudson, R. J., Christophersen, M. K., and Arelis, C.,** Effect of posture, feeding, low temperature, and wind on energy expenditures of moose calves, *Proc. North Am. Moose Conf. Workshop*, 14, 126, 1978.
87. **Wesley, D. E., Knox, K. L., and Nagy, J. G.,** Energy metabolism of pronghorn antelopes, *J. Wildl. Manage.*, 37, 563, 1973.
88. **Christopherson, R. J., Hudson, R. J., and Richmond, R. J.,** Comparative bioenergetics of American bison, yak, Scottish Highland, and Hereford calves, *Acta Theriol.*, 23, 49, 1978.
89. **Chappel, R. W. and Hudson, R. J.,** Winter bioenergetics of Rocky Mountain bighorn sheep, *Can. J. Zool.*, 56, 2388, 1978.
90. **Moen, A. N.,** personal communication in **Mount, L. E.,** The concept of thermoneutrality, in *Heat Loss from Animals and Man*, Monteith, J. L. and Mount, L. E., Eds., Butterworths, London, 1974, chap. 21.
91. **Christophersen, M. K.,** Metabolic response of moose to ambient temperature, in *Wildlife Productivity and Management Program*, Department of Animal Science Annual Report, University of Alberta, Edmonton, 1982, 78.
92. **Hales, J. R. S.,** Effects of heat stress on blood flow in respiratory and nonrespiratory muscles in the sheep, *Pflugers Arch.*, 345, 123, 1973.
93. **Gallivan, G. J. and Ronald, K.,** Temperature regulation in freely diving harp seals *(Phoca groenlandica), Can. J. Zool.*, 57, 2256, 1979.
94. **McArthur, A. J.,** Thermal resistance and sensible heat loss from animals, *J. Therm. Biol.*, 6, 43, 1981.
95. **Scholander, P. F., Walters, V., Hock, R., and Irving, L.,** Body insulation of some arctic and tropical mammals and birds, *Biol. Bull.*, 99, 225, 1950.
96. **Mahoney, S. A. and King, J. R.,** The use of the equivalent blackbody temperature in the thermal energetics of small birds, *J. Therm. Biol.*, 2, 115, 1977.
97. **Jacobsen, N. K.,** Differences of thermal properties of white-tailed deer pelage between seasons and body regions, *J. Therm. Biol.*, 5, 151, 1980.
98. **Gates, C. C. and Hudson, R. J.,** Effects of posture and activity on metabolic responses of wapiti to cold, *J. Wildl. Manage.*, 43, 564, 1979.
99. **Silver, H., Colovos, N. F., Holter, J. B., and Hayes, H. H.,** Fasting metabolism of white-tailed deer, *J. Wildl. Manage.*, 33, 490, 1969.

100. **White, R. G. and Yousef, M. K.,** Energy expenditure in reindeer walking on roads and on tundra, *Can. J. Zool.*, 56, 215, 1977.
101. **McNab, B. K.,** On estimating thermal conductance in endotherms, *Physiol. Zool.*, 53, 145, 1980.
102. **Schmidt-Nielsen, K.,** *Animal Physiology: Adaptation and Environment*, 2nd ed., Cambridge University Press, Cambridge, 1979, chap. 8.
103. **Casey, T. M., Withers, P. C., and Casey, K. K.,** . Metabolic and respiratory responses of arctic mammals to ambient temperature during summer, *Comp. Biochem. Physiol.*, 64A, 331, 1979.
104. **Bradley, S. R. and Deavers, D. R.,** A re-examination of the relationship between thermal conductance and body weight in mammals, *Comp. Biochem. Physiol.*, 65A, 465, 1980.
105. **Holmes, C. W.,** A note on the protection provided by the hair coat or fleece of the animal against the thermal effects of simulated rain, *Anim. Prod.*, 32, 225, 1981.
106. **Chappel, M. A. and Bartholomew, G. A.,** Standard operative temperatures and thermal energetics of the antelope ground squirrel *(Ammospermophilus leucurus)*, *Physiol. Zool.*, 54, 81, 1981.
107. **Bakken, G. S. and Gates, D. M.,** Heat transfer analysis of animals: some implications for field ecology, physiology, and evolution, in *Perspectives of Biophysical Ecology*, Gates, D. M. and Schmerl, R. B., Eds., Springer-Verlag, New York, 1975, chap. 16.
108. **Ozoga, J. J. and Gysel, L. W.,** Response of white-tailed deer to winter weather, *J. Wildl. Manage.*, 36, 892, 1972.
109. **Piltiers-Gauthier, H.,** Observations sur l'ecologie du dromadaire dans le Sahara nord occidential, *Mammalia*, 25, 195, 1961.
110. **Nilssen, K. J., Johnsen, H. K., Rognmo, A., and Blix, A. S.,** Heart rate and energy expenditure in resting and running Svalbard and Norwegian reindeer, *Am. J. Physiol.*, in press.
111. **Beall, R. C.,** Winter Habitat Selection and Use by a Western Montana Elk Herd, Ph.D. thesis, University of Montana, Missoula, 1974.
112. **Leckenby, D. A.,** Management of mule deer and their habitats: applying concepts of behavior, physiology, and microclimate, *West. Proc. Annu. Conf. West. Assoc. State Game and Fish Comm.*, 57, 206, 1977.
113. **Black, H., Scherzinger, R. J., and Thomas, J. W.,** Relationships of Rocky Mountain elk and Rocky Mountain mule deer habitat to timber management in the Blue Mountains of Oregon and Washington, in *Proc. Elk-Logging-Roads Symposium*, Hieb, S. R., Ed., University of Idaho, Moscow, 11.

Chapter 9

GROWTH AND DEVELOPMENT

M. A. Price and R. G. White

TABLE OF CONTENTS

I. INTRODUCTION

Although this chapter is concerned mainly with ungulates, much of the basic biology of growth which will be discussed applies to all members of the animal kingdom. All animals grow, and in doing so, most of them change shape. These processes are familiar to all of us, but properly understood by none.

A. Growth

While crystals and clouds can certainly grow, and towns and continental topographies develop, there are aspects of biological growth which set it apart from any other. The basic difference is that living things begin their existence carrying a plan, not only of their development but also of the maximum rate at which it will occur regardless of available "growth substrate". In birds and mammals, this blueprint even specifies a maximum size to which the body will grow.

Growth, in biological terms, is difficult to define, and exceptions to almost any suggested definition can be found. It cannot be defined as an increase in weight or even in dry matter since some things, germinating seeds for example, can be shown to be losing dry matter while gaining weight and size, and immature animals on submaintenance nutrition may continue skeletal growth while losing live weight. Here, growth is defined as an increase in size, though to be completely accurate a far longer, and inevitably more cumbersome, definition would be needed. It is not accurate to assume that weight is an infallible measure of size, though it is the best measure for most purposes. Linear and girth measurements often yield information not available from weight measurements, particularly on skeletal remains, road- and hunter-killed animals. The repeatability of linear measurements in animals, however, can be surprisingly low.[1]

However defined, the growth of an animal must be recognized as the sum of the growth of all its parts. Thus, when a human increases in height, that increase is simply the sum of the increases in length of the leg bones, the vertebrae, the skull, and the associated joint tissues. Similarly, an increase in weight represents the sum of the increases in weight of all the parts of the body.

Growth may be defined as positive, static, or negative. It is important, too, to differentiate positive growth following a period of negative growth from continuous positive growth. The former is best referred to as catch-up growth[2,3] though it is often called compensatory growth. Growth is not simply the result of adding new material to already existing material; it results from the interaction of anabolic and catabolic processes jointly referred to as turnover. If anabolism in a particular tissue proceeds faster than catabolism, positive growth will result; conversely, if the rate of catabolism is greater, negative growth will result. If the two proceed at equal rates, weight will be static, but the tissue will be in a state of dynamic equilibrium.

B. Development

If a growing deer increased in length by 10%, it cannot be assumed that the parts contributing to that length each increased by 10%. Similarly, if its weight increased by 10% it cannot be assumed that the weight of every tissue and organ in its body increased by 10%. Indeed, they would not. For reasons which form the basis of this chapter, different parts of the body grow at a variety of rates relative to each other and relative to the body as a whole. The overall result is a gradual change in shape which is referred to as development. Development is in no way a haphazard event; it follows genetically defined lines towards an adult shape which phylogeny has indicated to be the most suitable for the environmental niche which the animal expects to occupy. Most development in mammalian herbivores occurs in the prenatal and immediate postnatal period. Barring secondary sexual characteristics, the changes in shape which occur after puberty are quite minor.

C. Homeostasis

Although the environment, particularly the nutritional environment, can strongly influence the rate at which an animal grows and develops, it has remarkably little influence on the sequences involved. At a particular body weight, an animal's composition is largely dictated by its genotype. The variations which occur are usually either minor or the result of some extreme environmental insult. In either case, removal of the insult is normally followed by a rapid return to normal body composition. This remarkable ability of growing animals to defy their environment is referred to as homeostasis. Animals which are fully adapted to a particular environmental niche will have developed homeostatic mechanisms to combat the extremes of conditions encountered in that niche. However, if the niche changes or populations migrate to a new niche, their homeostatic mechanisms may be defeated and permanent growth impairment may result. The intrusion of man into many ungulate niches has resulted in environmental changes and migrations of the resident populations to new niches. Unless the species has sufficient genetic diversity and an appropriately short generation interval, it may be unable to evolve quickly enough to the new conditions and become extinct.

The homeostatic mechanisms of the body appear to be more weight- than time-oriented. Maximum positive growth rates are dictated by the maximum rate at which the body can assimilate nutrients and the maximum amount by which anabolism can exceed catabolism. Maximum negative growth rates are dictated by the maximum amount by which catabolism can exceed anabolism. Between these extremes, the body mechanisms attempt to ensure that each increment in body weight contains an appropriate proportion of each body component, regardless of how long it takes to achieve that increment. In practice, time can be of great importance because of its association with season and the changes including photoperiod that accompany changes in season in many locations. Since many ungulates are seasonal breeders and have seasonal changes in body composition, failure to achieve a particular body weight or condition in a particular time may be fatal. This places strong evolutionary pressure on appropriate adaptation.

II. PRINCIPLES OF GROWTH

Growth and development are directed, but not totally controlled, by the genetic code. Some of the rules controlling their progress are known, many are not. It is even possible to develop quite simple mathematical expressions describing growth and development. Nevertheless, these mathematical models will not apply under all circumstances, since the genetic code includes built-in adaptability which allows the growing animal to adapt to some particular environmental changes by apparently switching to "plan B" which may be outside the scope of the model proposed for that phase of growth.

A. Organization

1. Cellular Growth

Growth can be studied at many levels, but a logical starting point is the cell. Cells have two ways of growing: they can increase in number (hyperplasia), or they can increase in size (hypertrophy). It is also possible for cells to accumulate noncellular material either around them (e.g., bone) or within themselves (e.g., fat). This process is called accretion.

2. Tissue Growth

Beds of cells make up tissues or organs, and each, depending on function and location, will exhibit a type of growth most appropriate to its needs. Most tissues increase in size interstitially (endogenously), meaning that the growth takes place throughout the tissue. Some tissues, bone for example, cannot grow this way, since the rigidity of the tissue itself prevents any internal growth. These tissues grow only at growth surfaces; this is called appositional (exogenous) growth.[4]

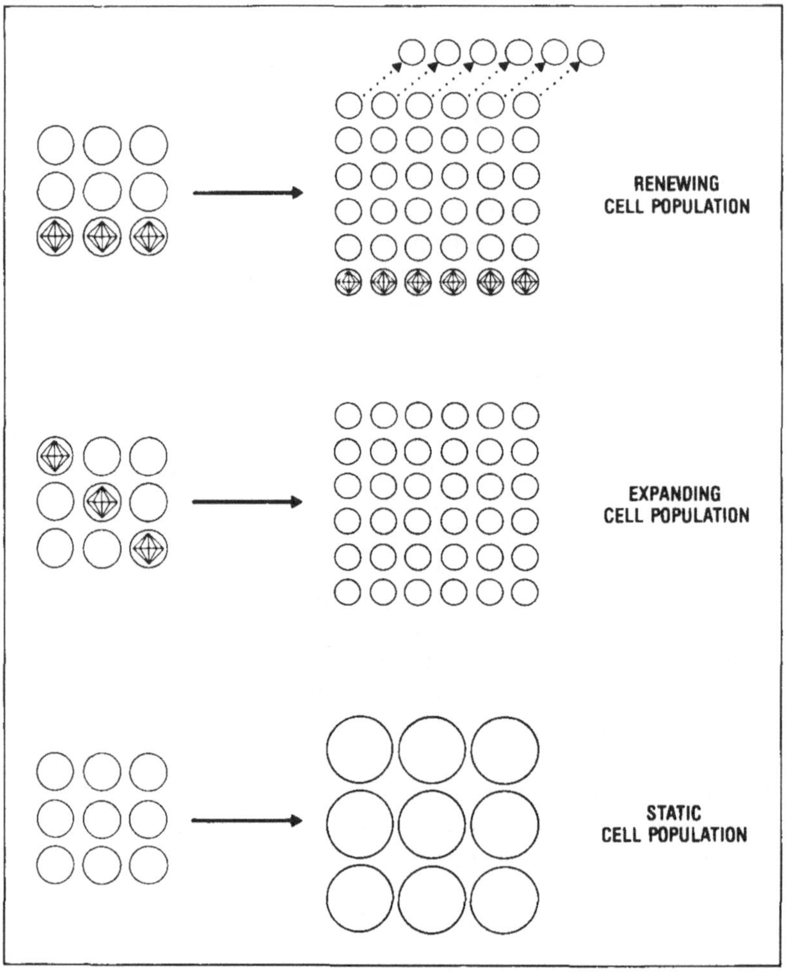

FIGURE 1. Modes of growth within a tissue. In renewing populations, cells are continually being lost or destroyed and replaced from mitotically active stem cells; in expanding cell populations, growth is by diffuse proliferation or cellular hyperplasia; in static cell populations, growth is by cellular hypertrophy. (From Goss, R. J., *Adaptive Growth*, Academic Press, New York, 1964. With permission.)

Few mammalian cells maintain mitotic ability and continue hyperplastic increase to maturity; most use hyperplasia to establish the cell complement of the tissue as early in life as possible and then use the process of hypertrophy to grow the tissue to its adult size. Tissues can be described as having static cell populations, where the cells have completed the hyperplastic phase and the number of cells is fixed, or they can have expanding cell populations where the cells retain their mitotic ability and grow hyperplastically (Figure 1). There is a third type called a renewing cell population where mitosis continues, but the number of cells is more or less fixed. This strategy is employed by tissues which continually lose cells such as the skin, epithelial tissues, and red blood cells. In each case, a population of mitotically active "stem" cells exists, which gives rise to daughter cells which lose their mitotic ability and perform their functions in the body for a limited period of time. Eventually they die and are lost, either by erosion or by reabsorption into the body. The rate of loss of mitotic division in the stem cells precisely matches the rate of loss of effete cells.

Tissues composed of cells which become mitotically inactive can only grow by hypertrophy. This effectively sets a ceiling to their ultimate size, since continued unchecked hypertrophy would eventually render any cell functionally incompetent. These tissues are said to exhibit determinate growth, as opposed to the indeterminate growth of an organ such as the liver where all cells retain their mitotic ability. The mammalian body contains both determinate and indeterminate tissues, and it can be reasonably generalized[3] that tissues made up of cells which excrete their products — liver cells, fibroblasts, and various exocrine glands — have an indeterminate growth pattern, whereas those which accumulate their products inside the cell — nerve, muscle, and fat cells — tend to lose mitotic ability when they are fully differentiated, and therefore have a determinate growth pattern.

3. Body Growth

Indeterminate growth carries with it the power of regeneration of lost tissue. A liver, for example, can be surgically reduced to 20% of its mass and will fully regenerate its original size, though certainly not its original shape. The rate of this restoration is remarkable: a two thirds hepatectomy in rats has resulted in a doubling of weight in the remaining one third in two or three days with complete regeneration of the original weight by the third week. In one published case, rat livers were repeatedly subjected to partial hepatectomy at monthly intervals for a whole year and continued to regenerate successfully.[5] Many "lower" animals have indeterminate growth of all organs and tissues, and as a result have no maximum or mature body size; they can often regenerate accidentally reduced limbs and tissues. It is interesting to speculate why "higher" animals, including wild herbivores, have chosen a determinate body growth pattern with a maximum size and very limited regenerative ability. Goss[3] has written extensively on the subject and speculates that the answer is twofold: on the one hand is the danger of growing too large for the environmental niche which the animal is trying to exploit, and on the other hand is the danger of living forever! Goss suggests that aging may not be built into the indeterminate growth pattern making it theoretically possible, in the absence of accident or disease, for animals such as some fish to live forever. As renewal and rejuvenation are the essence of evolution, this would not be in the best interest of the species as a whole. Thus, Goss suggests, indeterminate growth is an alternative to sensescence, ensuring that, ultimately, the individual will perish by simply growing too big, rendering it a victim of environmental limitation and similitude.

B. Growth Curves

If the progress of almost any growing system is plotted against time, the curve generated will be S-shaped or sigmoid. This will be found in plots of the lengths of a limb-bone, the area of a bacterial colony on an agar plate, or the total weight of a population of fish in a pond. It would probably also be found for the growth of a nonliving system like a rain cloud or a salt crystal. The S-shape can be divided into two parts, a self-accelerating part and a self-inhibiting part, joined by an inflection. Each part can be explained in biological terms and modeled mathematically. It is even possible to use single, quite simple models to describe the whole S-shaped curve.[6] However, biological systems being very complex, a mathematical model, particularly a simple one, is unlikely to fit any but the most primitive system. Nevertheless, a study of these generalized models can be very useful in understanding some of the factors influencing growth.

The early part of the self-accelerating period of growth represents a time when multiplication, whether of individuals, cells, or organelles, can proceed with no inhibition. Thus, if a bacterium were capable of mitotic division every hour and there were no limitations, it would proceed to double every hour in a geometric progression. At the end of 6 hr, it would have become a colony 64 times its original size. Six hours later, it would be over 4000 times the original size. Early in its second day, the number of individuals would exceed the current human population of the world. It is this process that causes the self-accelerating part of the sigmoid growth curve. Progress appears slow at first because it is visualized additively. Thus, in the first time period, one individual (or unit of length, weight, or whatever growth is being considered) is added; in the second, two more. In the tenth time period, 512 units are added and in the twentieth, over 1 million. In any real system, however, the environment quickly begins to offer resistance to this unbridled expansion. It may take the form of limited nutrients, or difficulty in getting nutrients to the individuals in the center of the colony, or space limitations caused either by the environment or the inability of individuals to move out of each other's space. The end result is a gradual slowing of the rate of expansion until a constant rate is achieved. This is the inflection of the curve. Continued resistance takes growth into the self-inhibiting phase which is asymptotic. At the asymptote, the population will be in balance with the environment. It is of interest to note that at least one of the currently used models for describing liveweight increases over time (the logistic equation) had its origin as a model for describing the increase in numbers of individuals in a human population.[7]

There are several ways of constructing mathematical models to describe growth, and it must be recognized that very similar curves can be created by very different models. The most acceptable method of fitting a model would be to study the growth organization of the animal concerned and use that as a guide to the best type of model. Thus, in the early stages of growth, the increase in cell numbers is exponential, and an exponential model would be appropriate for describing growth in volume or weight. Length growth, however, occurs in an additive fashion, with new cells being added at a growth surface. Stephenson[8] has argued from a logical point of view that the fetal period of growth in mammals conforms more closely to a cubic than to the more normally accepted multiplicative model. Certainly such a model would conform to a type of growth where new material is added to existing three-dimensional surfaces, though this would only represent an organ during hyperplasia.

The mammalian body is composed mostly of organs and tissues with determinate growth patterns, most of which have completed hyperplasia by birth or soon after. Postnatal growth of these animals, therefore, contains a fairly fixed proportion of hyperplasia and hypertrophy which is best described by an exponential relationship of the type $Y_t = A(1 - Be^{-kt})$ (where Y_t is body weight at time t, A is the asymptote, or mature weight, B and k are fitted constants, and e is the base of the natural log).[6] A number of variations on this basic model have been proposed, particularly the fitting of an inflection either as a fixed value or as a fourth parameter. Each model has its strengths and weaknesses, but none is considered to be uniformly applicable to the full range of prenatal and postnatal growth.[9]

A plot of the accumulation of weight over of period of time (i.e., a weight/time curve) is useful in visualizing the change in size of a growing animal. However, the same data can be analyzed in other ways to clarify other aspects of growth. Figure 2 illustrates the growth of an imaginary animal with a weight of 1 at time zero (e.g., birth) and a mature weight of 100 at time 20.

The plot of weight/time (Figure 2A) is the classic sigmoid shape discussed above. In the third row of data in Figure 2, the gain in each time period has been calculated, and plotting this against time gives the curve shown in Figure 2B. From this it can be seen that the animal increases its rate of gain to a maximum (corresponding to the inflection of the weight/time curve) where it may plateau for some time; it then decreases to zero at maturity. It has been suggested that this plateau in the rate curve corresponds to puberty.[10] In some tissues, notably bone, the change in hormonal balance at puberty can be shown to be the triggering mechanism for a progressive reduction in growth rate, though the relationships are far from simple.[11] The body growth curves shown in Figure 2, however, are the sum of all the growth curves for all the parts of the body, many of which show a decreased rate of gain considerably before puberty, and in some cases even before birth.

A third way of analyzing growth data is shown in the fourth row of data in Figure 2. The percentage increment is the gain during a time period expressed as a percentage of the weight

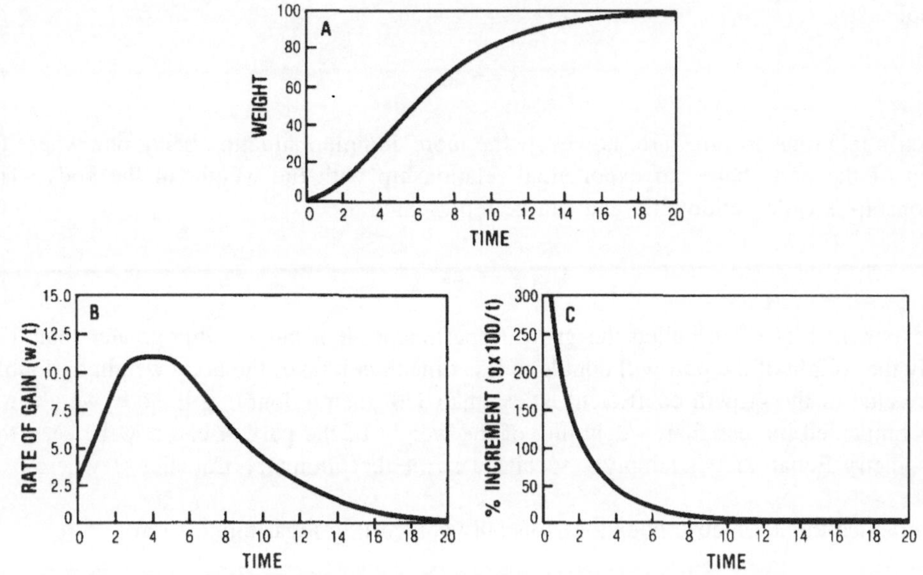

TIME, t	0	1	2	3	4	5	6	7	8	9	10	11	12	13	14	15	16	17	18	19	20
WEIGHT, w	1	4	10	19	30	41	52	62	70	77	82	86	90	93	95	97	98	99	99.5	99.8	100
GAIN, g		3	6	9	11	11	11	10	8	7	5	4	4	3	2	2	1	1	.5	.3	.2
% INCREMENT		300	150	90	58	37	27	19	13	10	6.5	5	4.5	3	2	2	1	1	.5	.3	.2

FIGURE 2. An increase in weight over a period of time can be represented graphically in a number of ways depending upon the aspect of growth to be illustrated.

at the beginning of that time period. It can be seen to be continually decreasing to zero. Its maximum value (the first time period) depends upon the length of the time period chosen. This type of curve is particularly useful in illustrating the efficiency of gain in a growing animal. It shows clearly that as the animal grows, each unit of weight increases more slowly. For example, a 100 kg deer gaining 100 g per day has twice as great a daily percentage increment (0.1%) as a 200 kg deer gaining at exactly the same daily rate (0.05%). Since the bulk of energy consumed by mammalian herbivores, even when growing rapidly, is used for maintenance, it can clearly be seen that increased size carries with it the cost of decreased growth per unit of input and hence reduced energy efficiency, unless rate of gain is appropriately augmented.

It has been noted[12] that the sigmoid liveweight/time curve applies more accurately to the mean of a population of individuals, than it does to any particular individual. It must also be noted that the body weight growth curve is the sum of the separate growth curves of each part of the body, and that the inflections of these curves are spread out from before birth to well after puberty. With few exceptions, the order of these inflections is fixed and an increase in growth rate simply brings them closer together in time, without changing their sequence.[13] Changes in body proportion result from the relationships of size and timing of the individual growth curves of the various parts of the body. It can be shown that, in general, the relationship between the weight of a part of the body (limb, tissue, organ, etc.) and the body as a whole is fixed.[14]

If the weight of a tissue or organ (y) remains a constant proportion (a) of the weight of the body (x) as it grows, then the relationship is said to be isometric and the weight of the part can be calculated at any time as a fraction of the weight of the body at that time. For example,

$$y = ax \quad (1)$$

Isometric relationships are rare, however, the more common situation being one where the weight of the part shares an exponential relationship with the weight of the body. This relationship is called allometric and can be represented as:

$$y = ax^b \quad (2)$$

The exponent b is often called the growth coefficient. If b has a value greater than 1.0, clearly the weight of the part will double in less time than it takes the body weight to double. If the value of the growth coefficient is less than 1.0, then a doubling in body weight will be accompanied by less than a doubling of the weight of the part. It can now be seen that the isometry Equation 1 is simply a special case of the allometry Equation 2 where b = 1.0

In practice the allometric equation is usually solved in a logarithmic form:

$$\log y = \log a + b.\log x \quad (3)$$

which is rectilinear with a slope b and an intercept on the log y axis at log a. The use of allometry to study relative growth has revealed remarkable constancy in relationships within the body and has greatly improved our understanding of the factors which influence growth and development.

C. Regulation of Growth

The study of relative growth has made it clear that most parts of the body are programmed as to their size relative to the sizes of all the other parts of the body. The mechanisms which allow them to achieve this are unknown, but several hypotheses exist. It may be, for example, that organs produce inhibitors to their own growth in proportion to their own size. This self inhibition would act as a negative feedback ensuring that growth was increasingly inhibited and finally switched off when the "right size" was reached.[3] Proof of this mechanism would require identification of the inhibitors as well as of the receptor sites. Variations on this suggestion include the possibility of organs producing stimulators to encourage the growth of other organs, which continue to respond to this stimulation until it is equalled by their own production of growth inhibiting substance.

It has also be suggested[15] that there are two aspects to size control: a genetic aspect, utilizing mechnisms such as those suggested above to grow the organ to "basic" size; and an environmental aspect, in which functional stimuli modify the size of the organ depending upon the demands placed upon it. This conforms to the overuse hypertrophy and disuse atrophy seen in most organs and tissues of the body, and also to the observation that disuse seldom results in the complete disappearance of the atrophying tissue.

D. Size Adjustments and Similitude

Increasing size presents the animal with the problems of what Thompson[16] and others have called similitude. The problem is basically mathematical: increasing the size but not the shape of any object will change the ratios among its volume, its surface area, and any of its linear or cross-sectional dimensions. Thus, if a cube of side length two units is compared with one of side length four units (Figure 3), the larger one will be found to have four times the surface area (2^2) and eight times the volume (2^3). Also, the pressure

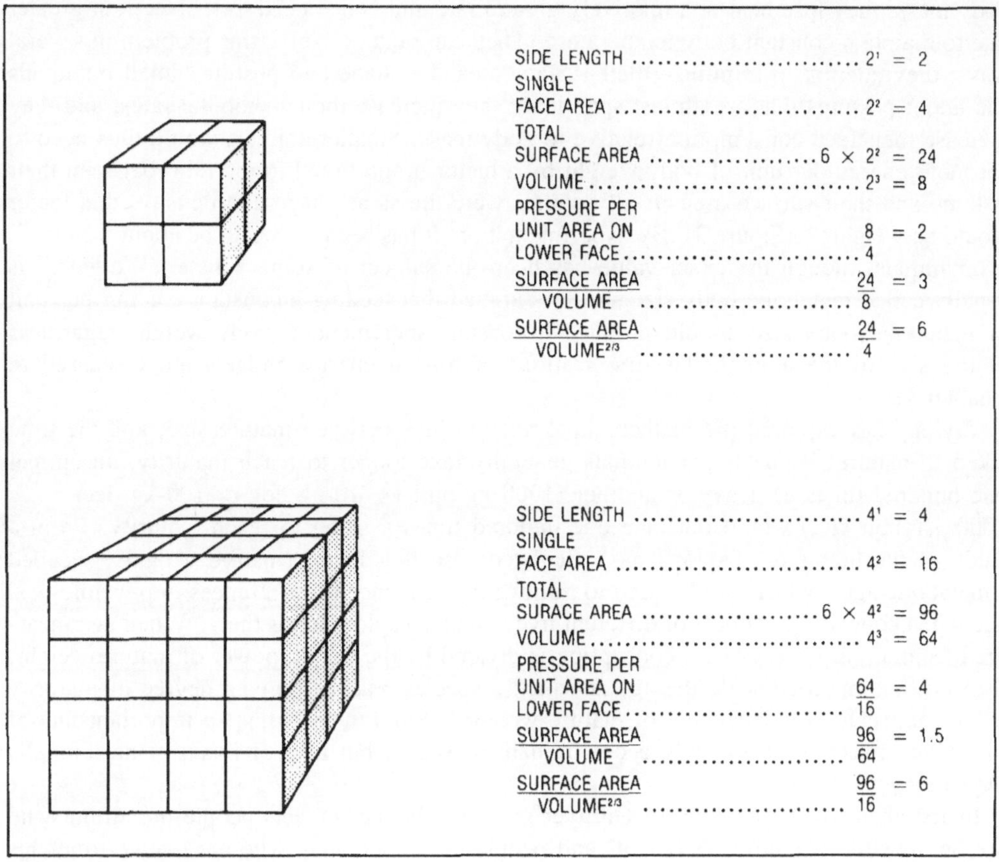

FIGURE 3. The effects of similitude on various size relationships in two bodies of constant shape (cubic) and density (1.0) but different size.

per unit area which it exerts on its nether face will be twice as great as that exerted by the smaller cube as will its surface to volume ratio. Continued increase in the size of the cube, with no change in the density of the material, would eventually lead to its collapse because of an inability of the lowest layers to sustain the pressure from above. For these reasons, no land animal has ever achieved the dimensions of the largest aquatic animal. Similarly, the smallest sized mammal is of the same order of magnitude as the largest insect. Similitude prevents the insect-type body from being larger, or the mammal-type body from being smaller. Although increasing the size of the body without changing the shape distorts the ratio of volume (v) to surface area (s) they still retain an exponential relationship ($s = kv^{0.66}$); k is a function of the shape of the body. For a cube, $k = 6.0$ (Figure 3); for a sphere $k = 4.836$.

Because of similitude, some organs must be of a different shape in larger animals. Thus, the kidneys of small animals are "kidney-shaped" and consist of a single renal papilla collecting into a single renal pelvis. Simply increasing the size of papilla and pelvis in increasingly larger animals results in problems of similitude. These are solved by increasing the number of papillae in medium sized animals (sheep, man) and ultimately increasing the number of pelvises to give a multilobular kidney in large animals (cattle, whales).

Similitude also has important implications in energy balance: heat is generated in the mammalian body in proportion to its mass. The same heat passes out of the body through its surface. Thus, since by similitude smaller animals have a larger surface area per unit of body mass, they lose heat at a relatively greater rate and hence need to replace it at greater rate to maintain constant body temperature. They can, and do, solve this problem in several ways: they attempt to minimize their surface areas by shape and posture (small mammals can adopt postures that are almost spherical); they increase their metabolic rates; and they increase their feed consumption relative to body mass. Smaller animals would thus need to eat more energy per unit of body weight by a factor proportional to the ratio between their volume and their surface area. If all animals were the same shape and density, this factor would be weight$^{0.66}$ (Figure 3). By experimentation, it has been shown to be about weight$^{0.75}$ in mammals, though the exact value has been the subject of some debate.[6] Weight$^{0.75}$ is usually called metabolic body size and it is argued that feeding a constant amount per unit of metabolic body size should produce a constant increment in body weight regardless of the size of the animal. The implications of this in energy budgets are discussed in Chapter 1.

Taylor[17] has explored the mathematical relationships between mature size, and the time taken to mature. While larger animals generally take longer to reach maturity, an animal one hundred times as heavy as another (1000 kg bull vs. 10-kg dog or 100-kg deer vs. a 1000-g rabbit etc.) would not take one hundred times as long to reach maturity. From a study of the literature, Taylor[17] has calculated a scaling factor (mature weight$^{0.27}$) called "metabolic age" which can be used to remove the influence of differences in physiological age when comparing animals of different liveweights. Taylor claims that: "When age measured from an origin at or near conception is divided by the 0.27th power of mature weight, then on this modified scale the age at which a species reaches a given degree of maturity has an expectation independent of mature weight." Similitude is thus an important fact of life which makes size not only a determinant of shape, but also of much of an animal's physiology.

In Jonathan Swift's account of Gulliver's travels, his hero meets people in Lilliput who are one twelfth of Gulliver's height, and people in Brobdingnag who are twelve times his height. Without regard to the principles of similitude, Swift made these mythical people the same shape as Gulliver. In reality, the Brobdingnagians would weigh 1728 times as much as Gulliver, and carry this weight on bones of only 144 times the cross-sectional area. It is likely that they would crush their bones by merely standing still; they certainly would if they attempted to run. Their muscles, as well, would be exerted beyond endurance. The Lilliputians on the other hand, would need to spend their whole day running to get sufficient food. (They would need approximately ten times as much food per unit of body weight as Gulliver). Unfortunately, with mouthparts and gut surface area proportional to Gulliver's they would be unable to consume or assimilate this quantity of food. Pigmy shrews, the smallest of mammals, have been known to starve to death overnight after being caught in a trap.

III. STAGES OF GROWTH

Animal life begins with the union of two gametes to form a fertilized egg. From that time until maturity, the animal must establish, develop, and grow all the organs and tissues which it needs, and yet remain living and functioning through every change. This remarkable series of events is somewhat akin to a huge production plant beginning operation the day the first sod is turned, installing its power generators immediately, and then never turning them down until it has reached full size. The whole process is genetically defined, with the code not only indicating the size, location, and anatomy of every part, but also how they should develop, both individually, and in relation to each other.

A. Prenatal Growth

In mammals, the most vulnerable anabolic period is spent inside the mother's body, warm and amply supplied with nutrients. This prenatal growth period has historically been divided into three trimesters, originally the first, middle, and the last 3-month periods of human gestation. In more general terms, these trimesters correspond to the period of the blastocyst, the embryo, and the fetus, respectively. They are not equal in length in all mammals, and in any case blend into each other rather than being sharply divided.

At conception, the genetic admixture from sire and dam is established, and much of the animal's future is fixed. Physically, at least, the environment is unlikely to cause any but minor deviations from the established plan. The study of animal behavior, and particularly its genetic basis, is rapidly developing, but the relative contributions of genetics and environment to animal behavior remain an area of debate.

1. Blastocyst

The period of the blastocyst precedes the establishment of a placental exchange between the individual and its dam. During this period, it is free in the uterus and bathed in uterine secretions (deutoplasm or uterine milk) which act as a nutrient. It grows by hyperplasia, the first few divisions being synchronous and referred to as cleavage. The rate of cell division varies with species, but in the domestic sheep and cow the 32 cell stage is generally reached 3 to 4 days after fertilization. During the cleavage period, there is little change in the mass of the blastocyst, the individual cells becoming smaller with each division. At this stage, there is no specialization among the cells, and each one is capable of developing into a clone of the others. This occasionally happens in all species and in some it is the norm. (The nine-banded armadillo *Dasypus novemcinctus* usually gives birth to monozygotic quadruplets.[18]) By about the 64 cell stage, depending on species, the ball of cells is too large for the efficient diffusion of nutrients into the middle cells, and the first sign of organization occurs: a small hollow (the blastocelic cavity) develops in the body of the ball of cells into which the nutrient medium diffuses. Specialization and differential rates of cell division continue until the blastocyst is a hollow ball containing a small mass of cells at one end. This mass is the presumptive animal, the remainder the presumptive placenta. While this is happening, the blastocyst is free floating in the uterus, seeking a position of attachment. This period is particularly important in multiple pregnancies to avoid the overcrowding which would occur if the blastocysts were unable to move apart. In the domestic pig, blastocysts are even known to migrate as far as the opposite uterine horn. Once a site of attachment is occupied, the blastocyst will begin to form a placental contact with the dam. In some animals, this involves passing partly or wholly through the epithelial layer of the endometrium — a process called implantation. In many ungulates, implantation in this sense does not occur; rather, the blastocyst becomes lodged by its increasing size in the lumen of the uterus. Placentation takes several forms, but in all cases it consists of a fetal portion and a maternal portion which grow villi into each other. Both sides are highly vascular, but at no time does maternal blood come in contact with fetal blood, though nutrients and waste products are exchanged. At about the same time, the inner cell mass (presumptive animal) of the blastocyst "gastrulates" forming three separate layers of cells: the endo-, meso-, and ectoderm.

2. Embryo

The individual is generally referred to as an embryo at this stage and its basic function is to establish the tissues and organs of the body. An entire branch of biology (embryology) is devoted to the study of these changes and details should be obtained from appropriate textbooks. Briefly, by cell multiplication, differentiation, and migration (swirling, inrolling, invagination, budding, etc.), the endoderm becomes the primitive gut from which arises most of the internal organs of assimilation and excretion; the mesoderm forms the somites from which arise the muscles, the major connective tissues (tendons, ligaments, cartilage, bone, and adipose tissue), and the vascular and lymphatic systems; the ectoderm becomes the skin, with its many specialized parts, such as hair, hooves, horns, various secretary glands including the mammary system, and also the central nervous system consisting of the brain and spinal cord.

The nutritional requirements of the embryo are tiny compared to the size of the dam, and it is heavily shielded from the external environment, physically by amniotic fluid and its dam's bulk, and chemically by the placenta which acts like a liver modifying, detoxifying, and resynthesizing much of what passes through it. In this heavily protected condition, the individual passes its most vulnerable phase. Interference with development in the embryo stage can lead to monstrous deformities which may only prove fatal after birth, "wasting" the pregnancy. A particular risk for wild herbivores in this context is the ingestion of weeds which contain teratogenic ("monster-forming") substances which can cross the placental barrier and interfere with normal embryology, producing characteristic monstrosities depending upon the exact stage of gestation when the weed is eaten.[19] Interference with development in the blastocyst stage is more likely to result in embryonic death or failure to establish placental contact. This is less catastrophic to the species since the dam can then simply return to estrus and have a second chance to become pregnant in the same breeding season. In species with closely defined breeding seasons, however, failure to establish placental contact would mean the loss of reproduction for an entire year. Early embryonic death is thought to be a significant cause of low reproductive efficiency in farm animals,[20] but its importance in wild species is difficult to establish. Environmental stressors (high temperature, disease, poor nutrition) have all been linked to "preimplantation" losses.[20]

3. Fetus

Once all the tissues and organs of the body are established, the individual can be referred to as a fetus. The goal in the fetal stage is to prepare the animal for birth. Phylogeny has taught the species what to expect in the postnatal environment, and development in the fetal stage will favor those organs most necessary immediately postnatally:

- A new-born kangaroo has particular need of a mouth and jaws to suck, and front claws to pull itself, after birth, into its mother's pouch. These parts are well developed at birth, but there is little evidence of the massive hind legs and tail since they are unnecessary for life in the pouch.
- Ruminants have no need of a rumen at birth and so leave its full development until later.
- Most herbivores are subject to carnivorous predation, and so place a high priority on prenatal limb development (to run away from predators and to stay near the grazing herd).
- Many ungulates are subject to cold-stress at birth; they place a high priority on development of pads or sheaths of brown adipose tissue which can be metabolized to maintain the temperature of vital organs (nonshivering thermogenesis) and to warm blood returning from the extremities.
- Neonates of carnivorous species (dogs, cats) tend to place high priority on subcutaneous fat, with little emphasis on mobility.

Much of fetal development concerns enlargement of organs and tissues, initially by hyperplasia but increasingly, as birth approaches, by hypertrophy. In most wild herbivores, the increase in cell numbers which occurs after birth is quite small, and many tissues have almost or completely lost mitotic ability by the time the animal's birth weight has doubled (nervous tissue, skeletal and cardiac muscle, renal nephrons, adipose tissue). Body weight increases exponentially until near the end of gestation (depending upon species) when an inflection occurs.

B. Birth

Final preparation for birth involves increasing, and increasingly complex, fetal movements, which are assumed to be both calisthenic exercise, and a need to correctly align the body for birth.[21-23] This is followed by a brief quiescent period, allowing the storage of energy in the form of fat and glycogen. The fetus initiates birth, but most of the work is done by the dam, involving massive abdominal and uterine muscle contractions. Once born, the neonatal herbivore must quickly get to its feet (since, with few exceptions, herbivores nurse in a standing position), locate the mammary gland, and suck. These tasks must be performed promptly since the neonatal herbivore usually has a limited energy supply. In addition, early suckling maximizes dam-offspring bonding and the passage of immunoglobulins from the dam. The placenta, in a number of mammals, blocks the transfer of immunoglobulins before birth, and the permeability of the gut to immunoglobulins decreases rapidly postnatally (in bovines, it reaches a minimum as little as 12 hr after birth). From an evolutionary point of view, the risk involved in a system such as this is probably minimal since a wild herbivore which is not sound or active enough to obtain an early suck is probably doomed anyway.

Prenatally, the animal is very heavily insulated from the vagaries of the outside environment. Under all but the most extreme environmental insult, an animal will be born with its full complement of cells and at a stage of development appropriate for its species.

The ideal size for a newborn animal is as large as possible without causing dystocia (difficult birth). Large size has the double advantage of reducing the surface area to volume ratio, and of increasing the body's total store of nutrients. In practice, maternal size is the chief determinant of neonatal size; the relationships are discussed in detail in Chapter 10. The dam's contribution is in two parts: she shares with the sire an equal contribution to the genotype of the offspring; she also provides and delivers nutrients through the maternal part of the placenta, and provides the physical environment in which the fetus grows. The importance of this "maternal influence" was shown in the classic work of Walton and Hammond,[24] who reciprocally crossed small Shetland and large Shire breeds of horses. They found that although the foals of both crosses were genetically similar (half Shire, half Shetland), the offspring of the Shire dam was considerably larger than the offspring of the Shetland dam, and that this difference persisted to maturity. Maternal influence has also been studied by ovum transfers between large (Lincoln) and small (Welsh Mountain) sheep breeds.[25] The conclusion was that while the influence of the recipient ewe on the birth weight of the lamb was greater than that of the donor ewe, the lamb's own genotype had a still greater effect. In addition, while smaller ewes were able, to some extent, to restrict the size of their lambs, large ewes were not able to increase lamb size much above the size which would have been expected had the lamb been carried by its natural mother. Older females tend to have larger offspring, and the offspring of later parities tend to be heavier than earlier ones. However, in both these latter cases, the influence plateaus after two or three pregnancies and there may be a senescent decline in birth weights from old females. The sex of the offspring commonly influences birth weight, the male being slightly larger in most ungulates. The number of offspring in the litter will also affect birth weight. As the number increases, the total weight of offspring also increases, but the weights of the individuals decrease, and they are born increasingly prematurely. Most of the larger wild herbivores typically give birth to a single offspring, and multiple births can pose serious problems for survival.

The effect of maternal nutrition on birth weight is complex, and is further complicated by the fact that variations in birthweight do not necessarily indicate variations in birth size or development, since the most vulnerable aspect of prenatal development is energy storage. Under most circumstances, the dam will maintain a balance of metabolites in her placental blood supply, even if her own nutrition is in imbalance. Thus, maternal nutrition, particularly during the blastocyst and embryo stages, has little effect on the individual's growth and development. Towards the end of the fetal stage, when protein synthesis is very rapid, it is probably normal for the dam to be unable to fully supply fetal needs. In anticipation of this, the dam in some species uses the early part of pregnancy, when the demands of the conceptus are small, to build up her own protein and energy status, a process known as pregnancy anabolism. In the later stages of pregnancy, she enters a catabolic phase, particularly mobilizing nitrogen for use in fetal protein synthesis.[26] The energy stored during pregnancy anabolism appears not to be so readily available to the fetus; instead, it tends to be retained by the dam to supplement the energy cost of lactation.[27]

Clearly, although the greatest nutritional demands are placed on the dam late in pregnancy, body condition and feed availability early in pregnancy can also influence the birth weight by affecting the dam's body stores. The most easily influenced aspect of birth weight, however, is energy storage, particularly body fat, buffering the other body tissues and organs.

The detailed effects of undernutrition vary with species, but there is ample evidence in many wild herbivores of a significant relationship between nutrition during pregnancy and birth weight and subsequent survival.[28,29] There is also a large body of literature relating fertility to body composition in females during the breeding season. This relationship would serve to avoid the potentially disastrous consequences of a large proportion of females giving birth when feed is already scarce, and they themselves are in poor body condition. If the breeding season is preceded by good nutritional conditions, then the females are likely to have sufficient energy and protein stored to make the risk of pregnancy worthwhile. It then seems eminently sensible following conception to continue to store as much protein and energy as possible, in case the feed supplies diminish. Making the stored protein available for fetal growth late in pregnancy ensures a good chance of the offspring being born in a state of development appropriate to its species and hence able to function normally. Carrying this teleological argument one stage further, the observed reluctance of dams to deplete their energy stores late in pregnancy would also have an evolutionary advantage. If the dam chose to "mine" her own energy stores to ensure ample energy in the neonate, she would leave herself vulnerable by having insufficient reserves to face the huge energy drain of lactation should the feed supply be poor. Under these conditions, both she and her offspring could die. If, on the other hand, she retained the energy stores herself and the feed supply became critical, her offspring would die, but she would have an improved chance of surviving to breed again in a better season.

In the domestic pig, the inability of the fetus to store fat even when the dam is in good body condition and positive energy balance may constitute a serious problem to producers.

Thus, the major influence of maternal nutrition on the offspring appears to be in its effect on their immediate vitality and hence survival. If they get over this stage, there will be little residual effect on their growth and development, but in nutritionally difficult times neonatal deaths will spare the herd the difficulties of coping with a large increase in population.

C. Suckling Period

For the first weeks in most ungulate species, the milk supply of all but the most severely impaired dams will be more than adequate for the neonate's needs. Early in lactation, milk quantity and quality will be largely a reflection of the dam's genotype and her nutritional status during pregnancy (see Chapter 10). Hyperplasia continues in mammals after birth, but at a rapidly decreasing rate (it has been estimated in rats that the daily addition of new cells equals 60% of existing cell numbers 10 days before birth, 20% at birth, and 2% at puberty[30]). An abundant milk supply soon after birth is therefore necessary to ensure as little interference with hyperplasia as possible. Under conditions of abundant nutrition, the dam will match her milk supply to the demands of her offspring after this initial period. However, as the offspring grows, and particularly when feed is limited, her supply will fall short of its needs and she will increasingly reject its efforts to suck. The offspring, therefore, looks elsewhere, slowly becoming less dependent on the dam, until it is completely able to supply its own needs. The rapidity with which the young animal becomes herbivorous depends on the species and the influence of nutrition on the milk production; it is also strongly influenced, in red deer *(Cervus elaphus)* at least, by the pregnancy status of the dam. Clutton-Brock et al.[31] indicated that, in this species, pregnant dams often weaned their calves by the time they were 5 to 7 months old, before the end of November. However, nonpregnant dams commonly allowed their offspring to suck until the following September or October. The suckling period fulfills the dual role of providing the offspring with adequate feed in a readily digestible form when it is most needed and of keeping it near its dam who can protect it and teach it self-defense.

It is, however, mutually advantageous to break this nutritional dependence as early as reasonably possible. A dam who continues to feed her offspring will have little chance to replace her own body stores in preparation for mating and the possibility of an unfavorable season. She will also increase the interval between offspring which, unless this increases her life expectancy, is contrary to the evolutionary argument that reproductive strategies must maximize the number of offspring which survive to reproduce. The offspring which doesn't wean itself is also placed at a disadvantage if this is a sign that it is inadequately foraging for its own feed. When the milk supply dries up, it may starve. Extreme examples of this are sometimes seen in farm livestock which have had many of their basic survival mechanisms severely curtailed by domestication; it is not unusual in areas where weaning is not imposed on domestic cattle herds to see grown animals still sucking their mothers, sometimes while their own calf sucks them (Figure 4). Time spent trying to milk a dry dam is clearly not time spent grazing.

D. Puberty

Puberty is a series of events culminating in the ability of the individual to breed successfully. The mechanisms which trigger these events are not fully understood, and vary among species, as well as between the sexes within a species. Mammals do not wait until they are fully grown before becoming reproductively competent, and the main trigger to puberty appears to be the achievement of a minimum proportion of mature weight. Age, in most species, is less important. Puberty is associated with the inflection in the weight/age growth curve, and it can be shown that the surge of gonadal hormones (androgens and estrogens) at puberty leads ultimately to the cessation of bone growth. Bones do not, however, all stop growing at the same time, and the order in which they stop growing leads to the final refinement of adult shape.

Female puberty in wild herbivores has received considerably more research attention than male puberty because it has a far greater influence on population dynamics. In most wildlife species, young males reach puberty long before they are permitted to sire offspring. Females on the other hand are likely to become pregnant immediately following puberty, so that unlike their brothers, the time at which females reach puberty can have a profound influence on the number of offspring born to the herd. Puberty in females (often described for convenience rather than accuracy, as first estrus) is also easier to observe than puberty in males (e.g., first ejaculate containing live sperm).

Development of secondary sexual characteristics is part of the puberty process and begins before full reproductive competence is achieved. In males the complete development of these traits is not normally achieved until close to maximum size, placing them at a competitive disadvantage versus other males until they are mature. Females, particularly in herding species, have less need of secondary sexual characteristics to find a mate; they are likely to do so as soon as they begin to cycle and present themselves to a male.

Many species of herbivores have restricted breeding seasons, and the study of puberty, particularly of females, in these species is complicated by the fact that first estrus cannot

FIGURE 4. Failure to wean is detrimental to both dam and offspring. It is assumed to be heavily selected against in the wild. (Photographs from Yeates, N. T. M. and Schmidt, P. J., *Beef Cattle Production*, Butterworths, London, 1974. With permission.)

occur outside this season. The onset of estrous cycles at the beginning of the breeding season is a photoperiodic response in most cases,[32] though obviously not in tropical latitudes.

A number of factors other than weight, age, and photoperiod are involved in puberty, often in a permissive manner. Thus, if body fat content is below some threshold level, females of most species will delay puberty with its attendant "risk" of pregnancy.[33] The absence of males will also delay female puberty in a number of species, though excessive male attention before puberty will delay first estrus.[34] In some species, environmental stressors have been shown to initiate first estrus in peripubertal females. In humans, it has been shown that blind, deaf, and diabetic girls all tend to reach puberty earlier than "normal" girls.

Thus puberty, defined as first estrus (which, from a population dynamics viewpoint, is a useful definition), results from an interaction of many factors, all of which must reach some threshold level. The relative importance of these factors is a function of species, but weight and age, generally in that order, are the most important, provided that requirements of breeding season (a photoperiodic effect), body condition, and social environment, are satisfied.

E. Adulthood

Full mature size is not achieved in a single year by any of the larger mammalian herbivores, and some, such as the African elephant *(Loxodonta africana),* may take 20 or more years to reach it.[35] Most environments include peaks and troughs in the feed supply: the troughs may be annual such as winter in arctic and cool temperate regions or the dry season in areas of seasonal precipitation; they may also occur at irregular intervals such as the droughts, often exacerbated by bushfires, common in parts of Africa and Australia. The severity of feed shortage in these times also varies widely, and species which successfully colonize niches in these environments must be able to tolerate moderate to severe weight loss during postnatal growth without any major influence on their adult body size or shape. Very few wild herbivores grow to maturity along their sigmoid "growth potential" curves; instead, live weight follows a saw-tooth pattern over time (Figure 5), with the troughs corresponding in timing and severity to feed shortages. Wood et al.[36] have proposed a growth model consisting of four curves to describe this periodic growth to maturity: one to describe prepubertal growth; a second, the actual weight change in an annual cycle; and the third and fourth, the annual progression of maximum and minimum weights reached respectively. It is probable that all animals fail, to a greater or lesser extent, to reach their potential size, since the environment will always be something of a limiting factor. Human populations show secular trends in mature body size, and age at puberty which can probably be explained, at least in part, by environment.[37]

Season influences both the amount and the quality of feed available and these will have separate as well as combined influences on growth. Very few large herbivores have adapted to withstand severe water shortage, and none — not even the legendary camel — is able to store water (though the fat in its hump will yield more than its own weight of metabolic water during catabolism and it can tolerate an extraordinary level of tissue dehydration without concentrating its blood). Some mainly smaller mammals (desert rats, koalas) choose never to drink, and many others are able to survive in the complete absence of drinking water provided their feed is reasonably lush or dew is available. The study of water economy is well covered in Schmidt-Nielsen's classic text.[38] For the present discussion, it is sufficient to say that the absence of water is so serious that its influence on growth is dwarfed by its influence on survival.

Mineral and, in some instances, vitamin imbalances can also have serious and long-term effects on growth. A species will not successfully colonize an environment to which it cannot adapt, and there are many examples of habitats which depend entirely upon the presence of natural mineral licks. As long as the animals can continue to find and utilize the licks, they can continue to inhabit an otherwise hostile environment.

The most common short-term nutritional problems with which herbivores must cope are the supplies of protein and energy. These are normally limiting in the cold or dry season and abundant in the growing season. Arctic and temperate region herbivores generally adopt strategies which allow them to build up nutrient (particularly energy) stores in preparation

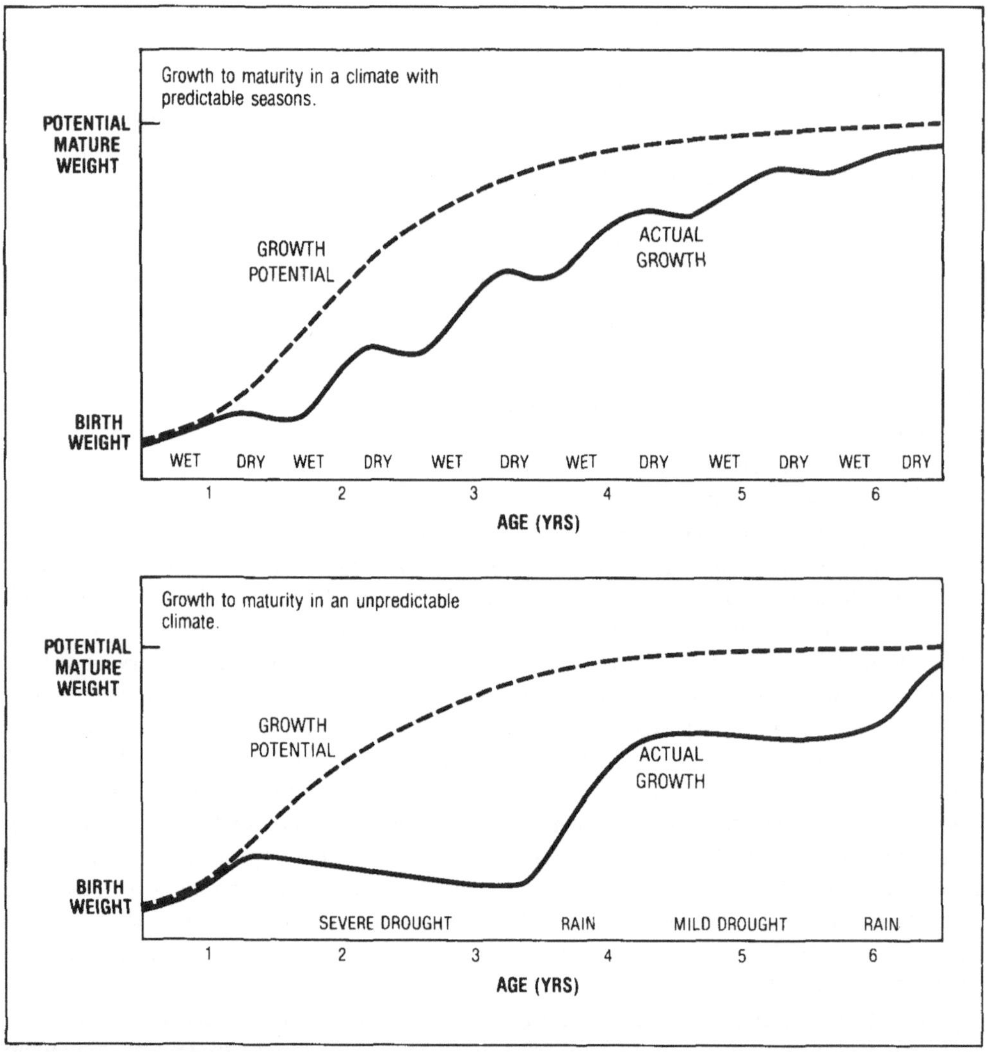

FIGURE 5. Wild animals rarely grow at their maximum genetic potential; instead, their growth occurs in periodic spurts corresponding to the availability of nutrition in their habitat.

for winter and to reduce energy expenditure during the winter. Body fat is a highly efficient form of energy storage. It is particularly energy dense, having over twice the dry matter energy content of other body components, and being stored in association with very little water. In addition, the synthesis of fat is a more energy efficient process than the synthesis of other body tissues. Many cold temperate and arctic animals therefore adjust their autumn metabolism to maximize the storage of fat in readiness for winter.[39]

A spring-summer increase in voluntary feed intake and an autumn increase in lipogenesis occur in a number of ungulates, regardless of body condition. Captive white-tailed deer have been shown to continue lipogenesis in the autumn, even when in negative energy balance,[40] though other workers using a more severe restriction in the same species also in the autumn have indicated a loss of fat during a period when the deer managed to maintain body weight and continue protein deposition.[41] Clearly the mechanisms involved are complex but a lipogenic period is normal in the autumn followed by a lipolytic period in the winter. During winter, there is a reduction in voluntary feed intake, even in captive animals amply provided with feed. Feeding times are concentrated in the warmest part of the day during the depths of winter. Metabolic rates are also lower in some species during the winter[42] and this may be accompanied by a reduction in body temperature. In most cases, these responses are photoperiodically induced and will occur regardless of body composition, nutrient availability, or ambient temperature. However, the magnitude of the response is influenced by those factors. The animal, therefore, will expect to lose weight during winter, but in spring will enter a catch-up growth phase which is particularly active for protein synthesis.

The prime function of fat in the animal body is to act as an energy reserve, but it also has an important secondary role as a thermal and mechanical insulator. To this end, the animal will tend to accumulate its energy stores in positions where they can best perform these secondary functions. In cold adapted animals, this will tend to be an evenly distributed subcutaneous layer to minimize heat loss. In some hot- and desert-adapted animals, fat is deposited in dorsal depots (e.g., the hump of camels and *Bos indicus* cattle, fat rumps and tails of some sheep, and the fat buttocks of the Hottentots and Kalahari bushmen of southern Africa). It has been argued[43] that the secondary function of these energy stores is to melt in direct solar radiation, thereby absorbing the latent heat of fusion of fat. When ambient temperatures drop, this heat is released again to the body as the fat solidifies, thus carrying some of the excessive heat of the day over into the cool of the night.

Energy conservation strategies are particularly important in females which must take the additional risk of pregnancy at some stage during the year. It is normally found that the fertility of the herd as a whole is closely correlated to the body condition of the females, which in turn is a function of the feed supply.[44]

In species occupying niches with definite reductions in winter feed supply, the normal reproductive strategy involves giving birth at, or shortly before, the beginning of spring in order to allow maximum time for dam and offspring to build up body stores for the following winter. Therefore, the breeding season, where it exists, usually precedes spring by the length of the gestation period.

Corresponding strategies are adopted by animals living in other climates involving an annual "difficult" season. In many environments, seasonal and non-seasonal breeders live side by side. In Zimbabwe, for example, zebra and buffalo breed year round while impala and kudu living in the same area have a closely defined breeding season.[45] Whether this indicates that the impala and kudu know something that the zebra and buffalo don't, or that the species are occupying markedly different niches in the same habitat, is not clear.

A further adaptation by female herbivores to periods of nutritional stress is their greater tendency than males to accumulate body fat. This is then catabolized for energy during weight loss. Females of reproductive age survive longer than males or juveniles of either sex under conditions of negative energy balance.[46]

Male herbivores, particularly those of species with pronounced breeding seasons are placed at considerable risk during the winter. They typically build up muscle and fat for the "rut" and may expend a great deal of their reserves during this period. Even in a long autumn, the male has a greater propensity than the female to put on muscle and a lower propensity to store fat which means that even though he may be heavier than the female, his gross energy content may be considerably less. Actively growing juveniles are at a similar disadvantage.

Adult size and shape is not totally controlled by genotype; adverse conditions can result in modifications to both. Bergmann's rule[47] states that warm blooded animals from colder climates tend to be larger, and hence by similitude, have a lower surface area to body weight ratio than animals of the same species from warmer climates. A refinement of this, known as Allen's rule,[47] states that there is a tendency for appendages such as ears and limbs to be smaller in colder climates. The extent to which these changes are a result of environment as opposed to genetic adaptations to local conditions is not clear, but there are some examples in the literature of these effects being induced in the laboratory.[48]

F. Senescence

In nature, few animals die of "old age", but the determinate growth pattern seen in all ungulates carries with it a process of incomplete repair of the accumulated depredations of living. As a juvenile animal becomes older, it is increasingly capable of resisting "decimating factors"[45] in its environment, until it reaches a peak of fitness in middle age. After that, it becomes decreasingly fit, a process referred to as senescence. A grazing animal which wears down or loses its incisor teeth has a much reduced chance of survival. Similarly, gradual erosion of articular cartilage will lead to impaired movement, as will a reduction in total muscle mass. Attrition of irreplaceable functional units in the internal organs (kidney nephrons, cardiac muscle fibers, pulmonary alveoli, intestinal villi, etc.) will also add to the burden of senescence, making the aging animal easy prey to predators or adverse conditions.

In summary, an ungulate will not be conceived unless its dam's body condition offers a reasonable chance of carrying it to term. From conception, the stages of growth through which it passes will ensure maximum protection from the environment while the animal is most vulnerable to permanent damage. Under all but the most adverse conditions, the animal will be born with all its tissues and organs (and hence growth potential) intact, and at the appropriate level of development for its species. The amount of energy which it has stored, however, will be subject to wide variation, depending upon its dam's state of nutrition. The period immediately following birth is a high risk period, and it is argued teleologically that this may represent a safety valve for the herd, causing offspring to die if conditions are inclement rather than adding to the population density, placing the whole herd at risk. If the neonate survives, the dam's milk supply will normally meet its immediate needs, during the period of rapidly decreasing hyperplasia. Once it has approximately doubled its birth weight, the amount of hyperplasia taking place is negligible and it is unlikely that any environmental insult short of causing death would cause permanent stunting. From then on, the animal will grow by hypertrophy in all but a few organs and tissues. During negative energy or protein balance, it will slow down or reverse growth but this, in itself, will not affect the animal's ultimate size or shape. Having reached adult weight, it will increase and decrease its body weight depending on environmental conditions. Ultimately, the animal will die and be replaced in the herd by a new model.

IV. BODY COMPOSITION

In order to understand the factors which influence total body composition and growth, it is necessary to look briefly at the growth of each of the major body components. Few organs consist of a single tissue, and most have aspects that arise from two or more embryological tissue-beds. As the organ grows, development occurs by allometric growth of the various parts. For a detailed discussion of individual organ growth, see Goss.[3]

A. Internal Organs

By similitude, animals eat less feed per unit of body weight as they grow larger. Since the weights of the organs of assimilation and digestion, including the liver, are closely related to the weight of feed which the animal eats, they tend to show a general negative allometry relative to body weight. However, since the actual weight of these organs, particularly the liver, is so strongly influenced by recent feed consumption, the relationship with body weight tends not to be a close one.[49,50]

Other internal organs have functions which are closely related to body size (heart, kidneys, lungs, spleen), and in these cases, allometric relationships will tend to hold not only in animals of different size within a species, but also in animals of different species. Thus, Prothero[51] has shown that in 104 mammalian species, covering the full range from shrew to blue whale, heart weight varies as the 0.98th power of body weight, with a very high coefficient of correlation (0.99). Kidney weight varies as the 0.67th power of body weight[3] (i.e., is related to the surface area of the body), and again this relationship tends to hold not only within but also among species.

Some organs, notably neural (eyes, brain, etc.) and to a lesser extent skeletal, have functions which are not closely related to body size or the nutritional environment. Their growth tends to be a function of time, though it would clearly be impossible for them to grow in size independent of the structures which contain and support them. Nevertheless, the sizes of these organs (eye lens, auditory ossicles, skull length, brain weight) have often been used as a measure of the age of the animal.[52] The characteristic eruption patterns of teeth can also be used as a measure of physiological age, and within a species can serve as a highly repeatable, easily observed indicator of chronological age,[53] though the extent to which it can be influenced by environment is not known.

B. Epidermis

The epidermis consists of a variety of structures whose major functions are to interact with the physical environment. Clearly, the weight of the skin will tend to be a function of the surface area of the body (i.e., approximately the two thirds power of body weight), but the relationship is imprecise. Skin can be used as a nitrogen and energy store, and so can be affected by short-term fluctuations in nutrition. It can also be affected by environmental conditions, particularly temperature, as can associated structures such as sweat and sebacious glands, pellage, and hooves. The mammary glands are epidermal structures, and their size will be greatly affected by sex and reproductive status.

The growth of antlers, ivories, and horn tissue is highly related to body size under normal habitat conditions[31,35,54-58] even though they represent tissues approximating bone, teeth, and nail, respectively. All these tissues make seasonal growth, with antlers being shed annually at some variable time following their social and behavioral role during rut.[31,56,59,60] An exception to this is the genus *Rangifer* (reindeer and caribou) in which females also grow antlers and the pregnant female retains her antlers until a few days after calving.[61,62] Ivories and horns grow seasonally but are not shed. Their roles in fighting and social status have been reviewed by Geist[63] and Guthrie and Petocz.[64] The size of ivories in elephants[35] and horns in sheep and muskoxen develop in relation to the general level of habitat quality.[55,65] Under conditions of extremely severe under-nutrition, including mineral imbalances, the antlers of deer may be poorly developed.

C. Mesodermal Tissues

The main bulk of the body is derived from the embryonic mesoderm. The growth of the various mesodermally derived tissues has been studied in great detail in farm livestock because of the importance of these tissues in meat production.[66-68] The weights of bone, muscle, and fat tend to be very closely related, allometrically, to the weight of the body as a whole and their growth and development is important in the nitrogen and energy budgets of the animal.

1. Cartilage and Bone

Cartilage is a connective tissue which is present extensively in the early fetus, and is largely replaced by bone as the animal grows. At maturity, cartilage is only found in places where its particular qualities of resilience and elasticity are required (external ear, respiratory passages, articular surfaces of bone, etc.). Cartilage is nonvascular and so is incapable of turnover, repair, or remodeling. As such, it cannot be considered an available nitrogen store. During normal growth, it is calcified in preparation for the invasion of blood vessels, then eroded, removed, and replaced by bone. The appearance and texture of the cartilage and the progress of ossification (the replacement of cartilage with bone as the animal matures) can be used as indicators of physiological age. They are, however, imprecisely related to chronological age.

Bone, the connective tissue which forms most of the skeleton of adult mammals, is the body's major store of several minerals. Despite its stable appearance, bone is in an active state of turnover throughout life. Under conditions of negative mineral balance, the body can demineralize the bone to some extent, so that body weight loss, in the extreme, can be accompanied by bone weight loss.[69] There is commonly a confusion in the literature between loss of bone mineral and loss of bone marrow which is mainly an energy store.

Studies of skeletal development in domestic ungulates[70,71] have indicated that within species, neither breed nor sex has a strong influence on the distribution of bones in the skeleton, and that the weights of individual bones retain a close allometric relationship with each other, and with the total weight of the skeleton. This allows the development of prediction equations to estimate skeletal weight from the weights of individual bones or limited numbers of bones.[72]

2. *Skeletal Muscle*

The number and location of the individual skeletal muscles is fixed within a species, and does not change as the animal grows. The attachment sites of muscles may, however, "creep" along the bone as they develop. Hyperplasia of muscle fibers ceases soon after birth (probably by the time the animal has approximately doubled its birth weight in most ungulate species) and all subsequent muscle growth is by hypertrophy. This minimizes the risk of permanent stunting as a result of unfavorable nutrition during growth.

The total mass of muscle in an animal is a function of its live weight, since this is the load which it has to support and move. Stewart,[73] using data from 13 species ranging in size from a mouse (21.5 g) to a whale (5.9×10^7 g), showed that the relationship between total muscle mass (y) and body mass (x) was $\log y = -0.199 + 0.947 \log x$. Within any individual species, however, the relationship will be influenced by such factors as genotype, sex, age, and nutritional status.

Within a species, the distribution of muscle, like that of bone, is very predictable, though unlike bone, it will be influenced somewhat by sex. For a detailed discussion of the factors influencing muscle growth, see Stewart[73] and Berg and Butterfield.[74]

The body's muscle mass constitutes an important store of both nitrogen and energy. When an animal loses weight, it will catabolize its muscle as well as its fat stores, even in the early stages of weight loss. Price[69] has suggested that this loss of muscle weight is a response to the reduced work load resulting from the weight loss. In cattle, the weight of muscle lost during a period of submaintenance nutrition is approximately equal to the weight of fat lost.[69,74] It is possible that if submaintenance nutrition were imposed in a period of active muscle development (e.g., in prerutting males), the impetus to gain or maintain muscle mass could negate the impetus to lose muscle as a consequence of weight loss. In this case, however, the greater activity and muscle tone of the animal would be offering the same stimulation to muscle hypertrophy as would an increase in live weight.

3. *Fat*

The major energy store of the body is fat. Although wild ungulates would rarely, if ever, have a fat mass equal to their muscle mass, the energy available from fat is far greater. Not only is fat very energy dense, it is also stored in association with very little water, and the animal can, under certain conditions, mobilize all but a tiny portion of its fat to maintain life. Fat is stored as triglyceride in adipose connective tissue cells called adipocytes. There are two types of adipocyte: brown and white. Considerably less is known about fat metabolism in brown adipose tissue (BAT) than in white (WAT).

BAT is found in a number of depots in neonates of many mammalian species. In most, it is rapidly replaced by WAT postnatally, but in rodents, hibernators, and possibly man, it persists in detectable amounts into adulthood. BAT cells differ histologically from WAT cells in having fat contained in a number of discrete droplets, and in their greater vascularity and content of cytochrome.[75] BAT is a highly metabolic organ whose major function is the local production of heat perhaps by oxidation without ATP synthesis through the proton conductance pathway,[76] a process known as nonshivering thermogenesis. The need to produce heat without wasteful muscular work is particularly acute in mammalian neonates and awakening hibernators. Newborn domestic piglets, which typically have little or no BAT, are particularly vulnerable to hypothermia. Little is known of the BAT content of neonatal or adult wild herbivores.

Recent work[77] has suggested that thermoregulation, and particularly thermogenesis in BAT, may play an important role in energy balance, at least in some species. The size of the BAT depot is known to increase or decrease in response to the need for thermogenesis (presumably by triglyceride accretion since differentiated adipocytes are mitotically inactive), and it is thought to play an important role in cold acclimation and even in "excreting" energy in cases of chronic overeating.[77] Much of the evidence for these functions comes from work with rodents, particularly the genetically obese strains. The application of this work to other species is a matter of controversy. It seems likely, however, that brown adipocytes, perhaps dispersed through WAT depots, are more common particularly in cold-adapted wildlife species than has been reported, and that the contribution of BAT to energy metabolism may be substantial.

White adipocytes differ from brown in their histology as well as their metabolism. The fat is coalesced into a single vacuole taking up most of the cell and pushing the cytoplasm and nucleus to the periphery, giving a typical cygnet ring appearance. Once an adipocyte has differentiated from the mesenchyme, it becomes incapable of de-differentiation or of mitosis. However, there is strong evidence that growing and even adult animals can recruit undifferentiated mesenchymal cells, which exist extensively throughout the connective tissue, to become adipocytes[78] though the ability to recruit in this way decreases markedly with age. The ability to store fat in adipocytes is prodigious (adipocytes can expand up to 10 diameters through fat accretion) and, combined with the capacity to "recruit" cells, this makes the animal's ability to store fat effectively limitless; certainly, gross interference with normal body functioning would occur long before the fat depots could be considered "full". Unlike the protein and mineral stores of the body, the fat store can be totally depleted, though in fact most animals will die before this can happen. Depleted adipocytes take on a wrinkled appearance under microscopy, and a characteristic "slimy" appearance macroscopically.

D. Estimation and Prediction of Body Composition

The assessment of energy reserves of animals provides a powerful tool for the study of bioenergetics and energy balance of animals.[79-83] However, the complete chemical analysis of the animal is costly and likely to leave little saleable material. It is also time consuming and generally impossible in field situations. Therefore, a large amount of research has centered on predicting directly the morphological and chemical composition from easily measurable predictors such as derivatives of body weight and carcass weight and measures (usually mass) of indicator bone, muscle, and fat tissues.[74] In addition, the repeated prediction of body composition of the same individual in vivo has been made possible using an array of techniques which estimate the whole body pool of water (or a mineral such as potassium) from which the body lean, fat, and energy contents can be calculated.[74,84,85] Under some circumstances, an index of fatness can be made from an ultrasonic estimate of back-fat thickness.[86]

1. Differentiation of Genetic and Environmental Influences

From the biological viewpoint, the morphological or chemical components of the gain can be partitioned into genetic and environmental components. This is necessary to make comparisons among populations in natural systems; i.e., are populations reaching their growth and condition potentials? To what extent do the observations reflect natural selection? Based on findings that the metatarsus is a bone on which environmental factors have a minor influence on linear growth,[87-89] Klein[90] suggested that the length ratio of femur/metatarsus would provide a sensitive index of the relative completion of growth. Therefore, these indexes have been compared with body composition estimates to differentiate environmental effects on cervid populations.[90-93] Other ratios of bones (e.g., two mandibular measurements)[94,95] and muscle groups could also be used to add to this genetic-environmental partition as the principle used is the comparison of the ratio of slow growth (femur) with fast growth (metatarsus) entities (a principle currently advocated by wildlife biologist Langvatn[92]). However, current views of growth based on tissue weights show that the rate of growth has little effect on the relative amount of muscle to bone or on the relative weights of individual muscles when compared at the same *total muscle weight* in cattle.[74] The explanation lies in the observation that when compared at the same *age*, the slow/fast growth ratios reveal differences in muscle and bone weight distribution. Therefore, the complete comparison is made by sampling at both equal body weight, to determine genetic effects, and equal age to emphasize environmental effects.

Under conditions of body weight loss, muscles most vital for survival undergo least weight loss[96] and there is a tendency towards a muscle weight distribution more typical of immature animals.[74] Therefore, conclusions on environmental-genetic differences based on weight ratios may not be valid following seasonal or chronic weight loss, and ratios based on bone lengths may be the only appropriate measures under this effect.

2. Body Composition

The estimation of body composition can be based on either the morphology of the animal (e.g., bone, muscle, fat, individual organs, etc.), commercial marketing cuts, or chemical components (e.g., energy, N, C, macro- and microminerals). Also, the carcass can be halved and one half analyzed morphologically and the other chemically.[97] Alternately, use of morphological analyses to predict the chemical composition is based on the chemical composition of the components,[84,98-101] but for many species, actual measurements are few and the biologist most usually resorts to analyses made on domestic species. A major limitation of this conversion is the fat content of muscle which may change markedly with general body condition in some species,[102] and this will greatly alter the energy content of muscle in such a conversion.

Following dissection, weights of tissues are usually expressed as functions of the body weight or carcass weight[101,103-106] for the interpretation of growth and fattening priorities.[87-89] Such ratios change with age and, although they give insight into growth priorities, they have limited application for the prediction of body composition for bioenergetic and nutritional balance studies. In recent years, it has been found expedient to regress body constituents, morphological or chemical, on a convenient independent variable (e.g., carcass weight, empty body weight, or live weight).[74,97,107-111] In many species, carcass weight is very highly correlated with live weight.[39,101,104,105,112,113] However, the measurement of body weight is extremely difficult for large animals and a correlate of body weight, e.g., heart girth, shoulder height, length, or leg length,[35,114,115] may be used. However, the error of prediction is very high and weights of body components, e.g., hind leg weight, may be a better estimator of body weight.[116]

3. Prediction of Body Composition

Extensive detailed morphological and chemical analyses of the body composition of animals indicate that age is not a good predictor[101,117] but that body weight and indicator muscles/bones and fat reserves (Figure 6) are variably good to excellent predictors of body composition.[58,72,84,101,102,107,118-120] This is a considerable advance over index of fatness (back-fat and kidney fat index, abdominal fat, and femur marrow fat, reviewed by Riney[121]). In a number of species, some of these indexes are variably correlated, either separately or as multiple variables, with total dissectable or chemical fat for growing animals or older animals in a general positive energy balance.[122-125] Bone marrow fat[122] is correlated with the fat reserves of animals during chronic under-nutrition but is not related to fat levels during growth and fattening. However, the curvilinear relation of the kidney fat index and fat content in deer[124,126] limits its application as Robbins[58] emphasizes; the index must have a linear or only a slightly curvilinear relation to the independent variable to be suitable for predictive purposes. Prediction equations are estimated over a wide range of values whereas biologists frequently must make predictions over a narrow weight change. Under these conditions, the correlation coefficient may be a poor indicator of the accuracy of the estimates and Norton[127] suggests that the individual variable between the predicted and observed value, the standard error of the estimate, should be used.[74] Berg and Butterfield,[74] in reviewing these techniques, concluded that the individual muscle and bone weights and backfat thickness give the best prediction of morphological composition, particularly when combined with the specific gravity of the carcass.

Other predictors of the morphological or chemical composition include those related to the prediction of fat, from specific gravity of carcasses,[128] or from the total body water space as estimated with isotopes (see review of accuracy by Sheng and Huggins[129]). Both techniques rely on treating the carcass, or the animal, as a simple two-compartment system, fat and fat-free, and that the fat-free compartment contains a constant water content.[130] This two-compartment model works reasonably well for carcass analyses. However, for the in vivo estimation, alimentary water constitutes an important source of error.[74,84,131-134] Therefore, in experiments where body composition, particularly the energy and protein reserves, are to be estimated sequentially, prediction equations relating the isotopically estimated water pool size (using tritiated water or deuterium oxide) is regressed against the body constituents.[97,110,133-137] Such relations have not been developed for wildlife species. Currently, an index of fatness is calculated by applying the Pace and Rathbun[130] formula to the water space, estimated with tritiated water or deuterium oxide, following correction for water loss before mixing, and alimentary water.[138,139] This isotopic space variably overestimates tissue water volume[129,132,138] and mean correction factors have been derived to correct the space estimate. The final complete body compositional analyses involves estimating the fat-free dry matter (100 − (% fat + % water)) and computing protein and ash-free dry matter from the ratio 80.3:19.7.[84] Clearly, this series of calculations carries with it error terms and biases which are of an unknown magnitude. More research is needed before a generalized calculation protocol (including the error terms) can be recommended. A final potential limitation to applying the Pace and Rathbun equation is that the water content of body lean is not constant at 73.2% but varies with age and body condition including prolonged under-nutrition.[58,134] Sheng and Huggins,[129] in reviewing the errors in fat estimation for a range from 70 to 76% (means for many species), found it to be significant; however, the largest error lies in the isotopic estimation of total body water.

Of prime importance to biologists and to the formulation of simulation models is the estimation of the energy content and composition of normal gain[58,140] and that gain during recovery following body weight loss. Few studies have investigated this phenomenon in wildlife species. Berg and Butterfield[74] concluded that the pattern of fattening reflects the normal pattern at the reinitiation phase and depends on the current level of fatness rather than being dependent on age, muscle weight, or sex. Therefore, the energy content of gain could be calculated from fattening curves appropriate to the level of fatness and from the

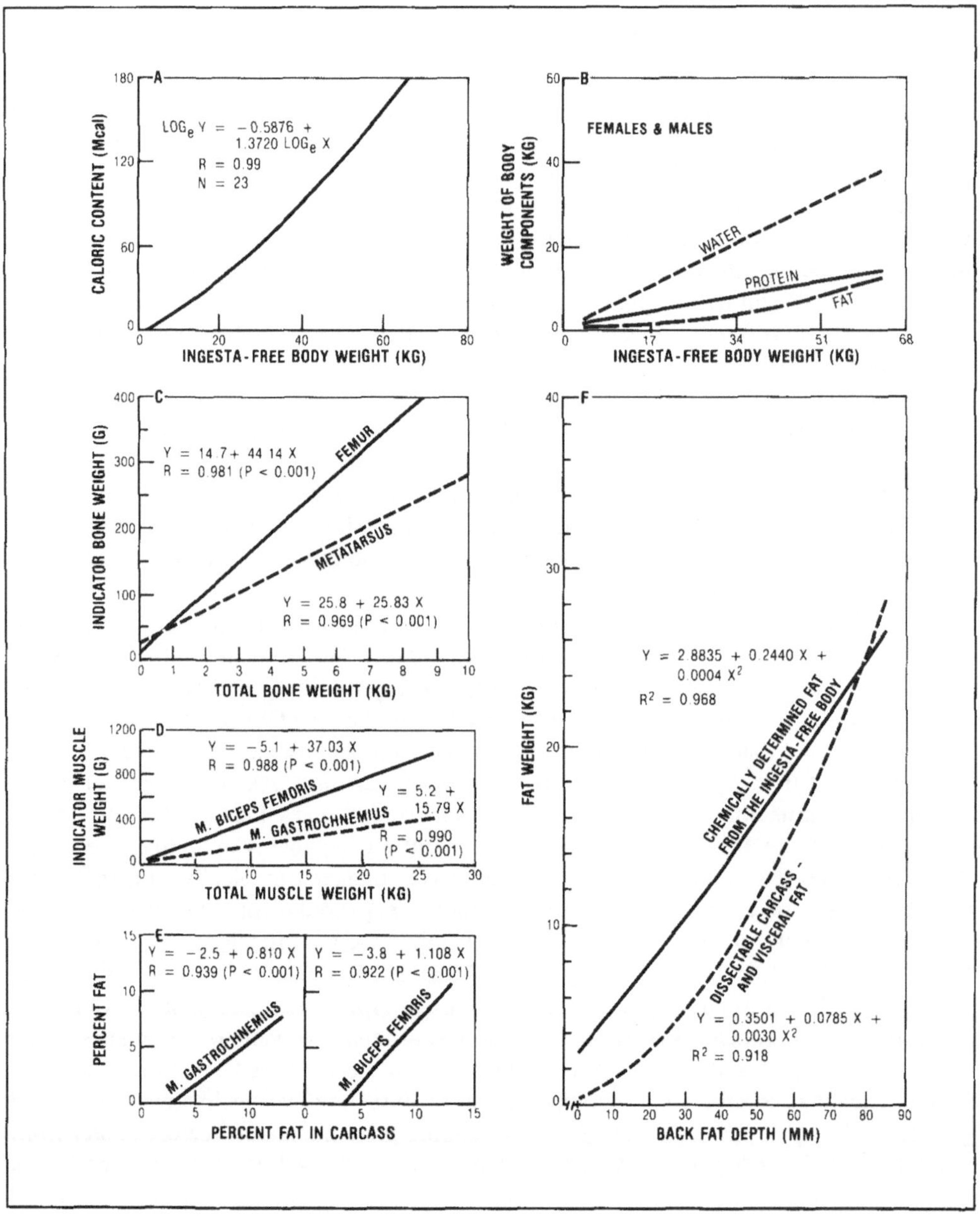

FIGURE 6. Some published relations for the prediction of body components from ingesta-free body weight in white-tailed deer (A,B) and indexes of bone (C), muscle (D), and fat in domestic and Svalbard reindeer (E,F). (Relations (A) and (B) redrawn from Robbins, C. T., Moen, A. N., and Reid, J. T., *J. Anim. Sci.*, 38, 1974; (C), (D), and (E) redrawn from Ringberg, T. M., White, R. G., Holleman, D. F., and Luick, J. R., *Can. J. Zool.*, 59, 1981; and (F) redrawn from Reimers, E. and Ringberg, T., *Acta Zool. Fenn.*, 175, 1983.)

energy content of protein (22.65 KJ/g) and fat (39.71 KJ/g).[84,98,101] In white-tailed deer, the energy content of normal gain increased from 4.8 to 14.7 KJ/g for a body weight range of 3 to 60 kg. Robbins[58] listed regression equations for a range of bird and mammalian species that can be used to approximate the energy and protein requirements for body gain. During prolonged under-nutrition, the relation between tissue weight and empty body or carcass weight is slightly below the gain relations except for fat which is lost much faster than it is gained.[141] The importance of each of the fat reserves is species dependent as the bulk of reserves can be variably distributed (some species showing a widely different propensity to deposit fat subcutaneously, viscerally, or intramuscularly). Much of this deposition is dictated by selection for agility and shape associated with reproductive behavior and strategies for predator avoidance[74,142] rather than physics of fat mobilization.

REFERENCES

1. **Green, W. W.,** Evaluation of the accuracy of instruments and data when measuring beef cattle, University of Maryland Agric. Exp. Stn. Bull. A186, College Park, Md., 1976.
2. **Williams, J. P. G., Tanner, J. M., and Hughes, P. C. R.,** Catch-up growth in male rats after growth retardation during the suckling period, *Pediatr. Res.*, 8, 149, 1974.
3. **Goss, R. J.,** *The Physiology of Growth,* Academic Press, New York, 1978.
4. **Leeson, T. S. and Leeson, C. R.,** *Histology,* 4th ed., W. B. Saunders, Philadelphia, 1981.
5. **Ingle, D. J. and Baker, B. L.,** Histology and regenerative capacity of liver following multiple partial hepatectomies, *Proc. Soc. Exp. Biol. Med.*, 95, 813, 1957.
6. **Brody, S.,** *Bioenergetics and Growth,* Reinhold, New York, 1945.
7. **Pearl, R. and Reed, L. J.,** On the rate of growth of the population of the United States since 1790 and its mathematical representation, *Proc. Natl. Acad. Sci. U.S.A.*, 6, 275, 1920.
8. **Stephenson, S. K.,** Growth measurements and the biological interpretation of mammalian growth, *Nature (London),* 196, 1070, 1962.
9. **Goonewardene, L. A., Berg, R. T., and Hardin, R. T.,** A growth study of beef cattle, *Can. J. Anim. Sci.*, 61, 1041, 1981.
10. **Brody, S.,** Growth and development with special reference to domestic animals. III. Growth rates, their evaluation and significance, *Res. Bull. Mo. Agric. Exp. Stn.*, 97, 1927.
11. **Short, R. V.,** The hormonal control of growth at puberty, in *Growth in Animals,* Lawrence, T. L. J., Ed., Butterworths, London, 1980, 25.
12. **Finney, D. J.,** Growth curves: their nature, uses and estimation, in *Patterns of Growth and Development in Cattle,* DeBoer, H. and Martin, J., Eds., Martinus Nijhoff, The Hague, 1978, 658.
13. **Pàlsson, H.,** Conformation and body composition, in *Progress in the Physiology of Farm Animals,* Hammond, J., Ed., Butterworths, London, 1955, 430.
14. **Huxley, J. S.,** *Problems of Relative Growth,* Methuen, London, 1932.
15. **Goss, R. J.,** *Adaptive Growth,* Academic Press, New York, 1964.
16. **Thompson, D'A. W.,** *On Growth and Form,* Cambridge University Press, U.K., 1961.
17. **Taylor, St.C. S.,** A relation between mature weight and time taken to mature in mammals, *Anim. Prod.*, 1, 203, 1975.
18. **Burton, M. and Burton, R.,** *The International Wildlife Encyclopedia,* B.P.C., London, 1969.
19. **Binns, W., James, L. F., Keeler, R. F., and Van Kampen, K. R.,** Developmental abnormalities, in *Animal Growth and Nutrition,* Hafez, E. S. E. and Dyer, I. A., Eds., Lea & Febiger, Philadelphia, 1969, 156.
20. **Edey, T. N.,** Nutritional stress and preimplantation mortality in Merino sheep (1964—67). General discussion and conclusions, *J. Agric. Sci.* (Cambridge), 74, 199, 1970.
21. **Fraser, A. F., Hastie, H., Callicott, R. B., and Brownlie, S.,** An exploratory ultrasonic study on quantitative foetal kinesis in the horse, *Appl. Anim. Ethol.*, 1, 395, 1975.
22. **Fraser, A. F. and Terhune, M.,** Radiographic studies of postural behaviour in the sheep fetus. I. Simple fetal movements, *Appl. Anim. Ethol.*, 3, 221, 1977.
23. **Fraser, A. F. and Terhune, M.,** Radiographic studies of postural behaviour in the sheep fetus. I. Complex fetal movements, *Appl. Anim. Ethol.*, 3, 235, 1977.
24. **Walton, A. and Hammond, J.,** The maternal effect on growth and conformation in Shire horse — Shetland pony crosses, *Proc. R. Soc. London B,* 125, 311, 1938.
25. **Dickinson, A. G., Hancock, J. L., Hovell, G. J. R., Taylor, St. C. S., and Wiener, G.,** The size of lambs at birth — a study involving egg transfer, *Anim. Prod.*, 4, 64, 1962.
26. **Naismith, D. J. and Morgan, B. L. G.,** The biphasic nature of protein metabolism during pregnancy in the rat, *Br. J. Nutr.*, 36, 563, 1976.
27. **Naismith, D. J., Richardson, D. P., and Pritchard, A. E.,** The utilisation of protein and energy during lactation in the rat, with particular regard to the use of fat accumulated during pregnancy, *Br. J. Nutr.*, 48, 433, 1982.

28. **Thorne, E. T., Dean, R. E., and Hepworth, W. G.,** Nutrition during gestation in relation to successful reproduction in elk, *J. Wildl. Manage.*, 40, 330, 1976.
29. **Blaxter, K. L. and Hamilton, W. J.,** Reproduction in farmed red deer. II. Calf growth and mortality, *J. Agric. Sci.* (Cambridge), 95, 275, 1980.
30. **Cheek, D.,** *Human Growth*, Lea & Febiger, Philadelphia, 1968.
31. **Clutton-Brock, T. H., Guinness, F. E., and Albon, S. D.,** *Red Deer: Behavior and Ecology of Two Sexes*, University of Chicago Press, Chicago, 1982.
32. **Yeates, N. T. M.,** The breeding season of the sheep with particular reference to its modification by artificial means using light, *J. Agric. Sci.*, 39, 1, 1949.
33. **Thomas, D. C.,** The relationship between fertility and fat reserves of Peary caribou, *Can. J. Zool.*, 60, 597, 1982.
34. **Brooks, P. H. and Cole, D. J. A.,** The effect of the presence of a boar on the attainment of puberty in gilts, *J. Reprod. Fert.*, 23, 435, 1970.
35. **Laws, R. M., Parker, I. S. C., and Johnstone, R. C. B.,** *Elephants and Their Habitats*, Claredon Press, Oxford, 1975.
36. **Wood, A. J., Cowan, I. McT., and Nordan, H. C.,** Periodicity of growth in ungulates as shown by deer of the genus *Odocoileus*, *Can. J. Zool.*, 40, 593, 1962.
37. **Damon, A.,** Larger body size and earlier menarche: the end may be in sight, *Soc. Biol.*, 21, 8, 1974.
38. **Schmidt-Nielsen, K.,** *Desert Animals. Physiological Problems of Heat and Water*, Oxford University Press, London, 1964.
39. **Reimers, E. and Ringberg, T.,** Body composition of Svalbard reindeer, *Can. J. Zool.*, 60, 1812, 1982.
40. **Verme, L. J. and Ozoga, J. J.,** Effects of diet on growth and lipogenesis in deer fawns, *J. Wildl. Manage.*, 44, 315, 1980.
41. **Holter, J. B. and Hayes, H. H.,** Growth in white-tailed deer fawns fed varying energy and constant protein, *J. Wildl. Manage.*, 41, 506, 1977.
42. **Ryg, M. and Jacobsen, E.,** Seasonal changes in growth rate, feed intake, growth hormone, and thyroid hormones in young male reindeer *(Rangifer tarandus tarandus)*, *Can. J. Zool.*, 60, 15, 1982.
43. **Yeates, N. T. M., Edey, T. N., and Hill, M. K.,** *Animal Science: Reproduction, Climate, Meat, Wool*, Pergamon, Sydney, 1975.
44. **Reimers, E.,** Reproduction in wild reindeer in Norway, *Can. J. Zool.*, 61, 211, 1983.
45. **Dasmann, R. F.,** *Wildlife Biology*, 2nd ed., Wiley-Interscience, New York, 1981, 86.
46. **Barrett, M. W.,** Distribution, behaviour and mortality of pronghorns during a severe winter in Alberta, *J. Wildl. Manage.*, 46, 991, 1982.
47. **Allee, W. C., Emerson, A. E., Park, O., Park, T., and Schmidt, K. P.,** *Principles of Animal Ecology*, W. B. Saunders, Philadelphia, 1949.
48. **McBride, G. M. and Christopherson, R. J.,** Effects of cold exposure on young growing lambs, *Can. J. Anim. Sci.*, 64, 403, 1984.
49. **Leche, T. F.,** Proportions of carcass and offal components in Jersey and Friesian bulls in relation to plane of nutrition, *Aust. J. Agric. Res.*, 24, 623, 1973.
50. **Murray, D. M. and Slezacek, O.,** Growth rate effects on some offal components of sheep, *J. Agric. Sci.* (Cambridge), 95, 241, 1980.
51. **Prothero, J. P.,** Heart weight as a function of body weight in mammals, *Growth*, 43, 139, 1979.
52. **Larson, J. S. and Taber, R. D.,** Criteria of sex and age, in *Wildlife Management Techniques Manual*, 4th ed., Schemnitz, S. D. and Toschik, I., Eds., The Wildlife Society, Washington, D.C., 1980, chap. 11.
53. **Graham, W. C. and Price, M. A.,** Dentition as a measure of physiological age in cows of different breed types, *Can. J. Anim. Sci.*, 62, 745, 1982.
54. **Huxley, J. S.,** The relative size of antlers in deer, *Proc. Zool. Soc. London*, 819, 1931.
55. **Geist, V.,** *Mountain Sheep: A Study in Behavior and Evolution*, University of Chicago Press, Chicago, 1971.
56. **Chapman, D. I.,** Antlers — bones of contention, *Mammal Rev.*, 5, 121, 1975.
57. **Hyvarinen, H., Kay, R. N. B., and Hamilton, W. J.,** Variation in the weight, specific gravity and composition of the antlers of red deer *(Cervus elaphus* L.), *Br. J. Nutr.*, 38, 301, 1977.
58. **Robbins, C. T.,** *Wildlife Feeding and Nutrition*, Academic Press, New York, 1983, 343.
59. **Goss, R. J.,** The deciduous nature of deer antlers, in *Mechanisms of Hard Tissue Destruction*, Publ. No. 75, Am. Assoc. Adv. Sci., Washington, D.C., 1963.
60. **Geist, V.,** The evolution of horn-like organs, *Behavior*, 27, 175, 1966.
61. **Espmark, Y.,** Studies in dominance-subordination relationships in a group of semi-domestic reindeer *(Rangifer tarandus L.)*, *Anim. Behav.*, 12, 420, 1964.
62. **Henshaw, J.,** Antlers — The bones of contention, *Nature (London)*, 224, 1036, 1969.
63. **Geist, V.,** Horn-like structures as rank symbols, guards and weapons, *Nature (London)*, 220, 813, 1968.
64. **Guthrie, R. D. and Petocz, R. G.,** Weapon automimicry among mammals, *Am. Nat.*, 104, 585, 1970.

65. **Ullrey, D. E.,** Nutrition and antler development in white-tailed deer, in *Antler Development in Cervidae*, Brown, R. D., Ed., Caesar Kleberg Wildlife Research Institute, Kingsville, Texas, 1982, 49.
66. **Hafez, E. S. E. and Dyer, I. A.,** *Animal Growth and Nutrition*, Lea & Febiger, Philadelphia, 1969.
67. **Lawrence, T. L. J.,** *Growth in Animals*, Butterworths, London, 1980.
68. **Lister, D., Rhodes, D. N., Fowler, V. R., and Fuller, M. F.,** *Meat Animals: Growth and Productivity*, Plenum Press, New York, 1976.
69. **Price, M. A.,** The effects of severe feed restriction on bulls and steers. II. Carcass composition, *Aust. J. Agric. Res.*, 28, 529, 1977.
70. **Richmond, R. J., Jones, S. D. M., Price, M. A., and Berg, R. T.,** Effects of breed and sex on the relative growth and distribution of bone in pigs, *Can. J. Anim. Sci.*, 59, 471, 1979.
71. **Jones, S. D. M., Price, M. A., and Berg, R. T.,** Effects of breed and sex on the relative growth and distribution of bone in cattle, *Can. J. Anim. Sci.*, 58, 157, 1978.
72. **Price, M. A. and Berg, R. T.,** The prediction of bone weight in beef carcasses by partial dissection, *Can. J. Anim. Sci.*, 57 (Abstr.), 816, 1977.
73. **Stewart, D. M.,** The role of tension in muscle growth, in *Regulation of Organ and Tissue Growth*, Goss, R. J., Ed., Academic Press, New York, 1972, 77.
74. **Berg, R. T. and Butterfield, R. M.,** *New Concepts of Cattle Growth*, Sydney University Press, Sydney, 1976, 240.
75. **Leat, W. M. F. and Cox, R. W.,** Fundamental aspects of adipose tissue growth, in *Growth in Animals*, Lawrence, T. L. J., Ed., Butterworths, London, 1980, 137.
76. **Nicholls, D. G.,** Brown adipose tissue mitochondria, *Biochem. Biophys. Acta*, 549, 1, 1979.
77. **Trayhurn, P., Goodbody, A. E., and James, W. P. T.,** A role for the brown adipose tissue in the genesis of obesity? Studies on experimental animals, *Proc. Nutr. Soc.*, 41, 127, 1982.
78. **Enser, M. and Wood, J. D.,** The development of adipose tissue in cattle, in *Patterns of Growth and Development in Cattle*, De Boer, H. and Martin, J., Eds., Martinus Nijhoff, The Hague, 1978, 243.
79. **Blaxter, K. L.,** *The Energy Metabolism of Ruminants*, Charles C Thomas, Springfield, Ill., 1962.
80. **Searle, T. W. and Graham, N. McC.,** Comparisons of body composition and energy utilization between Merino and fixed halfbreed (Border Leicester × Merino) wethers, *Aust. J. Agric. Res.*, 23, 339, 1972.
81. **Graham, N. McC. and Searle, T. W.,** Balances of energy and matter in growing sheep at several ages, body weights and planes of nutrition, *Aust. J. Agric. Res.*, 23, 97, 1972.
82. **Graham, N. McC. and Searle, T. W.,** Growth in sheep. II. Efficiency of energy and nitrogen utilization from birth to 2 years, *J. Agric. Sci.*, 79, 383, 1972.
83. **ARC (Agricultural Research Council),** The nutrient requirements of ruminant livestock, *Commonw. Agric. Bur.*, 351, 1980.
84. **Reid, J. T., Bensadoun, A., Bull, L. S., Burton, J. H., Gleeson, P. A., Han, I. K., Joo, Y. D., Johnson, D. E., McManus, W. R., Paladines, O. L., Stroud, J. W., Tyrell, H. F., Van Niekerk, B. D. H., and Wellington, G. W.,** Some peculiarities in the body composition of animals, in *Body Composition in Animals and Man*, National Academy of Science, Washington, D.C., 1968.
85. **Bergner, P.-E. E. and Lushbaugh, C. C.,** *Compartments Pools, and Spaces in Medical Physiology*, U.S. Atomic Energy Commission, Div. Technical Information, Springfield, 1967, 15.
86. **Temple, R. S., Stonaker, H. H., Howry, D., Posakony, G., and Hazaleus, M. H.,** *Proc. West. Soc. Am. Soc. Anim. Prod.*, 7, 477, 1956.
87. **Hammond, J.,** Some factors affecting the quality and composition of meat, *Chem. Ind. Rev.*, 59, 521, 1940.
88. **McMeekan, C. P.,** Growth and development in the pig, with special reference to carcass quality characters, Part I, II, III, *J. Agric. Sci.*, 30, pp. 276, 387, and 511, 1940.
89. **Pàlsson, H. and Verges, J. B.,** Effects of the plane of nutrition on growth and the development of carcass quality in lambs, Part I, II, *J. Agric. Sci.*, 42, pp. 1 and 93, 1952.
90. **Klein, D. R.,** Range-related differences in growth of deer reflected in skeletal ratios, *J. Mammal.*, 45, 266, 1964.
91. **Klein, D. R.,** Ecology of deer range in Alaska, *Ecol. Monogr.*, 35, 259, 1965.
92. **Langvatn, R., Ed.,** *Criteria of Physical Condition, Growth and Development in Cervidae-Suitable for Routine Studies*, Nordic Council for Wildlife Research, Stockholm, 1977, 27.
93. **Reimers, E., Klein, D. R., and Sørumgard, R.,** Growth in relation to habitat quality and time of calving in Norwegian wild reindeer, *Arc. Alp. Res.*, 15, 107, 1983.
94. **Lowe, V. P. W.,** Variation in mandible length and body weight of red deer *(Cervus elaphus)*, *J. Zool. London*, 166, 303, 1972.
95. **Reimers, E.,** Growth in domestic and wild reindeer in Norway, *J. Wildl. Manage.*, 36, 612, 1972.
96. **Butterfield, R. M.,** The relationship of carcass measurement and dissection data to beef carcass composition, *Res. Vet. Sci.*, 6, 24, 1965.
97. **Searle, T. W. and Hilmi, M.,** *In vivo* prediction with tritiated water of chemical and dissectible components of the dressed carcass of sheep growing at different rates, *Aust. J. Agric. Res.*, 28, 963, 1977.

98. **Blaxter, K. L. and Rook, J. A. F.,** The heat of combustion of the tissues of cattle in relation to their chemical composition, *Br. J. Nutr.*, 7, 83, 1953.
99. **Paladines, D. L., Reid, J. T., Bensadoun, A., and Van Niekerk, H.,** Heat of combustion values of the protein and fat in the body and wool of sheep, *J. Nutr.*, 82, 145, 1964.
100. **Jagusch, K. T., Norton, B. W., and Walker, D. M.,** Body composition studied with the milk fed lamb. I. Chemical composition and caloric content of the body and organs of newly born lambs, *J. Agric. Sci.* (Cambridge), 75, 273, 1970.
101. **Robbins, C. T., Moen, A. N., and Reid, J. T.,** Body composition of white-tailed deer, *J. Anim. Sci.*, 38, 871, 1974.
102. **Ringberg, T. M., White, R. G., Holleman, D. F., and Luick, J. R.,** Prediction of carcass composition in reindeer *(Rangifer tarandus tarandus* L.) by use of selected indicator bones and muscles, *Can. J. Zool.*, 59, 583, 1981.
103. **Ledger, H. P. and Smith, N. W.,** The carcass and body composition of the Uganda Kob, *J. Wildl. Manage.*, 28, 827, 1964.
104. **Ledger, H. P.,** Body composition as a basis for comparative study of some East African mammals, in *Comparative Nutrition of Wild Animals*, Vol. 21, Crawford, M. A., Ed., Symp. Zool. Soc. London, Academic Press, London, 1968, 289.
105. **Weiner, J.,** Dressing percentage, gross body composition and calorie value of the roe-deer, *Acta Theriol.*, 18, 209, 1973.
106. **Kay, R. N. B., Sharman, G. A. M., Hamilton, W. J., Goodall, E. D., Pennie, K., and Coutts, A. G. P.,** Carcass characteristics of young red deer farmed on hill pasture, *J. Agric. Sci.* (Cambridge), 96, 79, 1981.
107. **Tulloh, N. M.,** Relation between carcass composition and live weight of sheep, *Nature (London)*, 197, 809, 1963.
108. **Butterfield, R. M.,** Prediction of muscle content of steer carcasses, *Nature (London)*, 195, 193, 1962.
109. **Butterfield, R. M.,** Practical implications of anatomical research in beef cattle, *Proc. N. Z. Soc. Anim. Prod.*, 25, 152, 1965.
110. **Searle, T. W.,** Body composition in lambs and young sheep and its prediction *in vivo* from tritiated water space and body weight, *J. Agric. Sci.* (Cambridge), 74, 357, 1970.
111. **Searle, T. W., Graham, N. McC., and Smith, E.,** Studies of weaned lambs before, during and after a period of weight loss. II. Body composition, *Aust. J. Agric. Res.*, 30, 513, 1979.
112. **Ledger, H. P.,** Animal husbandry research and wildlife in East Africa: the description of a standardized technique for the comparison of the body and carcase composition of East Africa's wild and domestic ruminants with observation on its use, *East Afr. Wildl. J.*, 1. 17, 1963.
113. **Mitchell, G. J.,** Measurements, weights and carcass yields of pronghorns in Alberta, *J. Wildl. Manage.*, 35, 76, 1971.
114. **Bandy, B. J., Cowan, I. McT., Kitts, W. D., and Wood, A. J.,** A method for the assessment of the nutritional status of wild ungulates, *Can. J. Zool.*, 34, 48, 1956.
115. **Kirkpatrick, R. L.,** Physiological indices in wildlife management, in *Wildlife Management Techniques Manual*, 4th ed., Schemnitz, S. D. and Toshchik, L., Eds., The Wildlife Society, Washington, D.C., 1980, 99.
116. **Smith, N. S. and Ledger, H. P.,** A method of predicting liveweight from dissected leg weight, *J. Wildl. Manage.*, 29, 504, 1965.
117. **Anderson, A. E., Medin, D. E., and Bowden, D. C.,** Growth and morphometry of the carcass, selected bones, organs, and glands of mule deer, *Wildl. Monogr.*, 39, 122, 1974.
118. **Price, M. A. and Berg, R. T.,** Predicting side muscle weight in beef carcasses, *J. Anim. Sci.*, 43(Abstr.), 245, 1976.
119. **Ringberg, T. N., White, R. G., Holleman, D. F., and Luick, J. R.,** Lean body growth and carcass composition in semidomesticated reindeer *(Rangifer tarandus tarandus)* from birth to maturity, *Can. J. Zool.*, 59, 1040, 1981.
120. **Reimers, E., Ringberg, T. M., and Sørumgard, R.,** Body composition of Svalbard reindeer, *Can. J. Zool.*, 60, 1812, 1982.
121. **Riney, T.,** Evaluating condition of free ranging red deer *(Cervus elaphus)* with special reference to New Zealand, *N. Z. J. Sci. Technol.*, 36(B), 429, 1955.
122. **Ransom, A. B.,** Kidney and marrow fat as indicators of white-tailed deer condition, *J. Wildl. Manage.*, 29, 397, 1965.
123. **Dauphine, T. C.,** Physical variables as an index to condition in barren-ground caribou, *Trans. North West Section of the Wildl. Soc. (U.S.A.-Canada)*, 28, 91, 1971.
124. **Finger, S. E., Brisbin, I. L. Jr., Smith, M. H., and Urbston, D. F.,** Kidney fat as a predictor of body composition in white-tailed deer, *J. Wildl. Manage.*, 45, 964, 1981.
125. **Reimers, E. and Ringberg, T.,** Seasonal changes in body weights of Svalbard reindeer from birth to maturity, *Acta. Zool. Fenn.*, 175, 69, 1983.

126. **Bamford, J.**, Estimating fat reserves in the brush-tailed possum, *Trichosurus vulpecula* Kerr (Marsupialia:Phalangeridae), *Aust. J. Zool.*, 18, 415, 1970.
127. **Norton, H. W.**, Opportunities and pitfalls in the mathematics of body composition studies, in *Body Composition in Animals and Man*, National Academy of Science, Washington, D.C., 1968.
128. **Garrett, W. N.**, Experiences in the use of body density as an estimator of body composition of animals, in *Body Composition in Animals and Man*, National Academy of Science, Washington, D.C., 1968.
129. **Sheng, H.-P. and Huggins, R. A.**, A review of body composition studies with emphasis on total body water and fat, *Am. J. Clin. Nutr.*, 32, 630, 1979.
130. **Pace, N. and Rathbun, E. N.**, Studies on body composition. III. The body water and chemically combined nitrogen content in relation to fat content, *J. Biol. Chem.*, 158, 685, 1945.
131. **Panaretto, B. A. and Till, A. R.**, Body composition *in vivo*. II. The composition of mature goats and its relationship to the antipyrine, tritiated water, and N-acetyl-4-aminoantipyrine spaces, *Aust. J. Agric. Res.*, 14, 926, 1963.
132. **Carnegie, A. B. and Tulloh, N. M.**, The *in vivo* determination of body water space in cattle using the tritium dilution technique, *Proc. Aust. Soc. Anim. Prod.*, 7, 308, 1968.
133. **Searle, T. W.**, Prediction of body composition of sheep from tritiated water space and body weight tests of published equations, *J. Agric. Sci.* (Cambridge), 75, 497, 1970.
134. **Farrell, D. J. and Reardon, T. F.**, Undernutrition in grazing sheep. III. Body composition and its estimation *in vivo*, *Aust. J. Agric. Res.*, 23, 511, 1972.
135. **Keenan, D. M., McManus, W. R., and Freer, M.**, Changes in the body composition and efficiency of mature sheep during loss and regain of liveweight, *J. Agric. Sci.* (Cambridge), 72, 139, 1969.
136. **Foote, J. Z. and Greenhalgh, J. F. D.**, The use of deuterium oxide space to determine the amount of body fat in pregnant Blackface ewes, *Br. J. Nutr.*, 24, 815, 1970.
137. **Smith, B. S. W. and Sykes, A. R.**, The effect of route of dosing and method of estimation of tritiated water space for the determination of total body water and the prediction of body fat in sheep, *J. Agric. Sci.* (Cambridge), 82, 105, 1974.
138. **Till, A. R. and Downes, A. M.**, The measurement of total body water in the sheep, *Aust. J. Agric. Res.*, 13, 335, 1962.
139. **Holleman, D. F., White, R. G., and Luick, J. R.**, Application of the isotopic water method for measuring total body water, body composition and body water turnover, in *Use of Tritiated Water in Studies of Production and Adaptation in Ruminants*, International Atomic Energy Agency, Vienna, 1982, 9.
140. **Drew, K. R. and Reid, J. T.**, Compensatory growth in immature sheep. I. The effects of weight loss and realimentation on the whole body composition, *J. Agric. Sci.* (Cambridge), 85, 193, 1972.
141. **Tulloh, N. M.**, The carcass composition of sheep, cattle, and pigs as functions of body weight, in *Carcass Composition and Appraisal of Meat Animals*, CSIRO, Melbourne, 1963.
142. **Guthrie, R. D.**, Mammals of the mammoth-steppe as paleoenvironmental indicators, in *Paleoecology of Beringia*, Hopkins, D. M., Ed., (Symposium Wenner-Gren Foundation for Anthropological Research No. 81), Academic Press, New York, 1982.
143. **Yeates, N. T. M. and Schmidt, P. J.**, *Beef Cattle Production*, Butterworths, London, 1974.

Chapter 10

PREGNANCY AND LACTATION

O. T. Oftedal

TABLE OF CONTENTS

Production of offspring is energetically expensive. Both the growing fetus and the suckling neonate exert substantial energy demands on the mother. Needed energy must largely be obtained from increased food intake although mobilization of body reserves may contribute. Thus reproductive cycles must be matched to cycles in food availability and quality.[1,2] Timing becomes especially crucial for ungulates when short seasons of vegetative growth are followed by prolonged periods of scarcity. For these conditions, females must not only nourish offspring, but must rebuild energy reserves to permit future survival and subsequent reproduction.[3,4]

Energy requirements at various stages of life have been determined for domestic ruminants and horses. Metabolizable energy (ME) requirements increase from 420 to 545 $kJ/kg^{0.75}$ per day at maintenance to 585 to 710 $kJ/kg^{0.75}$ per day in late pregnancy,[5-9] with a further rise to 1000 to 1130 $kJ/kg^{0.75}$ per day in early lactation in nondairy animals.[5-7] If extrapolation from domestic species is valid, one might expect wild herbivores in similar conditions (e.g., in zoos) to exhibit a similar one third increase in energy requirements during late gestation, and a more than twofold rise over maintenance energy needs during intensive lactation. Several studies of captive cervids[10-12] indicate increases in food intake of this magnitude during early lactation, with additional mobilization of body reserves in some cases.

Among domestic ruminants, energy requirements for pregnancy and lactation are proportional to the productive processes involved, i.e., synthesis of fetal, placental, uterine, and mammary tissue in pregnancy, and milk production in lactation.[13] Assuming comparable energetic efficiencies of these processes between domestic and wild ungulates, it is possible to predict energy requirements for pregnancy and lactation in wild species from data on fetal development up to birth and on the composition and yield of milk. Although less accurate than direct measurement by energy balance studies, the estimates obtained allow examination of the scaling of requirements to body size across an array of species. This approach to the prediction of energy requirements is termed the factorial method since individual components or factors of energy requirements are estimated separately and then summed together. The factorial method is commonly used for estimating requirements of domestic animals.[13]

This chapter represents a step by step estimation of factors involved in energy requirements for pregnancy and lactation. Much emphasis is placed on the scaling of birth weight and milk output to body size as these scaling relationships form the basis for estimated energy requirements. Most data are from well-nourished animals and thus represent potential reproductive output. Under poor nutritional conditions, fetal growth and milk production may be impaired, with consequent reduction in energy demand on the female. Energy required for additional activities associated with reproduction, such as mate selection and mating, transport of a large fetus during locomotion, and increased foraging activity to obtain needed nutrients,[14] may be important to overall energy budgets of animals but will not be discussed here. The energy needed to support pregnancy and lactation, over and above maintenance needs, will be termed ''energy requirements'' for pregnancy and lactation rather than ''energy costs''. This is to avoid confusion with the expression used in evolutionary theory, ''costs of reproduction'', which is defined in terms of impact of current reproduction on future survival and fecundity.[14]

I. ENERGETICS OF PREGNANCY

Calculation and comparison of the energy requirements for pregnancy by the factorial method requires an evaluation of fetal growth patterns, birth weights, energy deposition in fetal and associated tissues, and the energetic efficiency of these proceses. Equations developed for these processes will be used to predict the relationship of energy requirements for pregnancy to maternal weight in ungulates.

A. Fetal Growth

As with many growth phenomena, fetal growth follows a sigmoid curve. An initial phase of slow growth yields to an exponential growth period; the terminal slowing is rather brief, however, since birth initiates rapid postnatal growth. Based on developmental data on sheep and several laboratory species, Huggett and Widdas[16] proposed a simple equation to represent fetal growth:

$$W^{0.33} = a\,(t - t_0) \tag{1}$$

where W is fetal weight, t is elapsed time since conception, and t_0 is a calculated initial time period that, in effect, compensates for the initial phase of slow growth. The regression coefficient (a) was termed the "specific fetal growth velocity" and was used to compare fetal growth among species. If an estimate of t_0 can be made, specific fetal growth velocity can be calculated from data on birth weight (W_b) and gestation time (t_g):

$$a = \frac{W_b^{0.33}}{t_g - t_0} \tag{2}$$

The exponent 1/3 appears to have been selected for theoretical reasons and has been validated empirically for only a small number of species.[16]

Several authors have examined calculated values of specific fetal growth velocity across a broad range of mammalian species.[16-20] In most cases, the specific fetal growth velocity has been calculated from birth weights and gestation lengths, with t_0 extrapolated from other species or assumed to be time of implantation in species with delayed implantation. Among terrestrial mammals, calculated values of specific fetal growth velocity range from 0.01 to 0.28. Proboscideans (0.08 to 0.09), perissodactyls (0.06 to 0.16), and artiodactyls (0.04 to 0.18) fall in the low to intermediate categories of fetal growth velocity.[18] By contrast, both bats and primates have slow fetal growth velocities (0.01 to 0.09), while large baleen whales exhibit remarkably high velocities (0.25 to 0.58).[18] This method of comparing fetal growth essentially ignores growth rate in the initial phase (when $t < t_0$), and may be biased in cases where a significant slowing of growth rate occurs just prior to birth, as in sheep with twins.[21] The values of t_0 assigned are estimates of variable and often unknown accuracy, yet they directly affect calculated growth velocities. Sacher and Staffeldt[19] argue that t_0 should be omitted from comparative calculations while Calder[22] points out that size-dependency of t_0 may introduce size-dependency into calculated growth velocities, and produce error in comparisons among species varying in body size. Clearly, estimates of fetal growth rates from birth weight and gestation lengths are a poor substitute for weight data on known-age fetuses over the course of gestation.

The initial slow growth phase and subsequent exponential increase in fetal weight imply that the preponderance of mass accumulation occurs in the latter part of pregnancy. In domestic cattle and sheep, about 60% and 80% of gestation have elapsed by the time fetal weights achieve 10% and 40% of birth weights, respectively.[13] Unfortunately, relatively few data are available on fetal weights at known gestation dates in wild herbivores, largely due to difficulties in determining exact conception dates in free-ranging animals. Fetal ages have been estimated by comparison to domestic or other species[23,24] by interpolation from the Huggett and Widdas fetal growth equation[25,26] or by reference to mean breeding dates in species with relatively synchronous breeding during a restricted seasonal period.[27-29] Given the potential error of each of these methods, graphic or other comparison of the shape of fetal growth curves[30] does not seem particularly useful. In white-tailed deer bearing fetal twins of known age, Robbins and Moen[31] found that fetal weights at about 50%, 75%, and 95% of the expected gestation period were comparable to those of domestic cattle and sheep at these gestation stages if expressed relative to expected birth weight. Before concluding that similar relationships hold for other wild herbivores, it is essential that additional data be obtained on known-age fetuses of other species especially those of very small and large size.

B. Scaling of Birth Weight

The relationship of birth weight to maternal weight has been of considerable interest since Leitch et al.[32] showed that litter weight at birth was proportional to maternal weight to the power 0.83 across a broad array (114 species) of mammals. Birth weight per se may scale to a different exponent of material weight than does litter weight, however.[33] If comparisons are restricted to related species within taxonomic groups, the scaling of either birth weight or litter weight to maternal weight may differ markedly from the exponent for an entire array of mammals.[34] From assembled data on 32 species of ungulates and subungulates, Robbins and Robbins[30] calculated that birth weight was proportional to maternal weight to the power 0.79. The suborder Suina (pigs, peccaries, and hippos) was excluded from this calculation, but domestic ruminants and several subspecies of wild ungulates were included.

Inasmuch as birth weight is central to calculations of the energetic costs of pregnancy, a critical reevaluation of the relationship of birth weight to maternal weight was deemed essential. Problems arise in matching data on neonatal weights to maternal weights. In many species, very few neonatal weights are available and it is not clear if they represent normal, full-term offspring on the day of birth. In captive animals, improved plane of nutrition may affect birth weights by leading to larger maternal size, earlier sexual maturity, and/or increased fetal growth rates. Thus, captive caribou produced newborn calves that were 10 to 15% larger than wild caribou neonates in the Northwest Territories of Canada.[37] Conversely, inbreeding depression may arise in captive herds leading to a greater proportion of premature and/or underweight neonates with reduced survival.[38] Birth weights may vary among populations as well as between males and females of the same subspecies, as in Scottish red deer.[39]

Maternal weights are subject to even greater variation. Substantial seasonal variation in maternal weight is characteristic of north temperate ungulates as fat stores are accumulated and exhausted. Weight increases from the late spring/early summer low ("lean weight") to the late fall/early winter peak may average as much as 28% in saiga antelope, 29% in caribou, 42% in moose and, at the extreme, 84% in reindeer in the far northern population on Svalbard.[40-43] By contrast, many warm climate ungulates show little seasonal variation in weight or fat stores.[25,26,44] Inclusion of animals in late pregnancy will bias mean female weight unless the weight of the gravid uterus is subtracted.[25] Unfortunately, this is not always done, or it is not clearly stated.[45-49] Maternal weights may differ among populations, especially at the subspecies level. For example, mean female weights in different subspecies of *Connochaetes taurinus* can differ by 32%, while in *Ovis canadensis* differences reach 64%.[50,51] The range of mean female weights of *Dama dama*, 35 to 45 kg, in different deer parks in Britain, is similar in magnitude to the seasonal variation in female weight at any one park.[52] Weights of captive ungulates will reflect both geographical origin of the founding stock as well as feeding and management programs. Animals weighed when immobilized for medical treatment or at necropsy may not be typical.

Several criteria are necessary for establishing representative birth and maternal weights in wild herbivores:

1. Maternal and birth weights should be from the same population.
2. Maternal weights in herbivores with seasonal fat accumulation should be "lean weights", preferably weights obtained 0 to 4 weeks post partum.
3. Maternal weights should be for nonpregnant animals; if pregnant females are included, the weight of the gravid uterus should be subtracted.

4. Sample size should be sufficient to reduce the effects of individual variability, including variation due to sex of neonate.

Unfortunately, available data on wild herbivores that comply with all criteria are sparse. An extensive but not exhaustive review of the literature is summarized in Table 1, encompassing 37 species and populations. Female weights of all north temperate species except *Dama dama* and perhaps *Ovibos moschatus* are lean weights. In a few species *(Antilope cervicapra, Antilocapra americana, Alces alces, Ovibos moschatus, Giraffa camelopardis, Loxodonta africana)*, female weights and birth weights are from different populations or are partly wild and partly captive data. In most warm climate species, it is not clear whether pregnant females have been excluded or whether females are in "lean" condition. Data for common duikers, *Sylvicapra grimmia*,[45] have been omitted since the reported birth weights (mean = 0.16 kg) are very low and obviously in error.[53] Sample sizes are small for some species and may not be representative.

When birth weights are plotted as a percentage of maternal weight (Figure 1), several trends become apparent:

1. Birth weight tends to decline as a percentage of maternal weight as maternal weight increases.
2. At any given maternal weight, birth weights are quite variable among species.
3. Species that twin on more than an occasional basis produce single offspring that are quite small (4.5 to 7.0%) relative to maternal weight (e.g., *Odocoileus hemionus, Oreamnos americanus, Ovis canadensis, Antilocapra americana, Alces alces*). Saiga antelope appear to be an exception (13.3%) although it is not clear if the reported birth weight is for single young, twins, or some combination thereof.[40]
4. Extreme north temperate species that twin very rarely (*Rangifer tarandus platyrhynchus, Rangifer tarandus groenlandicus, Ovibos moschatus*) also produce small offspring (6.6 to 7.9%) relative to maternal weight.
5. Pygmy hippos *(Choeropsis liberiensis)* produce very small young like other Suina[30] even though only a single offspring is born.
6. Very large herbivores produce especially small neonates.

Among herbivores, birth weight appears to reflect maternal size, reproductive strategy, environmental constraints, and perhaps phylogenetic origin. The decline in birth weight as a percentage of maternal weight with increasing maternal weight indicates that the scaling factor is less than 1.0. Since species with very low birth weights (twinning species with single young, extreme north temperate species, pygmy hippos) appear to exhibit a different scaling relationship, these species were excluded from the logarithmic regression of birth weight on maternal weight (Figure 2). Data exhibit a high degree of linearity up to a maternal weight of about 400 to 500 kg, above which the lower birth weights are reflected in a reduced slope (Figure 2). Thus the data may be described by two regression equations, one covering 4 to 400 kg ($r^2 = 0.981$; $n = 23$), the other 500 to 2700 kg ($r^2 = 0.992$; $n = 5$). The equations correspond to:

$$W_b = 0.166\ W_m^{0.90} \qquad (4\text{—}400\ \text{kg}) \tag{3}$$

$$W_b = 0.510\ W_m^{0.68} \qquad (500\text{—}2700\ \text{kg}) \tag{4}$$

The difference in exponents is sufficiently large as to suggest that different constraints operate in restricting reproductive output in gestation in very large herbivores. The 4 to 400 kg

Table 1
BIRTH AND MATERNAL WEIGHTS OF WILD HERBIVORES

Species	Maternal weight (MW) (kg)	N[a]	Lean, nonpregnant condition[b]	Birth weight[c] (BW) (kg)	N[a]	Birth percentage BW/MW × 100 (%)	Location	Ref.
Madoqua kirki (dik-dik)	4.5	5	?	0.67	16	14.9	Captive	54,55
Gazella dorcas (Dorcas gazelle)	13.8	12	Y	1.6	11	11.6	Captive	54
Saiga tatarica (Saiga antelope)	24	25	Y	3.2?	90	13.3	USSR	40
Antilope cervicapra (Blackbuck)	26.5	36	?	3.2	22	12.1	Feral, captive	44
Antidorcas marsupialis (Springbok)	27.9	A	?	3.3	M?	11.8	S. Africa	56,57
Rangifer tarandus platyrhynchus (Svalbard reindeer)	38	6	Y	3.0	3	7.9	Svalbard	43
Aepyceros melampus (Impala)	38.3	M	?	5.0	?	13.1	S. Africa	27,58
Dama dama (Fallow deer)	39	130	N	4.5	93	11.5	England	29
Lama vicugna (Vicuna)	43	2	?	6.3	10	14.7	Captive	59
Axis axis (Axis deer)	44	30	?	3.7	2?	8.4	Hawaii	23
Tragelaphus spekei gratus (Sitatunga)	45.5	5	?	4.1	4	9.0	Captive	60
Tragelaphus imberbis (Lesser kudu)	49	5	?	6.1	8	12.4	Captive	61
Odocoileus hemionus columbianus (Black-tailed deer)	49.8	6	Y	3.0s 5.2t	20 84	6.0 10.4	Captive Captive	62,63
Antilocapra americana (Pronghorn)	50	M	?	7.0t?	M	14.0	Alberta, Montana	64

Oreamnos americanus (Mountain goat)	56.5	M	?	3.0s	1	5.2	Montana	65
Ovis canadensis (Bighorn sheep)	63.4	4	Y	4.6s	4	7.3	Captive	66
Damaliscus dorcas (Blesbok)	68	?	?	7.0	4	10.3	S. Africa	67
Rangifer tarandus groenlandicus (Caribou)	75.5	26	Y	5.0	43	6.6	Canada	68
Cervus elaphus scoticus (Red deer)	84.2	17	Y	7.9	17	9.4	Captive	69
Connochaetes gnou (White-tailed gnu)	116	2	?	12	3	10.3	Captive	70
Connochaetes taurinus hecki (White-bearded wildebeest)	165	43	?	16.6	5?	10.1	E. Africa	46
Tragelaphus strepsiceros (Greater kudu)	170	29	?	15.1	9	8.9	Zambia	47,71
Ovibos moschatus (Musk ox)	172	6	?	11.3	2	6.6	Alaska; captive	72
Choeropsis liberiensis (Pygmy hippo)	230	5	?	5.7	31	2.5	Captive	73
Cervus elaphus nelsoni (N. American elk)	249	7	Y	21.2	7	8.5	Captive	11
Equus burchelli boehmi (Grant's zebra)	267	2	?	32.3	32	12.1	Captive	74
Equus zebra hartmannae (Hartmann mountain zebra)	276	23	?	25	5	9.1	S.W. Africa	24
Alces alces (Moose)	278	M	Y	12.4s	5	4.5	Alaska, Michigan	42,75
Taurotragus oryx oryx (Highveld eland)	316	A	N	25.8	15	8.2	S. Africa	49
Equus burchelli antiquorum (Burchelli's zebra)	322	51	Y	33.7	4	10.5	S. Africa	25
Taurotragus oryx livingstonei (Bushveld eland)	392	8	?	31.9	7	8.1	Captive	76
Syncerus caffer (African buffalo)	503	52	?	38.4	3	7.6	Tanzania	77

Table 1 (continued)
BIRTH AND MATERNAL WEIGHTS OF WILD HERBIVORES

Species	Maternal weight (MW) (kg)	N[a]	Lean, nonpregnant condition[b]	Birth weight[c] (BW) (kg)	N[a]	Birth percentage BW/MW × 100 (%)	Location	Ref.
Bubalus bubalus (Water buffalo)	530	60?	?	35.5	16	6.7	Captive	78
Syncerus caffer (African buffalo)	536	21	Y	37.2	8	6.9	S. Africa	79
Giraffa camelopardalis (Giraffe)	964	2	?	54.5	16	5.7	E. Africa; captive	80,81
Rhinoceros unicornis (Indian rhino)	1604	2	?	71.3	20	4.4	Captive	82
Loxodonta africana (African elephant)	2700	A	?	113	1	4.2	Zambia; captive	83,84

[a] Number of animals: M = mean, number not specified; A = asymptote of growth curve.
[b] Data on maternal weights that represent seasonal low weights on nonpregnant adult females, or pregnant females with gravid uteri removed, are rated Y (yes). Data that include pregnant animals with intact uteri or nonlean females are N (no). Many entries could not be fully evaluated and are rated ? (uncertain).
[c] Data refer to weights of single young if not specified. For species that twin regularly, s = single young, t = combined twin weights, ? = uncertain.

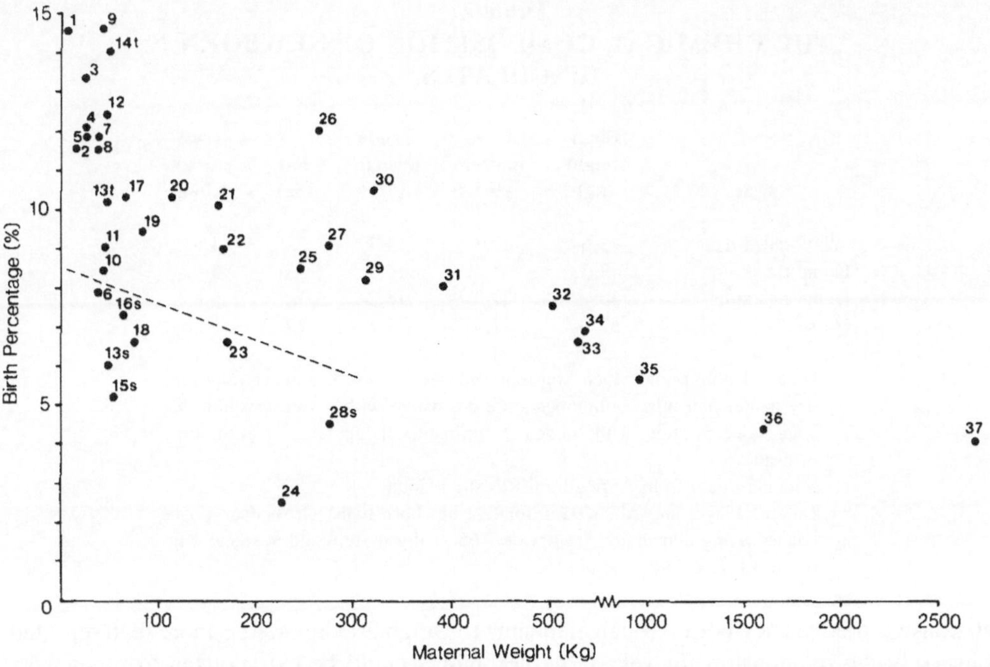

FIGURE 1. Scattergram of birth weight (as a percentage of maternal weight) vs. maternal weight (kg). Numbers refer to species as numbered in Table 1. In species which twin, single births are referred to with an "s" suffix, while combined weights of twins are referred to with a "t" suffix. Dotted line separates species considered to have very low birth weights and hence excluded from Figure 2. Note the change in scale between 500 and 1000 kg.

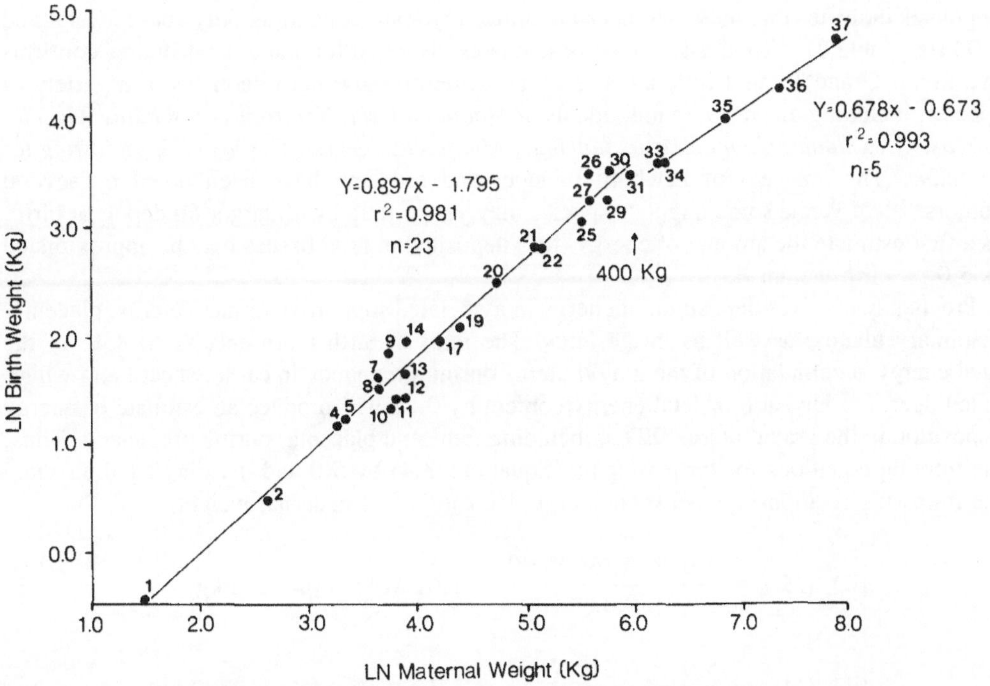

FIGURE 2. Logarithmic regression of birth weight vs. maternal weight. Two equations are displayed for the regression lines indicated, covering species 1-31 and 33-37, respectively. Numbers refer to species as numbered in Table 1.

Table 2
THE CHEMICAL COMPOSITION OF NEWBORN UNGULATES

Species	Birth weight (kg)	Dry matter (%)	Crude protein (%)	Fat (%)	Gross energy (kJ/kg)
White-tailed deer[a]	3.6	22.3	14.7	2.2	4.31
Domestic sheep[b]	3.5	—	16.4	3.0	5.02
Domestic cattle[b]	30	27.0	18.2	3.5	5.65
Horse[c]	52	27.3	17.1	2.6	5.06

[a] Data of Robbins and Moen[31] on four 190-day fetuses. The lower values for dry matter and other components are consistent with the observation that water concentration of the neonatal ruminant falls appreciably at the time of birth.[13]

[b] Summary data from Agricultural Research Council.[13]

[c] Data of Meyer and Ahlswede[85] for five newborn foals. Gross energy calculated using conversion factors of 5.65 kcal/g protein and 9.3 kcal/g fat where kJ = 4.184 × kcal.

relationship predicts newborn African elephants to weigh 203 kg or 80% more than reported; conversely the relationship for very large herbivores could be extrapolated to predict that newborn dik-dik should weigh 1.4 kg or 32% of maternal weight. Apparently, herbivores in different size classes need to be considered separately.

C. Energy Deposition in the Gravid Uterus

At birth, the neonate constitutes the end result of energy deposition in the fetus over the entire course of pregnancy. The few data available on the chemical composition of newborn ungulates indicates that these neonates are uniformly lean, containing only about 3% fat and 5.0 kJ/g (Table 2). Visual assessment of neonates dissected for anatomical studies confirms this trend. Grand[86] found little adipose tissue evident in zoo-born neonates of a variety of species, including one or more individuals of *Madoqua kirki, Neotragus moschatus, Gazella dorcas, Oryx dama, Connochaetus taurinus, Muntiacus reevesi, Cervus eldi, and Rangifer tarandus.* The leanness of newborn reindeer and red deer have been noted by several authors.[43,87,88] While some ungulate species may yet be found with larger fat depots at birth, as a first estimate the amount of energy (MJ) deposited in fetal tissues may be approximated as 5.0 × birth weight (kg).

Pregnancy involves deposition of energy in associated supportive tissues (uterus, placenta, mammary glands) as well as in the fetus. The fetus at birth represents 81 to 83% of the total energy accumulation of the gravid uterus during pregnancy in cattle, sheep, and white-tailed deer.[13,31] Division of fetal energy content by 0.82 will produce an estimate of energy deposition in the gravid uterus (ED_u), including fetus and placenta, during pregnancy. Thus, multiplying equations for birth weights (Equations 3,4) by 5.0 and dividing by 0.82 yield the following relationships between energy deposition and maternal weight:

$$ED_u\ (MJ) = \frac{5.0 \times 0.166\ W_m^{0.90}}{0.82} = 1.01\ W_m^{0.90} \quad (4\text{—}400\ kg) \tag{5}$$

$$ED_u\ (MJ) = \frac{5.0 \times 0.510\ W_m^{0.68}}{0.82} = 3.12\ W_m^{0.68} \quad (500\text{—}2700\ kg) \tag{6}$$

Species with low birth weights (twinners with single young, extreme north temperate species, hippos) will have proportionately lower ED_u.

Mammary development during pregnancy involves accumulation of an additional energy increment equivalent to about 25% of the energy content of the gravid uterus in domestic sheep.[21] An unknown portion of this accumulated energy constitutes initial secretions (colostrum) stored in the mammary glands prior to parturition. In most studies of the energetics of pregnancy or lactation, mammary energy accumulation is not factored out separately. Hence, it ends up included with other energetic processes that diminish calculated energy efficiencies for the production of the gravid uterus or milk.

D. Energy Requirements for Pregnancy

Thermodynamic inefficiencies in synthetic processes, the need to support the metabolic needs of the gravid uterus, and the requirement of energy for mammary development dictate that the energetic requirement for pregnancy is much greater than simply the amount of energy deposited in the gravid uterus. The energetic efficiency of pregnancy is usually measured as the proportion of the increase in metabolizable energy (ME) intake above maintenance that is incorporated into the gravid uterus. In most studies of domestic cattle and sheep, this efficiency has been estimated as 11 to 14% although higher and lower estimates have been obtained.[13,89] It is possible that efficiency improves as the energy density of maternal diet increases or as the proportion of energy derived from maternal stores rises.[89] The Agricultural Research Council in Great Britain[13] selected 13.3% as representative of available data for cattle and sheep that neither deposit nor deplete energy reserves during pregnancy. As a first estimate, this value can be applied to wild herbivores while recognizing that many species may draw on energy reserves to support the growth of the gravid uterus, especially in north temperate areas. Since estimates for cattle and sheep are similar despite differences in size, the assumption is made that the energetic efficiency of pregnancy is independent of body weight.

The ME required for pregnancy (E_p,MJ), over and above maintenance, can then be estimated by dividing energy deposition in the gravid uterus (ED_u; Equations 5, 6) by 0.133. The relationship of E_p to W_m (kg) for species of 4 to 400 kg maternal weight is then:

$$E_p = \frac{1.01\ W_m^{0.90}}{0.133} = 7.6\ W_m^{0.90} \tag{7}$$

While in species above 500 kg the relationship is

$$E_p = \frac{3.12\ W_m^{0.68}}{0.133} = 23.5\ W_m^{0.68} \tag{8}$$

Energetic costs will be reduced for species with relatively low birth weights, of course, such as twinning species with single young and extreme north temperate species.

The average daily energy requirement for pregnancy can be calculated as the total requirement (E_p) divided by gestation length (G). Across a wide spectrum of mammals gestation time has been reported to scale to maternal weight to the power 0.24 to 0.26.[19,33,90] Such broad comparison may be misleading because the relationship of gestation time to maternal weight differs among various mammalian orders, but the within-order scaling factors are all similar, 0.14 to 0.20.[91] For artiodactyls and perissodactyls, gestation time scales to the exponents 0.18 and 0.20, respectively.[91] Thus, the daily energy requirement for pregnancy scales to $W_m^{0.90} \div W_m^{0.19} = W_m^{0.71}$ in ungulates of 4 to 400 kg, and $W_m^{0.68} \div W_m^{0.19} = W_m^{0.49}$ in species above 500 kg. Although daily energy requirements for pregnancy may be proportional to maternal metabolic size in most ungulates of 4 to 400 kg, this is not true of larger species.

Calculation of average daily requirement does not address the energy needs of the pregnant female at different stages of pregnancy. Exponential growth of the fetus implies that energy needs will be greatest just prior to parturition since at that time both energy deposition and energy required for the metabolism of the gravid uterus will be large. In domestic cattle, the estimated daily energy required for pregnancy at 12 weeks, 8 weeks, and 4 weeks prior to parturition are only 19%, 33%, and 58% of the daily requirement at parturition.[13] At term, the energy requirement for pregnancy is equivalent to about 80% of the maintenance requirement in cattle. A similar pattern probably occurs with wild herbivores although exact predictions will depend on detailed studies of energy deposition in the fetus and associated structures as well as measurements of the energetic efficiency of pregnancy in nondomestic species.

II. ENERGETICS OF LACTATION

Parturition terminates the rising energetic costs of pregnancy, but initiates an even more energy-demanding phase, lactation. In well-nourished herbivores, milk production typically rises to a peak and then declines for a prolonged period as suckling young are gradually weaned to solid foods. Calculation of the energy requirements for lactation by the factorial method requires data on milk energy content and milk yield at various stages of lactation, as well as an estimate of the energetic efficiency of milk production. A discussion of milk composition and measurement of milk yield will therefore precede calculation of the energy required for lactation. The scaling of lactation energetics to maternal weight across species is the focus of this analysis, but brief comments on the effects of maternal nutrition and number of offspring on lactation performance are also included.

A. Milk Energy Content

Although sometimes measured directly, milk energy content is usually calculated from assayed levels of fat, protein, and carbohydrate in milk, using conversion factors developed for domestic species.[92] Sampling or analytical procedures which bias compositional data will produce error in energy calculations. The major energy component in most milks, fat, tends to be quite variable from sample to sample. In studies of horses and sheep at mid-lactation, for example, the coefficient of variation for fat (27 to 29%) was several fold greater than that for protein (7 to 11%) or sugar (4 to 6%).[93] Poor replication stems in part from the fact that a partial sample of mammary milk may not be representative of the whole. In many species, including cattle, sheep, water buffalo, eland, moose, bactrian camels, and horses, the fat content rises during the course of milk removal such that the last milk obtained may be three or more times higher in fat than the first milk drawn.[92] Milk samples taken shortly after the young have suckled are elevated in fat content and may not represent milk as consumed.[94] Careful and controlled sampling procedures and large numbers of samples are needed to avoid potential bias, but such criteria are not easily met in studies of wild herbivores.

Milk composition is a function of lactation stage. Dry matter and energy levels in milk bear an inverse relationship to yield in most herbivore species: Milk secreted at peak or mid-lactation tends to be lower in dry matter and energy than milk in early or late lactation.[92] Early lactation milk is uniformly elevated in protein content; fat levels may or may not be high. Artiodactyl and elephant milks exhibit a rise in fat, protein, and energy content in late lactation, but little compositional change is evident in rhino or equid milks at this lactation stage.[92] Thus, temporal trends do vary somewhat between phylogenetic groups.

Available data suggest that the Artiodactyla, Perissodactyla, and Proboscidea produce milks of low to moderate dry matter and gross energy content. At mid-lactation, rhino and equid milks contain 1.5 to 2.1 kJ/g, or about one half the energy in domestic bovid (2.9 to 4.6 kJ/g) and elephant (3.8 to 4.6 kJ/g) milks.[92] By contrast, many wild bovid and cervid milks are more energy dense (5.4 to 7.5 kJ/g). Increased energy density is achieved by an increase in the proportion of energy contributed by fat and a decrease in the proportion by sugar. Protein as a percentage of energy tends to be less variable (18 to 34% of milk energy) since protein levels rise with increases in energy density. In fact, the highest protein-energy percentages are found in milks of rather high (>4.6 kJ/g) energy content.[92]

B. Milk Yields for Peak and Total Lactation

Several methods have been used to quantify milk production in herbivores with suckling young; for purposes of the present discussion, studies on dairy animals are excluded. These procedures may be classified as: (1) weight-differential, (2) timed-milking, (3) isotope dilution, and (4) isotope transfer.[92] Weight-differential methods entail weighing suckling young before and after suckling bouts during a regimen of periodic, controlled access of young to mother. The cumulative weight gains during suckling are equated to milk intake. In timed-milking procedures the young are removed from their mothers and accumulated milk is collected by hand- or machine-milking at periodic intervals to estimate secretion rates. Isotope dilution techniques typically involve monitoring water kinetics in suckling young following administration of isotope-labeled water. Calculated water intakes of the young can be converted to milk intakes as long as water consumption from sources other than milk is negligible or quantifiable. Isotope transfer methods are usually coupled with dilution studies and allow water transferred from mother to young via milk to be distinguished from water from other sources.

All milk intake methods are based on assumptions and have potential errors.[92] Weight-differential methods require that the imposed regimen and procedures do not alter milk intakes even though suckling durations and intervals may be quite artificial. The importance of using trained animals has been emphasized in domestic animals,[95] and is even more vital with less docile species. Timed-milking methods are valid only if secretion rates are unaffected by separation and repeated milking, if a comparable and relatively complete evacuation of the mammary glands is accomplished at each milking, and if the volumes so obtained are representative of what the young would consume. Isotope dilution calculations may need to be adjusted for changing size of the water pool, isotope recycling, and metabolic water production, and will be in error if water consumption from nonmilk sources is not controlled or accounted for. Isotope transfer procedures require mathematical modeling and fairly frequent sampling of both mother and young, which may not be possible under free-ranging conditions.

Direct comparison of the various methods suggests that results may vary according to lactation stage. In sheep, yields measured by timed-milking are often higher than those obtained by weight differential or isotope dilution methods, especially in the first two weeks post partum.[95-98] It appears that young lambs do not empty the udders while suckling, such that volumes collected during timed-milking procedures overestimate lamb consumption.[95,99-101] If properly applied, all methods give comparable results at peak lactation.[93,95,98,103] In late lactation, isotope dilution methods overestimate milk intake due to extraneous water consumption, whereas weight-differential methods lose accuracy given the necessity of measuring small weight increments in large and active young.[95] In practice, milk yield estimates are rarely attempted in the prolonged tail of the lactation curve such that a small but unknown percentage of total milk yield is omitted.

Under conditions of adequate nutrition, milk yields of herbivores typically rise to a peak at about 3 to 6 weeks post partum (Table 3) although earlier peaks have been reported for sheep with twins[93] and later peaks for horses[109,110] and beef cattle[111] in some studies. Available

Table 3
PEAK MILK YIELDS OF VARIOUS HERBIVORES

Species[a]	Maternal weight (kg)	N[b]	Time of peak (days p.p.)	Daily milk yield[c] Amount (kg)	Percentage body weight (%)	Per unit metabolic wt. ($g/kg^{0.75}$)	Ref.
Dorcas gazelle	20.6	4	30—60	0.56	2.7	58	103
Bedouin goat	21.4	11	30—60	1.51	7.1	152	104
Ibex	30.3	6	30—60	0.64	2.1	49	103
Black-tailed deer	49.8	6	28	1.26	2.5	67	62
Sheep	52.6	5	17	2.47	4.7	126	93
Red deer	88.4	9	40	2.08	2.4	72	69,105
Reindeer	107	3	21—28	1.59	1.5	48	106
N. American elk	249	7	21	4.10	1.6	65	11
Beef cattle	340	30	14—28	7.42	2.2	94	107
Horse	515	5	39	17.6	3.4	163	108

[a] Refer to Table 1 for taxonomic binomials of nondomestic species; ibex is *Capra ibex*.
[b] N = number of females studied. Black-tailed deer and sheep had twins; all others had single young.
[c] All peak yields obtained by isotope dilution procedures except Bedouin goat, black-tailed deer, and red deer.

data for three small species (dorcas gazelle, Bedouin goat, ibex) may not represent peak yields per se since they are 1 month averages (Table 3). Maternal nutrition influences both the height and shape of the lactation curve. On poor pasture, sheep, red deer, and reindeer exhibit little or no post partum rise in milk yield; milk production begins to decline shortly after birth.[4,69,101,105,112] However, if nutritional conditions improve within the first month, post partum milk production will increase to more normal levels.[112]

Over the size range of 21 to 515 kg, milk yield at peak lactation increases from 0.6 to 17.6 kg/day in well-nourished animals (Table 3). Milk production is equivalent to 1.5 to 2.7% of body weight in most species although in horses (3.4%), sheep with twins (4.7%), and Bedouin goats (7.1%), peak levels are higher. The Bedouin goat yields are particularly remarkable but may reflect artificial selection for milk yield. As milk yields tend to vary in proportion to metabolic size rather than body weight,[92,113,114] yields have also been expressed as $g/kg^{0.75}$ (Table 3). Domestic species (Bedouin goat, sheep, beef cattle, horse) produce more milk per unit metabolic size (94 to 163 $g/kg^{0.75}$) than do nondomestic species (48 to 72 $g/kg^{0.75}$). The milks of domestic species are typically lower in dry matter and energy content, however, so yields of dry matter or energy per unit metabolic size are not necessarily elevated in domestic species.

Comparison of total milk yields across species is more difficult since the small daily amounts produced in the tail end of lactation are not easily measured. The nutritional significance of these small amounts at the end of lactation is probably minor.[115] One may define "effective lactation" as that period after birth in which yields are of sufficient magnitude to quantify by available methods. The length of effective lactation may increase with body size: in species of about 50 kg (e.g., black-tailed deer, sheep), effective lactation appears to be about 100 to 120 days, whereas animals of about 500 kg (e.g., beef cattle, horses) may have measurable yields for 150 to 180 days (Table 4). These data are consistent with the hypothesis that effective lactation length, like gestation length and other physiological cycles,[119] scales to weight to the power 0.25 or less. Rigorous testing of this hypothesis requires data on more species obtained by comparable methods and with the objective of measuring yields as late as possible in the lactation period.

Data on total yields over effective lactation are available for eight species (Table 4). These

Table 4
RELATIONSHIP OF TOTAL YIELD TO PEAK YIELD

Species	Number of animals[a]	Duration of study (days p.p.)	A. Total yield (kg)	B. Average yield (kg/day)	C. Peak yield (kg/day)	Yield ratio B/C (kg/day)	Ref.
Bedouin goat	12(1)[b]	120	173	1.4	1.6	0.88	103,104
Black-tailed deer	5(1)[b]	120	93	0.8	1.3	0.62	62
Sheep	28(1)	98	93	0.9	1.4	0.70	116
	6(2)	98	101	1.0	1.6	0.65	116
	6(2)	84	176	2.1	2.9	0.72	117
	6(3)	84	193	2.3	3.8	0.61	117
Red deer	3(1)[b]	150	160	1.1	1.6	0.68	10
	8(1)[c]	100	105	1.1	1.4	0.75	69,105
	9(1)	100	163	1.6	2.1	0.78	69,105
Reindeer[d]	?(1)[e]	98	118	1.2	2.0	0.60	4
	?(1)[f]	98	57	0.6	1.0	0.59	4
N. American elk[g]	7(1)[b]	150	410	2.7	4.1	0.66	11
Beef cattle	36(1)	180	825	4.6	6.4	0.72	118
Horse	8(1)	150	2260	15	18	0.84	110

[a] Values in parentheses indicate number of young per lactating female. All animals on good pasture unless otherwise indicated.
[b] Animals fed pelleted feed and hay.
[c] Animals on poor quality hill pasture.
[d] Yields estimated from graph of milk yield presented by White and Luick.[4]
[e] Animals on pasture supplemented with pelleted feed.
[f] Animals given restricted grazing.
[g] Data extrapolated to 150 days using equation developed for 20 to 80 days post partum by Robbins et. al.[11]

data have been obtained by various procedures: weight-differential (Bedouin goat, black-tailed deer, red deer, beef cattle, horse), timed-milking (sheep), isotope dilution (elk), and isotope transfer (reindeer). Differences between species could therefore be influenced by methodological bias. Total yields range from 93 kg to 2260 kg (Table 4). In most cases, average daily yield for the period of measurement is equal to 60% to 70% of peak yield (yield ratio of average/peak yield = 0.60 to 0.70). The Bedouin goat appears to be atypical (yield ratio = 0.88), perhaps because Bedouin goats have been selected by humans for sustained yields as a multipurpose milk and meat goat.[103,104] Dairy breeds of sheep also exhibit sustained yields.[120] Higher yield ratios for red deer on pasture (0.75, 0.78) than in pens (0.68) may reflect the shorter period of measurement (~100 vs. 150 days) with consequent omission of low yields in late lactation. Robbins et al.[11] provide equations relating yield to post partum time in elk such that yields can be extrapolated to 150 days to be comparable to red deer[10] (Table 4) even though originally measured only to 80 days. Poor grazing conditions reduce both peak and average yields to a comparable degree in red deer[69,105] and reindeer[4] so yield ratios are little affected (Table 4). Increasing number of offspring from one to two or from two to three stimulates peak milk yield more than average yield in sheep such that yield ratios fall (Table 4) but remain within the usual range, 0.61 to 0.72. Horses have a high yield ratio (0.84) since yields decline slowly after the lactation peak.[109,110] Sustained yields may be important to maintain energy transfer to suckling young in species with little increase in milk energy content in late lactation (e.g., rhinos, equids).

The relative constancy of yield ratio among species, despite nutritional influences and variation in numbers of young, suggests that peak yield is a useful indicator of overall lactation performance in herbivores.

Table 5
MILK ENERGY YIELDS FOR PEAK AND ENTIRE LACTATION

	Peak lactation[a]			Entire lactation[b]			
	Milk energy content	Daily energy yield		Milk energy content	Total energy yield		
Species[c]	(kJ/g)	(kJ)	($kJ/kg^{0.75}$)	(kJ/g)	(MJ)	($MJ/kg^{0.75}$)	Ref.
Dorcas gazelle	6.40	3,600	372	—	—	—	103
Bedouin goat	3.10	4,690	473	3.10	536	54.0	103,104
Ibex	6.78	4,350	335	—	—	—	103
Black-tailed deer	7.36	9,290	494	7.95	741	39.5	62,121
Sheep	4.64	11,500	586	5.40	950	44.8	93,116
Red deer	5.73	9,000	321	6.61	1079	37.4	10
Reindeer	6.95	11,050	331	8.20	967	29.1	4,106,122
N. American elk	4.60	18,900	301	5.10	2092	33.4	11
Beef cattle	2.97	22,050	280	3.24	2590	32.7	118, 123, 124
Horse	2.13	37,300	345	2.13	4812	38.9	108,110

[a] Based on peak milk yields in Table 3 except red deer data.[10] Milk energy content at mid-lactation as in Oftedal.[92]

[b] Based on total milk yields of well-nourished animals in Table 4; if possible, total yield corresponding to peak yield used. Milk energy content calculated as (milk energy content at mid-lactation + milk energy content at end of effective lactation) ÷ 2.

[c] All data for species with single young except sheep and black-tailed deer with twins.

C. Scaling of Milk Energy Output

The output of milk energy at any lactation stage is a function of both the yield and energy content of milk. Although variable in timing among studies and between species, peak lactation is easily defined and represents maximal energy demand on the female. Peak energy output in milk ranges from 3,600 to 37,300 kJ/day among 10 species studied (Table 5). Peak energy output scales to maternal metabolic size ($W^{0.75}$) across a broad array of mammals, but the level of output appears to differ between ungulates with single young and mammals with large litters.[125] Most of the herbivores in Table 5 produce 280 to 370 $kJ/kg^{0.75}$ per day at peak lactation, although higher estimates were obtained for Bedouin goats (473 $kJ/kg^{0.75}$·day), black-tailed deer with twins (494 $kJ/kg^{0.75}$·day), and sheep with twins (586 $kJ/kg^{0.75}$·day). The elevated energy yield of the Bedouin goat probably relates to its dairy function, whereas twinning appears to be responsible for high energy yields in black-tailed deer and sheep. Milk yields of single-suckled ewes are typically 59 to 71% (mean = 65%) of twin-suckled ewes.[93] Multiplication by 0.65 produces estimates of milk energy output for black-tailed deer (321 $kJ/kg^{0.75}$·day) and sheep (381 $kJ/kg^{0.75}$·day) within the normal range. The scaling of milk energy output at peak lactation to maternal weight can be determined for nine species of ungulates with single young by logarithmic regression (Figure 3); the Bedouin goat was excluded from calculations as atypical. The data are highly linear ($r^2 = 0.981$) and correspond to the power function:

$$EO_{PL} = 435\ W_m^{0.70} \tag{9}$$

where EO_{PL} is energy output at peak lactation (kJ/day) and W_m is maternal weight (kg). Ungulates with twins are at a higher production level.

Total energy yield for the entirety of effective lactation can be accurately determined only if data on milk energy content and milk yield can be combined for each point on the lactation curve. This was not feasible for most species given available data. A simpler but less accurate

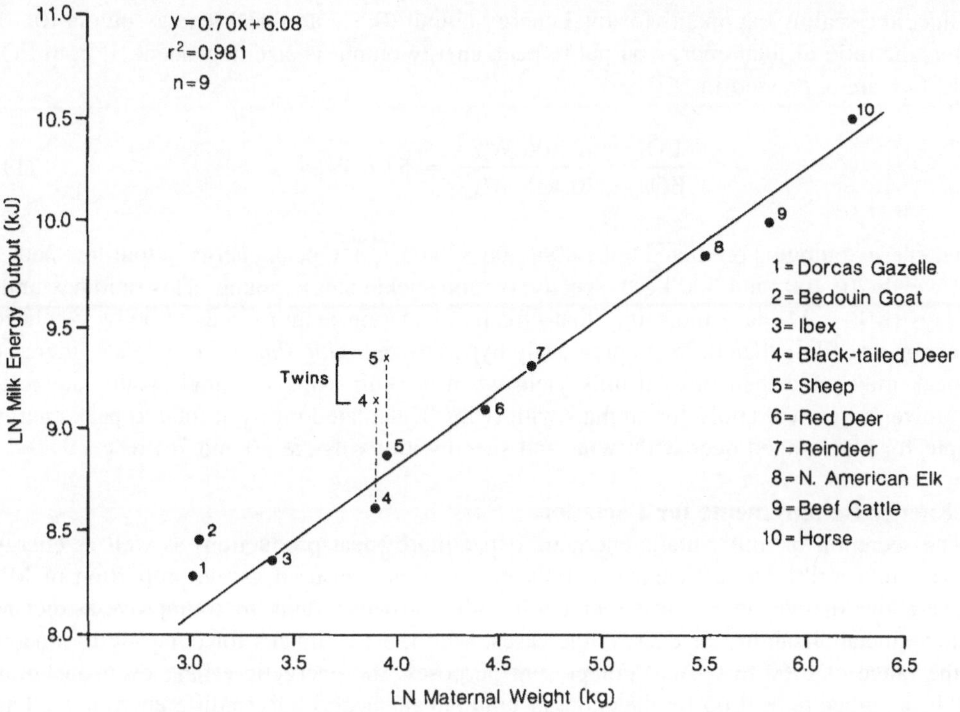

FIGURE 3. Logarithmic regression of milk energy output at peak lactation (kJ) on maternal weight (kg) for species with single young. Symbol "x" and dotted lines indicate females with twins. Bedouin goat excluded from calculations (see text).

approach involves multiplication of total milk yield by the average energy content of milk. The degree of change in milk energy content over the course of lactation varies among species: at the end of effective lactation, milk energy content has increased 6% above mid-lactation in sheep,[117] 17% in black-tailed deer,[121] 21% in elk,[11] 42% in red deer,[10] and 51% or more in reindeer.[122] A median energy value was calculated between the two extremes of mid- and end-lactation and used as an approximation of average energy content in calculation of total energy output in milk (Table 5). In view of declining yields in late lactation, this method may weight late lactation energy data too heavily, but it does make allowance for species-specific differences. The total energy yields so calculated range from 536 MJ in the Bedouin goat to 4812 MJ in the horse.

When expressed per unit metabolic size, total milk energy outputs of species with single young (other than the Bedouin goat) are remarkably close: 29.3 to 38.9 $MJ/kg^{0.75}$. The species with twins (black-tailed deer, sheep) have somewhat higher energy outputs (39.5 and 44.8 $MJ/kg^{0.75}$). As with peak yields, energy outputs of the twinning species can be corrected to a single young basis using sheep data. Ewes with singles appear to produce about 75% (range = 66 to 84%) of the total yields of ewes with twins.[117,120,126-128] The corrected energy outputs for ewes with single lambs (33.6 $MJ/kg^{0.75}$) and black-tailed does with single fawns (29.6 $MJ/kg^{0.75}$) are within the usual range. After logarithmic transformation, total milk energy output (EO_L, MJ) exhibits a linear relationship (r^2 = 0.985) to maternal weight (W_m, kg), corresponding to the power function:

$$EO_L = 25.06\ W_m^{0.81} \tag{10}$$

This relationship is based on seven species with single young (including black-tailed deer and sheep converted to a single young basis), but omits the Bedouin goat.

Since the scaling exponents for total energy output (0.81) and peak energy output (0.70) differ, the ratio of total energy output to peak energy output is size-dependent. If both EO_L and EO_{PL} are expressed in MJ:

$$\frac{EO_L}{EO_{PL}} = \frac{25.06\ W_m^{0.81}}{0.435\ W_m^{0.70}} = 57.6\ W_m^{0.11} \tag{11}$$

Total energy output is predicted to be 80×, 96×, and 114× peak energy output for species that weigh 20, 100, and 500 kg, respectively, and suckle single young. This ratio has units of days (MJ ÷ MJ/day) indicating that effective lactation length scales to body weight to an exponent of less than 0.25, as previously hypothesized (*vide supra*). The greater increase in peak milk yield than in total milk yield when nursing twins vs. single young suggests that lower ratios will apply for animals with twins. Calculated ratios of total to peak energy output for black-tailed deer with twins and sheep with twins are 80 and 76, respectively.

D. Energy Requirements for Lactation

The secretion of milk entails energetic expenditure (heat production) as well as energy transfer into milk. The efficiency of milk secretion is measured as the proportion of ME intake above maintenance that is secreted in milk. Efficiency appears to improve as dietary quality (metabolizability of energy) increases, with most estimates for domestic ruminants in the range of 0.60 to 0.70.[13] For present purposes, the energetic efficiency of lactation will be assumed to be 0.65 for the entire lactation in all species. Given differences in feeding habits, environmental factors, body size, etc., it is doubtful that any single value will apply to all species, but the error involved is likely 10% or less (i.e., actual value 0.60 to 0.70). Reproductive cycles in wild herbivores should be timed such that peak lactation occurs during periods of high dietary quality, thereby maximizing the energetic efficiency of milk production.

The metabolizable energy requirements for peak lactation (E_{PL}), over and above maintenance, can be calculated from Equation 9 for ungulates with single young as:

$$E_{PL}\ (\text{kJ/day}) = \frac{EO_{PL}}{0.65} = 669\ W_m^{0.70} \tag{12}$$

Somewhat higher values will be associated with nursing of twins (e.g., 920 $W^{0.75}$ for sheep with twins). ME requirements above maintenance for total lactation (E_{TL}) can be calculated from Equation 10 for ungulates with single young as:

$$E_L\ (\text{MJ}) = \frac{EO_L}{0.65} = 38.6\ W_m^{0.81} \tag{13}$$

This estimate omits energy requirements associated with the small milk yields occurring at the end of lactation. Once again, higher milk production levels for animals with two young imply higher energy requirements.

III. ENERGY REQUIREMENTS FOR FEMALE REPRODUCTION

The energy requirements for pregnancy and lactation have been calculated by a factorial method, using data on fetal growth, birth weight, energy deposition in the gravid uterus, milk energy content, milk production, and the energetic efficiencies of pregnancy and lactation. The scaling of energetic needs to body weight is influenced by number of young

Table 6
METABOLIZABLE ENERGY REQUIREMENTS FOR REPRODUCTION

Body weight (kg)	ME for pregnancy[a] E_P (MJ)	$E_P/E_R \times 100$ (%)	ME for lactation[b] E_L (MJ)	$E_L/E_R \times 100$ (%)	ME for reproduction[c] E_R (MJ)	$E_R/W^{0.75}$ (MJ/kg$^{0.75}$)	Multiples of maintenance[d]
4	26.6	18	118	82	145	51.3	107
20	113	21	436	79	549	58.0	120
100	482	23	1,607	77	2,089	66.1	137
400	1679	25	4,940	75	6,619	74.0	154
800	2211	20	8,660	80	10,870	72.3	150
1600	3543	19	15,190	81	18,730	74.0	154
2700	5056	18	23,200	82	28,250	75.4	157

[a] For species 4 to 400 kg, $E_P = 7.64\ W_M^{0.90}$; for species >500 kg, $E_P = 23.5\ W_M^{0.68}$.
[b] $E_L = 38.6\ W_M^{0.81}$.
[c] $E_R = E_P + E_L$.
[d] Maintenance assumed to be 0.481 MJ/kg$^{0.75}$·day.

reared and, at least in the case of pregnancy, the size class of ungulate. The total energy requirement for reproduction (E_R) will be the sum of requirements for pregnancy (E_P) and lactation (E_L). In species that always rear single young, the ME requirements for reproduction can be estimated as:

$$\text{(4—400 kg)} \qquad E_R\ (\text{MJ}) = E_P + E_L = 7.64\ W_m^{0.90} + 38.6\ W_m^{0.81} \qquad (14)$$

$$\text{(500+ kg)} \qquad E_R\ (\text{MJ}) = E_P + E_L = 23.5\ W_m^{0.68} + 38.6\ W_m^{0.81} \qquad (15)$$

The latter relationship assumes that milk energy output in very large herbivores can be extrapolated from species of 500 kg and less; unfortunately, there are no reliable data on lactation performance in species larger than a horse. These values of E_R as calculated do not include maintenance energy requirements for the reproductive animal itself.

Predicted energy requirements for reproduction over a size range of 4 to 2700 kg are presented in Table 6. Energy requirements increase from 145 to 28,250 MJ with increasing body size. Pregnancy needs account for only 18 to 25% of the total; lactation entails energy requirements of 3 to 4.5× those of pregnancy (Table 6). When energy requirements for reproduction are expressed per maternal metabolic size ($W_m^{0.75}$), calculated requirements increase about 40% (from 51 to 74 MJ/kg$^{0.75}$) as body size increases from 4 to 400 kg, but between 400 and 2700 kg there is little change (72 to 75 MJ/kg$^{0.75}$). The apparent constancy between 400 and 2700 kg is due to the declining proportion of E_R contributed by requirements for pregnancy (E_P) which in turn is a consequence of the proportionately smaller birth weights of very large herbivores. Energy requirements for reproduction can also be expressed as multiples of maintenance ME requirements. If 0.481 MJ/kg$^{0.75}$ is selected as representative of daily maintenance requirements, E_R is equivalent to 107 to 157× daily maintenance (Table 6). In other words, the metabolizable energy required to produce a single offspring would support a nonreproductive adult female for about 110 to 160 days. Reproduction is energetically expensive.

Species that twin retain more energetic options than those which always produce one offspring. When only a single young is produced by the twinning species, neonate size is small relative to non-twinner species so E_P is reduced. E_L remains more or less comparable to the non-twinner, however. When twins are born, combined birth weight is equivalent to the single young on non-twinners (i.e., E_P is comparable) but increased milk yields and hence increased E_L cause total energy requirements for reproduction to be higher. Thus with twins, twinners appear to require more energy for reproduction than non-twinners, but with a single young, E_R may be somewhat reduced in relation to species that do not twin.

In conclusion, the factorial approach allows an analysis of the scaling of the energy requirements of reproduction to body weight and examination of the effects of number of young. Many assumptions and extrapolations have been made so the predicted energy requirements must be considered approximate and in need of verification and refinement. There can be no doubt, however, that the energy requirements for reproduction represent a large and significant facet of the energetics of wild herbivores.

ACKNOWLEDGMENTS

I would like to thank K. Kranz for helping locate references on birth weights. A. Petric very kindly searched records at the Chicago Zoological Society (Brookfield Zoo) for both maternal and birth weights. K. Kranz and S. Thompson gave useful comments on the pregnancy section, while M. Allen reviewed the entire manuscript. Illustrations were prepared by J. Gradwohl. T. Cummings patiently typed several revisions of the manuscript. I am especially indebted to M. Allen for encouragement and moral support during the long and at times frustrating gestation of this manuscript.

REFERENCES

1. **Sadleir, R. M. F. S.,** The role of nutrition in the reproduction of wild mammals, *J. Reprod. Fertil., Suppl.* 6, 39, 1969.
2. **Sadleir, R. M. F. S.,** Reproduction in female cervids, in *Biology and Management of the Cervidae,* Wemmer, C., Ed., Smithsonian Inst. Press, Washington, D.C., in press.
3. **Thomas, D. C.,** The relationship between fertility and fat reserves of Peary caribou, *Can. J. Zool.,* 60, 597, 1982.
4. **White, R. G. and Luick, J. R.,** Plasticity and constraints in the lactational strategy of reindeer and caribou, *Symp. Zool. Soc. London,* 51, 215, 1984.
5. National Research Council, *Nutrient Requirements of Sheep,* National Academy of Sciences, Washington, D.C., 1975.
6. National Research Council, *Nutrient Requirements of Beef Cattle,* National Academy of Sciences, Washington, D.C., 1976.
7. National Research Council, *Nutrient Requirements of Horses,* National Academy of Sciences, Washington, D.C., 1978.
8. National Research Council, *Nutrient Requirements of Dairy Cattle,* National Academy of Sciences, Washington, D.C., 1978.
9. National Research Council, *Nutrient Requirements of Goats,* National Academy of Sciences, Washington, D.C., 1981.
10. **Arman, P., Kay, R. N. B., Goodall, E. D., and Sharman, G. A. M.,** The composition and yield of milk from captive red deer *(Cervus elaphus* L.), *J. Reprod. Fert.,* 37, 67, 1974.
11. **Robbins, C. T., Podbielancik, R. S., Wilson, D. L., and Mould, E. D.,** Growth and nutrient consumption of elk calves compared to other ungulate species, *J. Wildl. Manage.,* 45, 172, 1981.
12. **Sadleir, R. M. F. S.,** Energy consumption and subsequent partitioning in lactating black-tailed deer, *Can. J. Zool.,* 60, 382, 1982.
13. Agricultural Research Council, *The Nutrient Requirements of Ruminant Livestock,* Commonwealth Agricultural Bureaux, Farnham Royal, England, 1980.
14. **Clutton-Brock, T. H., Iason, G. R., Albon, S. D. and Guiness, F. E.,** Effects of lactation on feeding behaviour and habitat use in wild Red deer hinds, *J. Zool. London,* 198, 227,
15. **Bell, G.,** The costs of reproduction and their consequences, *Am. Nat.,* 116, 45, 1980.
16. **Huggett, A. St. G. and Widdas, W. F.,** The relationship between mammalian foetal weight and conception age, *J. Physiol. (London),* 114, 306, 1951.

17. **Payne, P. R. and Wheeler, E. F.**, Comparative nutrition in pregnancy, *Nature (London)*, 215, 1134, 1967.
18. **Frazer, J. F. D. and Huggett, A. St. G.**, Species variations in the foetal growth rates of eutherian mammals, *J. Zool. London*, 174, 481, 1974.
19. **Sacher, G. A. and Staffeldt, E. F.**, Relation of gestation time to brain weight for placental mammals: implications for the theory of vertebrate growth, *Am. Nat.*, 108, 593, 1974.
20. **Frazer, J. F. D.**, Growth of young vertebrates in the egg or uterus, *J. Zool. London*, 183, 189, 1977.
21. **Rattray, P. V., Garrett, W. N., East, N. E., and Hinman, N.**, Growth, development and composition of the ovine conceptus and mammary gland during pregnancy, *J. Anim. Sci.*, 38, 613, 1974.
22. **Calder, W. A.**, The pace of growth: an allometric approach to comparative embryonic and post-embryonic growth, *J. Zool. London*, 198, 215, 1982.
23. **Graf, W. and Nichols, L.**, The Axis deer in Hawaii, *J. Bombay Nat. Hist. Soc.*, 63, 629, 1966.
24. **Joubert, E.**, Size and growth as shown by pre- and post-natal development of the Hartmann zebra *Equus zebra hartmannae*, *Madoqua*, 1, 55, 1974.
25. **Smuts, G. L.**, Pre- and postnatal growth phenomena of Burchell's zebra *Equus burchelli antiquorum*, *Koedoe*, 18, 69, 1975.
26. **Spinage, C. A.**, *A Territorial Antelope: The Uganda Waterbuck*, Academic Press, New York, 1982, chap. 4.
27. **Fairall, N.**, Prenatal development of the impala *Aepyceros melampus* Licht., *Koedoe*, 12, 97, 1969.
28. **Markgren, G.**, Reproduction of moose in Sweden, *Viltrevy*, 6, 129, 1969.
29. **Chapman, D. and Chapman, N.**, *Fallow Deer: Their History, Distribution and Biology*, Terence Dalton Ltd., Lavenham, U.K., 1975.
30. **Robbins, C. T. and Robbins, B. L.**, Fetal and neonatal growth patterns and maternal reproductive effort in ungulates and subungulates, *Am. Nat.*, 114, 101, 1979.
31. **Robbins, C. T. and Moen, A. N.**, Uterine composition and growth in pregnant white-tailed deer, *J. Wildl. Manage.*, 39, 684, 1975.
32. **Leitch, I., Hytten, F. E., and Billewicz, W. Z.**, The maternal and neonatal weights of some mammalia, *Proc. Zool. Soc. London*, 133, 11, 1959.
33. **Blueweiss, L., Fox, H., Kudzma, V., Nakashima, D., Peteres, R., and Sams, S.**, Relationships between body size and some life history parameters, *Oecologia*, 37, 257, 1978.
34. **Leutenegger, W.**, Maternal-fetal weight relationships in primates, *Folia Primat.*, 20, 280, 1973.
35. **Millar, J. S.**, Adaptive features of mammalian reproduction, *Evolution*, 31, 370, 1977.
36. **Western, D.**, Size, life history and ecology in mammals, *Afr. J. Ecol.*, 17, 185, 1979.
37. **McEwan, E. H. and Whitehead, P. E.**, Reproduction in female reindeer and caribou, *Can. J. Zool.*, 50, 43, 1972.
38. **Templeton, A.R. and Read, B.**, The elimination of inbreeding depression in a captive herd of Speke's gazelle, in *Genetics and Conservation*, Schonewald-Cox, C. M., Chambers, S. M., MacBryde, B., and Thomas, W. L., Eds., Benjamin/Cummings Publ. Co., Menlo Park, Calif., chap. 15.
39. **Mitchell, B.**, The weights of new-born to one-day-old red deer calves in Scottish moorland habitats, *J. Zool. London*, 164, 250, 1971.
40. **Bannikov, A. G., Zhirnov, L. V., Lebedeva, L. S., and Fandeev, A. A.**, *Biology of the Saiga*, Israel Program for Scientific Translations, Jerusalem, 1967.
41. **Dauphine, T. C.**, Biology of the Kaminuriak population of barren-ground caribou. IV. Growth, reproduction and energy reserves, *Can. Wildl. Serv. Rep. Ser.*, 38, 1976.
42. **Franzmann, A. W., LeResche, R. E., Rausch, R. A., and Oldemeyer, J. L.**, Alaskan moose measurements and weights and measurement-weight relationships, *Can. J. Zool.*, 56, 298, 1978.
43. **Reimers, E. and Ringberg, T.**, Seasonal changes in body weights of Svalbard reindeer from birth to maturity, *Acta Zool. Fenn.*, 175, 69, 1983.
44. **Mungall, E. C.**, *The Indian Blackbuck Antelope: A Texas View*, Kleberg Studies in Natural Resources, Texas A & M University, College Station, Tex., 1978.
45. **Wilson, V. J. and Clarke, J. E.**, Observations on the common duiker *Sylvicapra grimmia* Linn., based on material collected from a tsetse control game elimination scheme, *Proc. Zool. Soc. London*, 138, 487, 1962.
46. **Talbot, L. M. and Talbot, M. H.**, The wildebeest in Western Masailand, East Africa, *Wildl. Monogr.*, 12, 1, 1963.
47. **Wilson, V. J.**, Observations of the greater kudu (*Tragelaphus strepsiceros* Pallas) from a tsetse control hunting scheme in Northern Rhodesia, *East Afr. Wildl. J.*, 3, 27, 1965.
48. **Howells, W. W. and Hanks, J.**, Body growth of the impala (*Aepyceros melampus*) in Wankie National Park, Rhodesia, *J. S. Afr. Wildl. Manage. Assoc.*, 5, 95, 1975.
49. **Jeffery, R. C. V. and Hanks, J.**, Body growth of captive eland *Taurotragus oryx* in Natal, *S. Afr. J. Zool.*, 16, 183, 1981.

50. **Blood, D.A., Flook, D. R., and Wishart, W. D.,** Weights and growth of Rocky Mountain bighorn sheep in Western Alberta, *J. Wildl. Manage.*, 34, 451, 1970.
51. **Attwell, C. A. M.,** Growth and condition of blue wildebeest in Zululand, *S. Afr. J. Wildl. Res.*, 12, 63, 1982.
52. **Pemberton, J. M. and Dancie, O.,** Liveweights of fallow deer *(Dama dama)* in British deer parks, *J. Zool. London*, 199, 171, 1983.
53. **Wilson, V. J., Schmidt, J. L., and Hanks, J.,** Age determination and body growth of the common duiker, *Sylvicapra grimmia, J. Zool. London*, 202, 283, 1984.
54. U.S. National Zoological Park, unpublished records.
55. **Dittrich, L. and Boer, M.,** *Verhalten und Fortpflanzung von Kirks Rüssel-Dikdiks (Madoqua(Rhynchotragus) kirki)* im Zoologischen Garten, Zoologischer Garten der Landeshauptstadt Hannover, Hannover, 1980.
56. **Penzhorn, B. L.,** Body growth of the springbok *Antidorcas marsupialis* in the Mountain Zebra National Park, *S. Afr. J. Wildl. Res.*, 8, 171, 1978.
57. **Skinner, J. D.,** An appraisal of the status of certain antelope for game farming in South Africa, *Z. Tierzüchtg. Züchtgsbiol.*, 90, 263, 1973.
58. **Fairall, N. and Braack, H. H.,** Growth and development of the impala *Aepyceros melampus*, *Koedoe*, 19, 83, 1976.
59. **Schmidt, C. R.,** Captive breeding of the vicuna, in *Breeding Endangered Species in Captivity*, Martin, R. D., Ed., Academic Press, New York, 1975, 271.
60. **Lang, E. M.,** Beobachtungen an der Sumpfantilope *(Tragelaphus spekei gratus)* im Zoologischen Garten Basel, *Zool. Gart., Lpz. (N.F.)*, 49, 8, 1979.
61. **Lang, E. M.,** Haltung und Zucht des kleinen Kudu *(Tragelaphus imberbis)*, *Zool. Gart., Lpz. (N.F.)*, 46, 3, 1976.
62. **Sadleir, R. M. F. S.,** Milk yield of black-tailed deer, *J. Wildl. Manage.*, 44, 472, 1980.
63. **Sadleir, R. M. F. S.,** Birth weights and early growth of captive mother-raised black-tailed deer, *J. Wildl. Manage.*, 44, 268, 1980.
64. **O'Gara, B.,** *Antilocapra americana, Mamm. Species*, 90, 1, 1978.
65. **Rideout, C. B.,** Mountain goat, in *Big Game of North America*, Schmidt, J. L. and Gilbert, D. L., Eds., Stackpole Books, Harrisburg, Pa., 1978, 149.
66. **Woolf, A.,** Influences of lambing and morbidity on weights of captive Rocky Mountain bighorns, *J. Mammal.*, 52, 242, 1971.
67. **du Plessis, S. S.,** Ecology of blesbok with special reference to productivity, *Wildl. Monog.*, 30, 1, 1972.
68. **McEwan, E. H.,** Growth and development of the barren-ground caribou. II. Postnatal growth rates, *Can. J. Zool.*, 46, 1023, 1968.
69. **Loudon, S. I., McNeilly, A. S., and Milne, J. A.,** Nutrition and lactational control of fertility in red deer, *Nature (London)*, 302, 145, 1983.
70. **von Richter, W.,** *Connochaetes gnou, Mamm. Species*, 50, 1, 1974.
71. Chicago Zoological Society (Brookfield Zoo), unpublished records.
72. **Tener, J. S.,** *Muskoxen in Canada. A Biological and Taxonomic Review*, Queen's Printer, Ottawa, Ontario, 1965.
73. **Lang, E. M.,** *Das Zwergflusspferd, Choeropsis liberiensis*, A. Ziemsen Verlag, Witenberg Lutherstadt, 1975.
74. **Wackernagel, H.,** Grant's zebra, *Equus burchelli boehmi*, at Basel Zoo. A contribution to breeding biology, *Int. Zoo Yrbk.*, 5, 38, 1965.
75. **Verme, L. J.,** Some characteristics of captive Michigan moose, *J. Mammal.*, 51, 403, 1970.
76. **Posselt, J.,** The domestication of the eland, *Rhod. J. Agric. Res.*, 1, 81, 1963.
77. **Sinclair, A. R. E.,** *The African Buffalo. A Study of Resource Limitation of Populations*, University of Chicago Press, Chicago, 1977.
78. **Camoens, J. K.,** . *The Buffalo in Malaysia*, Ministry of Agriculture, Malaysia, 1976.
79. **Pienaar, U. de V.,** Observations on developmental biology, growth and some aspects of the population ecology of African buffalo (*Syncerus caffer caffer* Sparrman) in the Kruger National Park, *Koedoe*, 12, 29, 1969.
80. **Dagg, A. I.,** *Giraffa camelopardis, Mamm. Species*, 5, 1, 1971.
81. **Dagg, A. I. and Foster, J. B.,** *The Giraffe. Its Biology, Behavior and Ecology*, Van Nostrand Reinhold Co., New York, 1976, chap. 6.
82. **Laurie, W. A., Lang, E. M., and Groves, C. P.,** *Rhinoceros unicornis, Mamm. Species*, 211, 1, 1983.
83. **Hanks, J.,** Growth in weight of the female African elephant in Zambia, *East Afr. Wildl. J.*, 7, 7, 1969.
84. **Hanks, J.,** Growth of the African elephant *(Loxodonta africana)*, *East Afr. Wildl. J.*, 10, 251, 1972.
85. **Meyer, H. and Ahlswede, L.,** Uber das intrauterine Wachstrum und die Körperzusammensetzung von Fohlen sowie den Nährstoffbedarf tragender Stuten, *Uebers. Tierernaehr.*, 4, 263, 1976.

86. **Grand, T.,** personal communication.
87. **Ringberg, T. M., White, R. G., Holleman, D. F., and Luick, J. R.,** Body growth and carcass composition of lean reindeer (*Rangifer tarandus tarandus* L.) from birth to sexual maturity, *Can. J. Zool.*, 59, 1040, 1981.
88. **Loudon, A. S. I. and Kay, R. N. B.,** Lactational constraints on a seasonally breeding mammal: the red deer, *Symp. Zool. Soc. London*, 51, 233, 1984.
89. **Robinson, J. J., McDonald, I., Fraser, C., and Gordon, J. G.,** Studies on reproduction in prolific ewes, *J. Agric. Sci.* (Cambridge), 94, 331, 1980.
90. **Zeveloff, S. I. and Boyce, M. S.,** Parental investment and mating systems in mammals, *Evolution*, 34, 973, 1980.
91. **Kihlstrom, J. E.,** Period of gestation and body weight in some placental mammals, *Comp. Biochem. Physiol.*, 43A, 673, 1972.
92. **Oftedal, O. T.,** Milk composition, milk yield and energy output at peak lactation: a comparative review, *Symp. Zool. Soc. London*, 51, 33, 1984.
93. **Oftedal, O. T.,** *Milk, Protein and Energy Intakes of Suckling Mammalian Young: A Comparative Study*, Ph.D. thesis, Cornell Universeity, Ithaca, N. Y., 1981.
94. **Erb, R. E., Sitarz, N. E., and Malvan, P. V.,** Blood plasma and milk prolactin, and effects of sampling technique on composition of milk from suckled ewes, *J. Dairy Sci.*, 60, 197, 1077.
95. **Doney, J. M., Peart, J. N., Smith, W. F., and Louda, F.,** A consideration of the techniques for estimation of milk yield by suckled sheep and a comparison of estimates obtained by two methods in relation to the effect of breed, level of production and stage of lactation, *J. Agric. Sci.* (Cambridge), 92, 123, 1979.
96. **Coombe, J. B., Wardrop, I. D., and Tribe, D. E.,** A study of milk production of the grazing ewe, with emphasis on the experimental technique employed, *J. Agric. Sci.* (Cambridge), 54, 353, 1960.
97. **Poulton, S. G. and Ashton, W. M.,** Studies on ewes milk. V. The effect of high cereal diets on ewes and on the yield of milk and milk constituents, *J. Agric. Sci.* (Cambridge), 78, 203, 1972.
98. **Wright, D. E., Jones, B. A., and Geenty, K. G.,** Measurement of milk consumption in young ruminants using tritiated water, *Proc. N.Z. Soc. Anim. Prod.*, 34, 145, 1974.
99. **Wallace, L. R.,** The growth of lambs before and after birth in relation to the level of nutrition, *J. Agric. Sci.* (Cambridge), 38, 93, 243, 1948.
100. **Barnicoat, C. R., Logan, A. G., and Grant, A. I.,** Milk secretion studies with New Zealand Romney ewes. I and II, *J. Agric. Sci.* (Cambridge), 39, 44, 1949.
101. **Barnicoat, C. R., Murray, P. F., Roberts, E. M., and Wilson, G. S.,** Milk secretion studies with New Zealand Romney ewes. V—XI, *J. Agric. Sci.* (Cambridge), 48, 9, 1956.
102. **LeDu, Y. L. P., MacDonald, A. J., and Peart, J. N.,** Comparison of two techniques for estimating the milk production of suckler cows, *Livestock Prod. Sci.*, 6, 277, 1979.
103. **Maltz, E.,** Productivity in the Desert: Bedouin Goat, Ibex, and Desert Gazelle, Ph.D. thesis, Tel Aviv University, Tel Aviv, 1979 (in Hebrew; English summary).
104. **Maltz, E. and Shkolnik, A.,** Milk production in the desert: lactation and water economy in the black Bedouin goat, *Physiol. Zool.*, 53, 12, 1980.
105. **Loudon, A. S. I., Darroch, A., and Milne, J. A.,** The lactation performance of red deer on hill and improved species pastures, *J. Agric. Sci.* (Cambridge), 102, 149, 1984.
106. **McEwan, E. H. and Whitehead, P. E.,** Measurement of the milk intake of reindeer and caribou calves using tritiated water, *Can. J. Zool.*, 49, 443, 1971.
107. **Yates, N. G., MacFarlane, W. V., and Ellis, R.,** The estimation of milk intake and growth of beef calves in the field by using tritiated water, *Aust. J. Agric. Res.*, 22, 291, 1971.
108. **Oftedal, O. T., Hintz, H. F., and Schryver, H. F.,** Lactation in the horse: milk composition and intake by foals, *J. Nutr.*, 113, 2096, 1983.
109. **Neseni, R., Flade, E., Heidler, G., and Steger, H.,** Milchleistung und Milchzusammensetzung von Stuten im Verlaufe der Laktation, *Archiv Tierz.*, 1, 91, 1958.
110. **Bouwman, H. and van der Schee, W.,** Composition and production of milk from Dutch warmblooded saddle horse mares, *Z. Tierphysiol. Tierernähr. Futtermittelk.*, 40, 39, 1978.
111. **Walker, D. E. K.,** Milk production in beef heifers, *N. Z. J. Agric.*, 107, 277, 1963.
112. **Peart, J. N.,** The influence of live weight and body condition on the subsequent milk production of Blackface ewes following a period of undernourishment in early lactation, *J. Agric. Sci.* (Cambridge) 75, 459, 1970.
113. **Linzell, J. L.,** Milk yield, energy loss in milk and mammary gland weight in different species, *Dairy Sci. Abstr.*, 34, 351, 1972.
114. **Hanwell, A. and Peaker, M.,** Physiological effects of lactation on the mother, *Symp. Zool. Soc. London*, 41, 297, 1977.
115. **Robbins, C. T., Oftedal, O. T., and O'Rourke, K. I.,** Lactation, early nutrition and hand-rearing of wild ungulates, with special reference to deer, in *Biology and Management of the Cervidae*, Wemmer, C., Ed., Smithsonian Inst. Press, Washington, D.C., in press.

116. **Corbett, J. L.,** Variation in the yield and composition of milk of grazing Merino ewes, *Aust. J. Agric. Res.*, 19, 283, 1968.
117. **Peart, J. N., Edwards, R. A., and Donaldson, E.,** The yield and composition of the milk of Finnish Landrace x Blackface ewes. II. Ewes and lambs grazed on pasture, *J. Agric. Sci.* (Cambridge), 85, 315, 1975.
118. **Boggs, D. L., Smith, E. F., Schalles, R. R., Brent, B. E., Corah, L. R., and Pruitt, R. J.,** Effects of milk and forage intake on calf performance, *J. Anim. Sci.*, 51, 550, 1980.
119. **Lindstedt, S. L. and Calder, W. A.,** Body size, physiological time, and longevity of homeothermic animals, *Q. Rev. Biol.*, 56, 1, 1981.
120. **Peart, J. N., Doney, J. M., and Smith, W. F.,** Lactation pattern in Scottish Blackface and East Friesland X Scottish Blackface cross-bred ewes, *J. Agric. Sci.* (Cambridge), 92, 133, 1979.
121. **Sadleir, R. M. F. S.,** Energy and protein intake in relation to growth of suckling black-tailed deer fawns, *Can. J. Zool.*, 58, 1347, 1980.
122. **Luick, J. R., White, R. G., Gau, A. M., and Jenness, R.,** Compositional changes in the milk secreted by grazing reindeer. I. Gross composition and ash, *J. Dairy Sci.*, 57, 1325, 1974.
123. **Macy, I. G., Kelly, H. J., and Sloan, R. E.,** *The Composition of Milks*, rev. ed., Publ. No. 254, National Research Council, Washington, D.C., 1953.
124. **Gleddie, V. M. and Berg, R. T.,** Milk production in range beef cows and its relationship to calf gains, *Can. J. Anim. Sci.*, 48, 323, 1968.
125. **Oftedal, O. T.,** Body size and reproductive strategy as correlates of milk energy output in lactating mammals, *Acta Zool. Fenn.*, 171, 183, 1984.
126. **Gardner, R. W. and Hogue, D. E.,** Effects of energy intake and number of lambs suckled on milk yield, milk composition and energetic efficiency of lactating ewes, *J. Anim. Sci.*, 23, 935, 1964.
127. **Peart, J. N., Edwards, R. A., and Donaldson, E.,** The yield and composition of the milk of Finnish Landrace x Blackface ewes. I. Ewes and lambs maintained indoors, *J. Agric. Sci.* (Cambridge), 79, 303, 1972.
128. **Maxwell, O. J., Doney, J. M., Milne, J. A., Peart, J. N., Russel, A. J. F., Sibald, A. R., and MacDonald, D.,** The effect of rearing type and prepartum nutrition on the intake and performance of lactating Greyface ewes at pasture, *J. Agric. Sci.* (Cambridge), 92, 165, 1979.

Chapter 11

ASSESSMENT OF NUTRITIONAL STATUS

A. W. Franzmann

TABLE OF CONTENTS

Application of the animal indicator concept to assess nutritional status of large herbivores is the focus of this chapter. Use of animals in populations as indicators to assess change in the animal's environment has, in recent years, received increased research effort. This effort is not in lieu of traditional direct monitoring of the environment (habitat analysis and evaluation) and vital statistics (sex and age ratios, natality, mortality, recruitment and survival) of a population, but an adjunct to them. Increased emphasis on the indicator concept was stimulated by a need for increased precision to better manage wildlife populations. Technological advances in animal capture (immobilization) and laboratory biochemical procedures (availability and cost) were important factors for recent gains in using the animal indicator.

Direct assessment of animals in a population has centered on morphology (form), physiology (function), and behavior (activity). This discussion will focus on form and function; behavior is covered in Chapter 3. Measurements of animal form and function in response to nutritional perturbations are designated direct measurements when compared to measurements of the environment and population statistics.

Animal condition is a vague term that may refer in a general way to the level of energy reserves of the animal with respect to internal and external demands,[1] or as expressed by Hanks,[2] the individual's chances of living or dying. The term health may better describe this state through measures of form, function, reproduction, and activity; nevertheless, the term condition has been widely used and generally accepted.

Requirements to use blood parameters to assess nutritional status include:[3] (1) determine boundary conditions or identify sources of variation altering function; (2) establish baseline values within boundary conditions; and (3) quantify physiologic changes related to environment. Hanks[2] provided a solid basis for adding; (4) assess the resilience of an animal population or some measurement for its ability to withstand further perturbations. These criteria may apply to assessments of animal form in a general way, but have been primarily used for blood parameters.

I. MORPHOLOGY (FORM)

A. Morphometrics

A review of early studies in North America noted that deer show differences in rates of growth and ultimate body size in relation to quality of food.[4] Measurements of both dead and live herbivores which may reflect this relationship include: live, dressed, and carcass weights; linear and circumferential body measurements; antler and horn weights and measurements; and various combinations. The selection of measurement was more often dictated by field circumstances than design.

Most studies on relationships of animals to their range focused on body weight or mass. Due to the difficulty in obtaining live weights and the ease of measurements on animals killed by hunters, carcass and dressed to live weight relationships were developed for white-tailed deer (*Odocoileus virginianus*),[5] bison (*Bison bison*),[6] pronghorns (*Antilocapra americana*),[7] Rocky Mountain and Roosevelt elk (*Cervus elephus*),[8-9] bighorn sheep (*Ovis canadensis*),[10] roe deer (*Capreolus capreolus*),[11] and moose (*Alces alces*).[12] Similarly, prediction equations were developed for linear and circumferential measurements to body weight for Colombian black-tailed deer (*Odocoileus hemionus columbianus*),[13] caribou and reindeer (*Rangifer tarandus*),[14,15] white-tailed deer,[16] moose,[17] and certain African ungulates.[18] Some authors noted that skeletal measurements were better than soft body measurements because they could be more accurately measured.[19,20]

Antlered herbivores provide an additional parameter, and antlers were reported to be more directly affected by limitation in forage quality than body size.[21] Moose populations in Alaska in good habitat developed larger antlers at an earlier age than those in poor habitats.[22]

Moose maintained at high densities at the Moose Research Center, Alaska, had extremely slow and poor antler development. Reindeer males on restricted diets for 1 year developed smaller, lighter antlers than those fed *ad libitum*.[23]

Studies assessing the effect of controlled nutritional intakes on morphometrics are limited, but those done have provided valuable information. Birth weights of white-tailed deer fawns from poorly fed does were significantly lighter than those from well-fed does.[24] Birth weights thus serve as an index of maternal nutritional status. White-tailed deer fawns which were restricted 40% in energy intake had a 76% reduction in body fat deposition, but had no significant decrease in body weight.[25] In another study, high energy intake of white-tailed deer fawns resulted in an increase in body weight, skeletal size, internal organ size, and fat deposits, but there was little effect of varying levels of protein intake.[26] In the extreme case, starved adult mule deer lost 37% and fawns 30% of prestarvation body weight. A 30% body weight loss was approximately the point at which the deer could not recover.[27]

Gut fill is an added source of variation related to body weight, especially in nonselective grazing ruminants. Weight variation can be significant, but standardization of procedures used for comparisons can minimize this influence, and is essential.

In spite of some interpretation and standardization problems with morphometrics, they have been used to assess range quality or nutrient status for some time by wildlife managers. The assessments were limited for determining "relative goodness" but valuable when differences in habitat and nutrient status were substantial.

B. Organs and Glands

Fetal development (weight; eyelens, thyroid, thymus, and adrenal weight) were related to maternal malnutrition caused by winter severity.[28] All measurements decreased in severe winters except adrenal weight. Adrenal weights were generally heavier in severe winter, but the trend was inconsistent.

Thymus weight in white-tailed deer fawns is strongly influenced by nutritional status.[29] Seasonal fluctuations were noted in free-ranging fawns, with the thymus reaching maximum size in autumn and minimum size in January and February. Animals fed restricted diets in autumn had smaller organs by December than those fed better diets. The authors devised a model for assessing the nutritional history of deer based upon their experiments.[29]

Enlargement of the adrenal gland in response to stress has not been as well documented in large as in small mammals.[2] The potential of the adrenal gland as an index of condition was suggested for mule deer (*Odocoileus hemionus*),[30] Sika deer (*Cervus nippon*),[31] and white-tailed deer.[32] However, in studies of ten East African ungulates, only one species had a significant correlation between adrenal weight and total fat.[33] A similar conclusion was made from studies of wildebeest (*Connochaetes taurinus*).[34] Seal et al.[35] found no evidence of increased adrenal activity indicative of stress in white-tailed deer at high densities. Hanks[2] noted that a variety of important factors influence adrenal weight and more studies are needed.

C. Fat Reserves

Fat reserves represent stored energy and measurements of this reserve have been applied to assessment of condition. The sequence of fat mobilization provides a basis for selecting anatomical sites to sample fat reserves, depending upon objectives. Harris[36] first described the sequence in white-tailed deer in order of earliest disappearance as (1) rump fat, (2) subcutaneous fat, (3) visceral fat, and (4) marrow fat. Subsequently, more detailed descriptions of this pattern have been published.[37-39] If a study objective were to evaluate early fat loss, a measure of rump fat may prove useful. Conversely, to assess starvation, bone marrow fat (BMF) may be preferred.

Kidney fat indexes[36]

$$\left[\frac{\text{KFI} = \text{perinephric fat}}{\text{kidney weight}} \times 100\right]$$

decrease to approximately 40 without appreciable change in marrow fat.[2]

Relative disappearance rates between other types of fat deposits are less obvious. Fat mobilization was noted to differ significantly between femur and metatarsus and humerus and metacarpus bones in white-tailed deer[40] and some African ungulates.[39] This difference in mobilization was not noted in moose, wherein fat was simultaneously mobilized from all limb bones.[41] Mandible and longbone marrow fat from moose and caribou were correlated,[42,43] but differences in fat mobilization by age classes were suggested.[43]

A characteristic difference in BMF between moose and impala (*Aepyceros melampus*) was noted. Marrow fat was considered of limited value for condition assessment in young impala (≤2 years old),[44] but it was very useful for moose calves.[45]

Other difficulties with interpreting BMF data were related to methodology. Visual estimates of color and consistency[40] were limited due to lack of quantification.[45,46,47] The compression method as described for wapiti[48] was an improvement, but was supplanted by the oven-dried[49] and reagent-dried methods,[50] particularly the former. Dessication may limit application of BMF indexes when assessments are attempted from bones which are not fresh[51] or those not frozen.[41,51]

The basic assumption for application of the KFI is that kidney weight is a constant function of body size.[36] This assumption was questioned[52] and later proven erroneous in caribou.[53] The KFI was distorted 70 to 80% due to seasonal kidney weight fluctuations in caribou. Adjustments must be made for interseasonal comparisons in caribou and further studies are needed in other species.

Total body fat (TBF) measurements are difficult and expensive to obtain,[2] and are generally derived from body composition studies. These studies provide invaluable baseline information for understanding and predicting energy and protein requirements for growth.[54] Body composition has been studied for several wild herbivores (Chapter 10), including: roe deer,[11] Uganda kob (*Kobus kob thomasi*),[55] white-tailed deer,[54,56] mule deer,[57] impala,[58] and reindeer.[59]

Body composition studies of impala were used to evaluate 38 condition indexes. Complete buttock dissection proved most useful followed by KFI. No other criteria were correlated with condition (body fat/ash × 100) or total body fat.[58] Another body composition study on white-tailed deer suggests that lipogenesis is an obligatory physiological event in autumn and proceeds despite undernutrition until a serious negative energy balance occurs.[56] They concluded that fat reserves may not reliably indicate relative range quality unless other physical parameters were also examined.

D. Rumen Morphology

Hofmann[60] described the changes in rumen papillation that corresponded to dietary fiber content. Rumen papillae can increase their surface enlargement factor (SEF) by growth of existing papillae and by increasing their density per unit surface or by both. This process occurs to optimize nutrient absorption during low fiber intake. Ruminants on poor quality roughage have a low SEF and may have papillae with keratinized epithelium and lowered absorption. Vascularization in the sub mucosa also increases with increases in SEF.[60]

Changes in the morphology of the rumen may be considered as an additional method for assessment of the quality of forage intake. Comparative studies would be beneficial, particularly when data on SEF for a species were not available. A limitation of the procedure is that the information must come from dead animals.

E. Condition Scores

Most wildlife managers and researchers have made subjective appraisals of animals such as good, fair, or poor condition. Characteristics associated with inanition were described to provide a basis for this system in red deer,[36] tsessebe (*Damaliscus lunatus*),[61] and elephant (*Loxodonta africana*).[62] A classification system from 1 to 10 (10 being best) based upon physical characteristics and activity patterns was devised for white-tailed deer[63] and was subsequently modified and applied to bighorn sheep[64] and moose.[65]

Most problems associated with such assessments are those associated with any subjective evaluations. In addition, it has been demonstrated that considerable change may occur in fat deposits without altering the external appearance of animals.[44,66] Nevertheless, there is considerable value in having this type of assessment in conjunction with other criteria.[67] They certainly are valid for identifying extremes in condition; their weakness is in differentiating "relative goodness".

Many subjective methods for evaluating condition by carcass quality have been used, but few were based on any type of system. Kistner et al.[68] outlined a technique that may be used for deer which has been widely tested in the field. This method is based upon fat reserve depletion and provides a numerical score for the carcass, but it is basically a subjective index.

II. PHYSIOLOGY (FUNCTION)

A. Hematology and Blood Chemistry

1. Boundary Conditions

Identifying boundary conditions or sources of variation for blood values is the first step in selecting blood parameters for nutritional assessment. Boundary conditions for free-ranging large herbivores have been identified as sex, age, location, season, year, habitat, rectal temperature, capture and handling methods, pregnancy, lactation, condition, sample time/post capture, weight, growth/weight, climatic factors, wild vs. captive, malnutrition, marrow or body fat, disease, and starvation (Table 1).

Individual blood parameters in the studies cited were evaluated to determine boundary conditions, but were summarized under classes of blood parameters (hematology, nonprotein nitrogen, proteins, lipids and fatty acids, glucose and lactate, electrolytes, minerals and trace elements, enzymes and hormones) to facilitate and simplify identification of those factors which must be considered in nutritional assessment using blood parameters (Table 1).

Several conclusions are obvious from Table 1: (1) most boundary conditions significantly alter one or more classes of blood parameters; (2) varied responses (some or none) were observed for the same perturbation; (3) representation from large ungulate species is poor (9 species); and (4) several important boundary conditions have been identified. Less obvious conclusions require checking the original publications to evaluate methodology and objectives. Most blood data come from studies with a great variety of objectives and methodology. This explains, in part, the inconsistency of blood parameter responses. Nevertheless, these studies provide data that identify important boundary conditions, the first requirement for using blood parameters for nutritional assessment.

Boundary conditions (Table 1) all had some influence on blood parameter classes. Ranking these boundary conditions in order of importance becomes difficult and even dangerous because each parameter has characteristics which make it unique, depending upon objectives. In some instances, we may want to measure short-term responses (minutes) and in others, long term responses (months). Different classes of blood parameters are thereby needed and their relative importance changes.

A conclusion from nearly all the studies on free-ranging herbivores is that to better evaluate influences on blood parameters, controlled studies using serial samples and/or captive animals are needed.

Table 1
RESPONSE OF BLOOD PARAMETER CLASSES TO BOUNDARY CONDITIONS TESTED IN NONCONFINED FREE-RANGING LARGE HERBIVORES

Classes of blood parameter	Species[a]	Ref.	Sex	Age	Location	Season	Year	Habitat	Rectal temperature	Handling and capture	Pregnancy	Lactation	Condition	Sample time/post capture	Growth/weight	Climate	Wild vs. captive	Malnutrition	Marrow or body fat	Disease	Starvation
Hematology	Aeme	58											0						0		
	Alal	65,67, 69,119	0	* 0	*	*	*		* 0	* 0	0	0	*								
	Anam	76,77,78	* 0	* 0	* 0		*	*	*	*				*	* 0	* 0					
	Bibi	74,75	0	* 0	*																
	Ceel	121,122, 123	* 0	0							* 0						* 0				
	Odhe	126,127,128	* 0	* 0		* 0	* 0	*					*		*	* 0		*	* 0		
	Odvi	79,80, 129,130	0	*	*	*		*													
	Ovca	64,88, 90,135, 136			*			* 0		* 0			* 0				* 0			* 0	
	Rata	81,138, 139,140, 141	0	*		* 0				* 0	* 0	*					* 0	*	0		
Nonprotein nitrogen	Aeme	58											0						0		
	Alal	65,67,69	0	* 0	*	* 0	0		0	0	0	0	0								
	Anam	76,78,120	0	* 0	*			*	0	* 0				* 0	* 0	0					
	Ceel	121,122, 123,124	0	*0										*			*				
	Odvi	79,80	* 0	*	*	*		*													
	Ovca	64,88,135			*			*		0			0				*				
	Rata	81,138,140	0	* 0		* 0				0								*			
Proteins	Aeme	58											0						0		
	Alal	65,67,69	* 0	* 0	* 0	*	* 0		* 0	* 0	0	0	* 0								
	Anam	76,78,120	* 0	* 0	* 0		* 0	* 0	* 0	0				* 0	* 0	* 0					
	Ceel	121,122, 123,124	* 0	* 0										0			* 0				
	Odvi	79,80, 131,132, 133	0	*	*	*		*												*	
	Ovca	64,135,137			*			*		*			0				0	*		*	
	Rata	81,138, 140	0	*		* 0				0		* 0									
Lipids and fatty acids	Aeme	58											0						0		
	Alal	65,67,69	* 0	0	*	*	0		0	0	0	*	0								
	Anam	76,78,120	* 0	0	*			*	0	0				0	0	0					
	Bibi	74	*																		
	Ceel	121,122, 123,124	0	0							0			0			* 0				

	Odvi	79,80, 131,134	* 0	* 0	*	* 0		*											
	Ovca	64,135			*			*		*				0			* 0		
	Rata	81,140,142	0	*								*						*	*
Carbohydrates	Alal	65,67, 69	* 0	0	*	*	0		*	*	0	0	0						
	Anam	76,120	0	* 0			*		0	0				*	0	0			
	Ceel	121,122, 123,124	* 0	0							0						* 0		
	Odvi	79,80,131	* 0	* 0	0	*		0											
	Ovca	64,135			*			*		*									
	Rata	81,138, 140,143	0	*						*									
Electrolytes, minerals and trace elements	Alal	65,67,69	* 0	* 0	* 0	*	* 0		0	0	0	0	0						
	Anam	76,120	* 0	*			* 0		0	0				*	0	0			
	Ceel	121,122, 123,124	0	* 0		*					0			*			0		
	Odvi	80,131	0	* 0	0	* 0		0											
	Ovca	64,88,135			*			* 0		* 0			0				* 0		
	Rata	81,138, 140,144	0	* 0		*				* 0	*						*	*	*
Enzymes	Alal	65,69	* 0	* 0	0	*	0		0	0	0	* 0	*						
	Anam	76,78,120	* 0	* 0	*		* 0	*	* 0	* 0				* 0	0	* 0			
	Bibi	74		*										*			* 0		
	Ceel	121,122, 123,124	* 0	*							* 0								
	Odvi	80,131	* 0	* 0	* 0	0		* 0				* 0				* 0			
	Ovca	88,135								*		*						*	
	Rata	81,140,144	0	* 0		* 0						0	0						
Hormones	Alal	82							*	*				* 0	0	* 0			
	Anam	78	0	0	*			*											
	Ceel	125								*				* 0			* 0		
	Odvi	80,131,136	* 0	0	* 0	0		* 0									*		
	Rata	81,138, 140,145	0	*		* 0				0		*						*	

Note: * = significant alteration, 0 = no significant change.

[a] Species listed by contracture of genus/species.

Abbreviations: Aeme = *Aepyceros melampus* (impala); Alal = *Alces alces* (moose); Anam = *Antilocapra americana* (pronghorn); Bibi = *Bison bison* (American bison); Ceel = *Cervus elephus* (American elk); Odhi = *Odocoileus hemionus* (mule deer); Odvi = *Odocoileus virginianus* (white-tailed deer); Ovca = *Ovis canadensis* (bighorn sheep); Rata = *Rangifer tarandus* (caribou/reindeer).

2. Baseline Values

Baseline values may be established once boundary conditions are identified. This may be done by eliminating boundary condition influences by improving study design and standardizing procedures. Many studies summarized (Table 1) produced a set of baseline data. For example, moose (*Alces alces*) blood data were sorted to provide listing of means and standard deviations for calves, subadults, and adults by season, sex, and location.[65,67,69] This classification resulted in sorting out some major boundary condition influences and provided a listing of expected values for those classifications. Additional sorting provided values related to condition[65,69] which provided an opportunity to compare the physiologic status of various moose populations.[67,70-72] Baseline values for white-tailed deer are the most extensive available.[73]

Population comparisons using blood data are the earliest applications reported for gross assessment of nutritional status. No quantifications were possible; the data simply made physiologic comparisons feasible. This was a vast improvement from taking selected data and drawing inferences from domestic animal values. Without quantification from controlled studies, these studies were limited to statements about exceptional populations at either end of the physiologic scale. They cannot be used as measures of "relative goodness" without further quantification, and are useful as supportive data combined with other population assessments. Studies comparing blood parameters from free-ranging populations other than moose have been reported for bighorn sheep,[64] bison,[74,75] pronghorn,[76,78] white-tailed deer,[79-80] and reindeer.[81]

3. Response to Perturbation

The third criterion needed to better refine blood assessments of nutritional status is difficult to attain because it requires maintaining captive animals for controlled studies or serially sampling from free-ranging animals. The only exceptions may be studies where animals are captured and handled under highly standardized procedures and quantification of the animals status is done based upon predetermined criteria.[58,65,67,69] Two groups of these quantified studies will be considered: First, those which are not closely related to nutritional influences (sex, age, immobilizing drugs, physical restraint, capture and handling, estrus, gestation, lactation, and sampling time after death) (Table 2); secondly, those studies which tested a boundary condition related to nutritional status (protein intake, energy intake, protein + energy intake, starvation, season, condition, and body fat) (Table 3).

Controlled studies have not added appreciably in identifying boundary conditions, but they provided a more reliable basis for our consideration of them in study design. Most effort was directed toward influences related to capture (immobilization, handling, and excitability[65,82-88]). The only values not affected by capture-related activity in any studies were BUN, sodium (Na), potassium (K), chloride (Cl), calcium (Ca), white blood cell count, and mean corpuscular values (Table 2). The mean corpuscular values may be disregarded because there are quantitative relationships between red blood cell counts, hemoglobin (Hb), and packed cell volume (PCV), and in all studies, these values were depressed; therefore, the relationship may not necessarily be affected. Sex appears to have little influence on several important parameters (Hb, PCV, MCHC, BUN, and cholesterol) based upon white-tailed deer studies,[89] whereas age was a factor for most parameters (Table 2).

Other than capture-related studies, limited information was provided by controlled studies (Table 2). This is partly because study design and standardized procedures were used to control many identified boundary conditions. This allowed studies to focus on more direct nutritional assessment (Table 3).

An evaluation of blood studies involving nutritional assessment (direct and indirect) does not provide the panacea many hoped for (Table 3). There remain many inconsistencies in response of blood parameters to nutritional perturbation. However, there were some useful findings. Eight controlled studies have identified the direct relationship of protein intake to BUN.[89-96] However, the response changed when intake dropped to starvation levels and body protein catabolism occurred.[27,95] This time varies, depending upon fat reserves of the animal going into a starvation period.[27] Starved animals also had elevated levels of amino acid nitrogen[95] and depressed levels of total serum protein (TSP)[95] and glucose.[27] Unfortunately, these were the only blood parameters tested in the two studies where starvation was controlled and monitored. Since BUN levels were not affected by capture procedures (Table 2), it is clear that this parameter is a valuable tool to monitor effects of protein intake. However, BUN is not a sufficient indicator of nutritional status since the energy level of the diet affects nitrogen reutilization.[92,94,97] Blood indicators of dietary energy as well as protein are needed.

A review of studies where energy intake was controlled did not provide an overwhelming choice of a blood parameter such as BUN for protein intake (Table 3). There is disagreement regarding response of those parameters which are most likely to be useful such as Hb, PCV, mean corpuscular hemoglobin concentrations (MCHC), TSP, albumin, cholesterol, nonesterified fatty acid (NEFA), and BUN. There was one study which evaluated responses of amino acids and triidothyronine (T_3) to energy intake, and they concluded that isoleucine, leucine, and T_3 were the best indexes.[92] They also indicated that NEFA was useful for monitoring energy intake; however, another study found no response of NEFA to energy intake.[89] Both studies were in white-tailed deer fawns.

Studies evaluating both protein and energy intake did not provide blood parameters that could be easily applied (Table 3). This boundary condition approaches assessment by use of a combination of inputs with complicated responses due to interactions. When we assess boundary conditions such as season and condition, the inputs are even greater and responses more difficult to evaluate. For this reason, blood value response was designated as fluctuating or not fluctuating (Table 3). An important finding from a study on seasonal influence was that all parameters investigated (BUN, amino acid, nitrogen, TSP, PCV) showed marked seasonal variation independent of dietary plane.[95] Bahnak et al.[99] reported that white-tailed deer that were semi-starved had significant drops in thyroxine (T_4) and triidothyronine (T_3) which suggested that decreasing thyroid hormones served as a mechanism to reduce energy requirements of malnourished deer. The thyroid hormone T_3 was proposed as diagnostic for dietary status and T_4 an indicator of relative well-being of deer.

Condition/blood relationships were identified for moose.[65,69] The parameters most useful for condition assessment were PCV, Hb, TSP, calcium, and phosphorus. A condition index (body fat/ash × 100) and body fat were tested in impala (*Aepyceros melampus*) and no relationships were detected for PCV, TSP, albumin, BUN, total lipids, free-fatty acids (FFA), and cholesterol.[58]

Confusion may arise on how to apply various parameters and which to use. Much of it arises because of differing objectives. Seal[100] used the different ranges of response to perturbations to demonstrate the importance of selecting proper blood parameters for an assessment. The responses ranged from seconds (glucose, catecholamines, cortisol, glucagon) to days (ketone bodies, NEFA, T_3, T_4, urea, amino acids) to weeks (proteins) to months (red blood cell morphology and Hb). He concluded that by measurement of the appropriate parameter and their correlation with growth, reproduction, and survival, it may be possible to assess the quantity and composition of the food supply available to an animal over time. In an assessment of blood values in white-tailed deer, it was concluded that the relationship of blood parameters to nutrition can be applied to wild deer.[73] Most blood data available today come from studies on white-tailed deer (Tables 1 to 3) which lends additional value to this conclusion.

Table 2
RESPONSE OF BLOOD PARAMETERS TO BOUNDARY CONDITIONS IN LARGE HERBIVORES TESTED IN CONTROLLED STUDIES

Blood parameter	Sex: Odvi(89), Rata(146)	Age: Odhe (147,148)	Age: Odvi (Fowns) (149)	Age: Rata (150)	Drug: Odvi (52,54,54,55): Short term	Drug: Odvi (52,54,54,55): Long-term	Physical restraint: Odvi (83, 84,85,86,87)	Excitability: Alal (65,82)	Estrus: Rata (150)	Gestation: Odvi (151)	Gestation: Rata (150)	Location: Odvi (152)	Location: Rata (153)	Time sampling after death: Odvi (154)
Hematology														
Red blood cell count					−	−	−					0		
Hemoglobin (Hb)	0	·	−	+	−	−	−				·	0		
Packed cell volume	0	−	−	+	−	·	−		−		·	0		
Mean corpuscular Hb			−		0		0							
Mean corpuscular volume			·		0	0	0							
Hb concentration	0		0		0	0	0							
White blood cell count				·	0	0	0							
Sedimentation rate		0												
Nonprotein nitrogen														
Nonprotein nitrogen		−												0
Urea nitrogen	0		0		0	0	0							
Proteins														
Total serum protein		· —	0	·	−		0		−	−	−			0
Albumin		·	0		0		0	·		−				
Globulin		−	0		0		0	−		+		0		
Gamma globulin			+					−		−		0		
Fibrogen														
Sialic acid					0									
Lipids and fatty acids														
Cholesterol	0		0			0	−							
Vietone TRB	0													
Carbohydrates														
Glucose			0		+	·	·	−					0	·
Electrolytes, minerals and trace elements														
Sodium			0		0	0	0							
Potassium			0		0	0	0							
Chloride			0		0	0	0							
Calcium			0		0	0	0							
Phosphorus			0		−	−	−							
Enzymes														
Alkaline phosphotase					+0	−0	−0							
Glutamic oxalycetic transaminare			0		+0	+0	+0							
Creative kinase					+0	+0	·							
Lactic pehydugenase			0		0	+0	−0	+						
Ceruloplasmin					0									
Hormones														
Thyroxine														
Cortisol	0				+	·	+	+						0
Projestin					+	−	+							0

Note: + = positive correlation, − negative correlation, 0 = no correlation.
Abbreviations: Alal = *Alces alces* (moose); Odhe − *Odocoileus hemionus* (mule deer); Odvi = *Odocoileus virginianus* (white-tailed deer); Rata - *Rangifer tarandus* (reindeer or caribou).

Table 3
RESPONSE OF BLOOD PARAMETERS TO NUTRITIONALLY INFLUENCED BOUNDARY CONDITIONS IN LARGE HERBIVORES TESTED IN CONTROLLED STUDIES

	Protein intake				Energy intake			Protein and energy		Starvation			Season (* indicates fluctuation)			Condition		Body fat:
	Bibi	Ceel	Odvi	Ovca	Bibi	Odvi	Odheco	Bibi	Odvi	Odvi	Odhe	Alal	Bibi	Odvi	Rata	Aema	Alal	Aeme
Blood parameters	(96)	(91)	(89,92,93,94)	(90)	(96)	(89,92,94,97)	(147,148)	(96,155)	(92,95,101,156)	(95,101)	(27)	(157)	(96,155)	(95,101,156,158)	(159,160,191)	(58)	(65,69)	(58)
Hematology																		
Red Blood Cells (RBC)			+			+			0					*				
Hemoglobin (HB)	+		+ 0	+	+	0	0	+ 0	0			*	*	*			+	
Packed cell volume	+		0	+	+	0	0	− 0	0			*	* 0	*		0	−	0
Mean corpuscular Hb	+		+		0	−		0	+			*	*	0				
Mean corpuscular volume	0		+		0	0		0	0				*	0				
Mean corpuscular Hb concentration	0		0		0	0		0	+				0					
White Blood Cells (WBC)	+		0		0	0		+ − 0	0				* 0	0				
WBC differential								+										
Sedimentation rate							0											
Nonprotein nitrogen																		
Urea nitrogen	+	+	+	+	+	− 0	0	+ 0	+ 0	+	+	*	*	*		0	0	0
Creatanine								− −					0					
Bilirubin						0						*					0	
Urea/Creatanine ratio			+															
Proteins																		
Total protein	−		0		+ −	0	0	0	0	−		*	* 0	*		0	·	0
Albumin	0		0		0	0	0	0	0			*	*	*		0	·	0
Globulin							0					*	*				0	
Alpha glubulin	+				+			0				*	0				0	
Beta globulin	+				0			+				*	*				+	
Gamma globulin	0		0		−	0	0			*	*				0			
Fibrinogen			0			0		0	0				0	*				
Sialic acid									+									
Lipids and fatty acids																		
Cholesterol	+		+		−	0		+	+			*	*	*		0	0	0
Triglyceride									0					0				0
Nonesterified fatty acid			0			− 0												
Ketones			+			0												
Carbohydrates																		
Glucose	−		0		−	0	−	+ 0	+ 0		·	*	*	0	*		+	
Electrolytes, minerals and trace elements																		
Sodium			+			+		0	0				* 0	0 *				
Potassium			0			0		0	0					*				
Chloride			+			+		0	+			*	0	0			+	
Calcium								0	0			*	0	*			+	
Phosphorous	+		0		−	−		0	0					0				
Magnesium									0			*	0				0	

Table 3 (continued)
RESPONSE OF BLOOD PARAMETERS TO NUTRITIONALLY INFLUENCED BOUNDARY CONDITIONS IN LARGE HERBIVORES TESTED IN CONTROLLED STUDIES

	Protein intake				Energy intake			Protein and energy		Starvation			Season (* indicates fluctuation)			Condition		Body fat:
Blood parameters	**Bibi (96)**	**Ceel (91)**	**Odvi (89,92,93,94)**	**Ovca (90)**	**Bibi (96)**	**Odvi (89,92,94,97)**	**Odheco (147,148)**	**Bibi (96,155)**	**Odvi (92,95,101,156)**	**Odvi (95,101)**	**Odhe (27)**	**Alal (157)**	**Bibi (96,155)**	**Odvi (95,101,156,158)**	**Rata (159,160,191)**	**Aema (58)**	**Alal (65,69)**	**Aeme (58)**
Enzymes																		
Alkaline phosphotase	+		0		+	+		0	0			0	*	0			0	
Glutamine oxalycetic transaminase	+		0		0		0	+0	0			*	0					
Glutamine pyruvic transaminase			0			+		+	0				*0	*				
Creatanine kinase	+		0		+	+		0	0				0				0	
Lactic dehydrogenase			0		+	+			0	0	0							
Hormones																		
Triidothyronine			0			+			+ 0	−			0	*	0			
Thryoxine			0				0	0	+ 0	−								
Growth Hormone									0					0	*			
Cortisol			0			+			+ 0									
Progestin			0			+			+									
Insulin			0			+			0									
Testosterone															*			
Amino acids																		
Isoleucine			0			+			0									
Leucine			0			+			0									
Phenylthalanine			0			+			0									
Histidine			0			+			0									
Theonine			0			+			0									
Glycine			0			+			0									
Valine			+			+			0									
Citrulline			0			0			+									
Taurine			0			0			+									
Glutamic acid			0			0			+									
Asportic acid			0			0			+									
Glutamine/Asparaguine			0			0			+									
Amino acid nitrogen										+					*			

Note: + = positive correlation, − = negative correlation, 0 = no correlation, * = significant change.

Abbreviations: Aeme = *Aepycerous melampus* (impala); Alal = *Alces alces* (moose); Bibi = *Bison bison* (bison); Ceel = *Cervus elaphus* (North American elk); Odhe = *Odocoileus hemionus* (mule deer); Odheco = *Odocoileus hemionus columbianus* (Columbia black-tailed deer); Odvi = *Odocoileus virginianus* (white-tailed deer); Rata = *Rangifer tarandus* (reindeer or caribou).

4. Resilience

The ability to utilize blood parameters to determine resilience of an animal to further perturbation can only be suggested by interpreting data from studies where animals are subjected to starvation.[27,95,99] The assumption may be made that when catabolism occurs, as identified by marked increases in BUN and/or amino acid nitrogen and depressed TSP and glucose, the animal approaches the point where resilience and response to further perturbation are not possible. Thyroid (T_3) levels may be valuable for monitoring resilience. More study is needed to evaluate resilience of populations using not only blood parameters but other physiological and morphological measurements. Resilence may be better assessed by the reproductive response of a population.

B. Urine

Urine nitrogen-based constituents (urine nitrogen (N), urea, creatinine, urea creatinine ratios, and ^{14}C-urea) were monitored in studies where diet was controlled in white-tailed deer,[25,89,93,98] mule deer,[60] and wapiti.[91] Factors identified as significantly influencing some measure of urine N were protein intake, energy intake, urea recycling, and tissue catabolism.[89] A review of urinary excretion of hydroxyproline expressed as the hydroxyproline-creatinine index (HCI) for nutritional assessment in mammals concluded that this promising technique requires additional study before its value can be assessed.[2]

An important conclusion regarding urinary/creatinine ratios was that additional nutritional indexes are required to differentiate the four main effects influencing this parameter.[89] This conclusion would hold for other measures of urinary N excretion.

C. Milk

Few data are available for making application of milk constituents to nutritional assessment in wild herbivores. Published data consist primarily of basic milk analyses taken on limited numbers of animals. A review of gross composition of milk from North America wild ruminants illustrated that there was a paucity of information available.[101]

The most extensive study was on reindeer which were sampled throughout the lactation cycle. The effect of nutritional factors on their milk composition was not clear, but no major nutritional effects were detected.[102] The difficulty of detecting differences in "relative goodness" may again be an important factor in this situation. Information from domestic animals exposed to nutritional alteration indicates the potential for using composition in wild herbivores.[103] Considerable study is required before this is feasible.

D. Rumen and Fecal Sampling

Most large herbivores are ruminants and sampling from the rumen to obtain information on nutrient intake has been done by analyzing volatile fatty acid (VFA) production in moose,[104] VFA concentration of rumen liquor in red deer,[105] white-tailed deer,[106] mule deer,[107] roe deer, and red deer,[108] nitrogen in black-tailed deer,[109] and crude protein in springbok (*Antidorcas marsupialis*).[110] Fecal nitrogen was used in East African ungulates for evaluating dietary quality.[111] Hungate[112] outlined the changes in concentration and species of rumen microbial populations associated with diet quality.

Volatile fatty acids are waste products of microbial fermentation and the amount present depends upon composition of intake and feeding regime.[112] Seasonal differences have been reported for both VFA production[104] and VFA concentration.[106] Gasaway and Coady[104] concluded that for studies of wild animals, VFA concentration can only be used as an approximate indicator of fermentation rates and not a measure of VFA production. Volatile fatty acid determinations are quite complex for field application and require fresh-killed or trocarized animals.

Analyses of nitrogenous compounds in the rumen are subject to error when attempts are made to relate them to protein content of forage due primarily to nitrogen recycling.[112] Additional problems (fermentation rates, gut surface to volume ratios, and feed passage rates) exist for fecal nitrogen determinations which reduces their usefulness as nutritional monitors.[113] Although measures of rumen contents are a step closer to monitoring what the animal ultimately does with nutrients than is forage analysis, there are complicating factors such as absorption and utilization. Measures that eliminate these sources of variation would be preferred.

Rumen microbial counts have not been used much in wildlife to assess quality of intake. Collection, storage, handling, and standardization problems have limited their use, but the potential remains.

E. Abomasal Parasite Count

The abomasal parasite count (APC) technique has been applied to white-tailed deer in southeastern United States as an indicator of status of deer density.[114] The association between haemonchosis (*Haemonchus contortus*) and malnutrition may be used as an indicator of overpopulation.[115] Where this parasite is endemic, this indirect assessment of malnutrition should be considered. This premise assumes that high density is synonymous with malnutrition which may not be true in all instances.

III. DISCUSSION

A plethora of potential measurements for nutritional assessment using the indicator animal were presented. The wildlife biologist must have a basis to select one or more of these measurements for application. The decision will be primarily influenced by management or research objectives, but considerations of cost, time, personnel, laboratory facilities, live or dead animal assessment, ease of collection and analyses, interpretation assistance, and ecological information enter the selection process.

Objectives may vary from a need to determine the response of short-term small perturbations such as protein intake to long-term population responses to vegetation successional processes. Each objective, and those falling in between, may benefit from nutritional assessment.

The use of multiple assessment measurement should not be overlooked,[38,100] but of greater importance is an understanding that these form and function assessments provide good supportive data, and are of less value when used alone. Supportive quantification is often required regarding changes in observed population statistics, behavior, or status. Nutritional assessment may be considered best as additional input to the decision-making process.

Nearly every parameter discussed has limitations, and many have little value for the manager. Most of the limitations are, of course, inherent in the nature of the parameter, such as being available only from the dead animal, but some may result from study objectives. For example, a study of condition using BMF indexes for animals in relatively good health may conclude that BMF indexes are of little value. However, for animals in extremely poor health, BMF indexes may prove quite useful. Similar problems occur with other parameters but may not be as obvious. The message is that proper selection of assessment parameters is critical in study design.

A serious practical problem in assessing condition via blood assays is that it is difficult to sample enough animals of each sex and age class to resolve factors of individual variation.

Perhaps the greatest limitation of most parameters is the inability to differentiate "relative goodness". The present ability to quantify the extremes is quite good, but the less dramatic differences are more difficult to assess. Perhaps more emphasis must be placed on measuring physiological entities, such as neuropeptides, that control and regulate homeostasis rather than the results of their action.

Future needs are not difficult to identify; more work is needed in the realm of nutritional assessment. Controlled studies using captive animals of a greater variety of wild species are needed. The entire continuum of nutrient intakes must be monitored with a greater diversity of measurements. The resilience of animals as outlined by Hanks[2] must be assessed. The use of remote-controlled blood collection systems would practically eliminate human interference in research animals. Several prototypes have been tested, and the one proposed by Bubenik[117] appears to be the furthest advanced in design. Hopefully, this technique will gain wider acceptance.

Statistical analyses for handling data vary as much as the objectives. Perhaps more standardization of analyses would be helpful. Greater use of techniques that can assimilate the complex multifaceted term condition, such as multivariate discriminant analysis,[118] should be applied. Standardization of techniques, in general, should be attempted in all facets of study from capture to analysis allowing greater application of the comparative value of these nutritional assessments.

Wildlife managers must be better informed on the use of the indicator animal in a manner that does not imply a panacea. As many of the heretofore listed techniques were described, they were perceived by some as entities in themselves and considerable disappointment followed when reality struck. Obviously, the researcher and manager must work closely to attain the most from the data, but not more than is warranted.

REFERENCES

1. **Taber, R. D., White, K. L., and Smith, N. S.,** The annual cycle of condition in the Rattlesnake, Montana, mule deer, in *Proc. Montana Acad. Sci.*, 72, 1979.
2. **Hanks, J.,** Characterization of population condition, in *Dynamics of Large Mammal Populations*, Fowler, C. W. and Smith, T. D., Eds., Wiley-Interscience, New York, 1981, 47.
3. **LeResche, R. E., Seal, U. S., Karns, P. D., and Franzmann, A. W.,** A review of blood chemistry of moose and other Cervidae with emphasis on nutritional assessment, *Nat. Can.*, 101, 263, 1974.
4. **Klein, D. R.,** Food selection by North American deer and their response to over-utilization of preferred plant species, in *Animal Populations in Relation to Their Food Resources*, Watson, A., Ed., Blackwell Scientific, Oxford, U.K., 1970, 25.
5. **Hamerstrom, R. and Camburn, F. L.,** Weight relationships in the George Reserve deer herd, *J. Mammal.*, 31, 5, 1950.
6. **Halloran, A. F.,** American bison weights and measurements from the Wichita Mountains Wildlife Refuge, in *Proc. Oklahoma Acad. Sci.*, 212, 1960.
7. **O'Gara, B. W.,** Derivation of whole weights for the pronghorn, *J. Wildl. Manage.*, 34, 470, 1970.
8. **Quimby, D. C. and Johnson, D. E.,** Weights and measurements of Rocky Mountain elk, *J. Wildl. Manage.*, 15, 57, 1951.
9. **Troyer, W. A.,** The Roosevelt elk on Afognak Island, Alaska, *J. Wildl. Manage.*, 24, 15, 1960.
10. **Aldous, M. C., Craighead, F. C., Jr., and Devan, G. A.,** Some weights and measurements of desert bighorn sheep *(Ovis canadensis nelsoni)*, *J. Wildl. Manage.*, 22, 444, 1958.
11. **Weiner, J.,** Dressing percentage, gross body composition and caloric value of roe deer, *Acta Theriol.*, 18, 209, 1973.
12. **Blood, D. A., McGillis, J. R., and Lovaas, A. L.,** Weights and measurements of moose in Elk Island National Park, *Can. Field Nat.*, 81, 263, 1967.
13. **Bandy, P. J., Cowan, I. McT., Kitts, W. D., and Wood, A. J.,** A method for assessment of the nutritional status of wild ungulates, *Can. J. Zool.*, 34, 48, 1956.
14. **McEwan, E. H. and Wood, A. J.,** Growth and development of barren-ground caribou. I. Heart girth, hind foot length, and body weight relationships, *Can. J. Zool.*, 44, 401, 1966.
15. **Aarak, E. and Lenvik, D.,** Weight development in young reindeer, in *Proc. 2nd Int. Reindeer/Caribou Symp.*, Reimers, E., Gaare, E., and Skjenneberg, S., Eds., Direktoratet for Vilg og Ferskvannafisk, Trondheim, Norway, 1980, 355.

16. **Smart, C. W., Giles, R. H., Jr., and Guynn, D. C.,** Weight tape for white-tailed deer in Virginia, *J. Wildl. Manage.*, 36, 553, 1973.
17. **Franzmann, A. W., LeResche, R. E., Rausch, P. A., and Oldemeyer, J. L.,** Alaskan moose measurements and weight and measurement-weight relationships, *Can. J. Zool.*, 56, 298, 1978.
18. **Talbot, L. M. and McCullough, J. S. G.,** Weight estimations for East African mammals from body measurements, *J. Wildl. Manage.*, 29, 84, 1965.
19. **Klein, D. R.,** Range-related differences in growth of deer reflected in skeletal ratios, *J. Mammal.*, 45, 226, 1964.
20. **Stains, B. W.,** The dynamics and performance of a declining population of red deer *(Cervus elaphus)*, *J. Zool.*, 184, 403, 1978.
21. **Cowan, R. L. and Long, T. A.,** Studies on antler and growth and nutrition of white-tailed deer, in *Proc. Nat. White-tailed Deer Disease Symp.*, 54, 1962.
22. **Gasaway, W.,** Moose antlers: how fast do they grow?, Alaska Department of Fish and Game, (pamphlet), Juneau, 1975.
23. **Wika, M.,** On growth of reindeer antlers, in *Proc. 2nd Int. Reindeer/Caribou Symp.*, Reimers, E., Gaare, E., and Skjenneberg, S., Eds., Direktoratet for Vilg og Ferskvannsfisk, Trondheim, Norway, 1980, 416.
24. **Verme, L. J.,** Effect of nutrition on growth of white-tailed deer fawns, in *Trans. 28th N. Am. Wildl. and Natural Resources Conf.*, Wildlife Management Institute, Washington, D.C., 1963, 433.
25. **Holter, J. B. and Hayes, H. H.,** Growth in white-tailed deer fawns fed varying energy and constant protein, *J. Wild. Manage.*, 41, 506, 1977.
26. **Verme, L. J. and Ozoga, J. J.,** Influence of protein-energy intake on deer fawns in autumn, *J. Wildl. Manage.*, 44, 305, 1980.
27. **DeCalesta, D. S., Nagy, J. G., and Bailey, J. A.,** Starving and refeeding mule deer, *J. Wildl. Manage.*, 39, 663, 1975.
28. **Verme, L. J.,** Influence of nutrition on fetal organ development in deer, *J. Wildl. Manage.*, 43, 791, 1979.
29. **Ozoga, J. J. and Verme, L. J.,** The thymus gland as a nutritional status indicator in deer, *J. Wildl. Manage.*, 42, 791, 1978.
30. **Hughes, E. and Mall, R.,** Relation of the adrenal cortex to condition of deer, *Calif. Fish and Game*, 44, 191, 1958.
31. **Christian, J. J., Flyger, V., and Davis, D. E.,** Factors in the mass mortality of a herd of Sika deer, *Cervus nippon*, *Chesapeake Sci.*, 1, 79, 1960.
32. **Welch, B. L.,** Adrenals of deer as indicators of population conditions for purposes of management, in *Proc. Nat. White-tailed Deer Symp.*, 94, 1962.
33. **Smith, N. S.,** Appraisal of condition estimation methods for East African ungulates, *East Afr. Wildl. J.*, 8, 123, 1970.
34. **Attwell, C. A. M.,** Growth and condition of blue wildebeest in Zululand, *S. Afr. J. Wildl. Res.*, 12, 64, 1982.
35. **Seal, U. S., Verme, L. J., Ozoga, J. J., and Plotka, E. D.,** Metabolic and endocrine responses of white-tailed deer to increased population density, *J. Wildl. Manage.*, 47, 451, 1983.
36. **Harris, D.,** Symptoms of malnutrition in deer, *J. Wildl. Manage.*, 9, 329, 1945.
37. **Riney, T.,** Evaluating condition of free-ranging red deer *(Cervus elaphus)*, with special reference to New Zealand, *N. Z. J. Sci. Technol.*, 36, 429, 1955.
38. **Ransom, A. B.,** Kidney and marrow fat as indicators of white-tailed deer condition, *J. Wildl. Manage.*, 29, 397, 1965.
39. **Brooks, P. M., Hanks, J., and Lundbrook, J. V.,** Bone marrow as an index of condition in African ungulates, *S. Afr. J. Wildl. Res.*, 7, 61, 1977.
40. **Cheatum, E. L.,** Bone marrow as an index of malnutrition in deer, *N. Y. State Conserv.*, 3, 19, 1949.
41. **Peterson, R. O., Allen, D. L., and Dietz, J. M.,** Depletion of bone marrow fat in moose and a correction for dehydration, *J. Wildl. Manage.*, 46, 547, 1982.
42. **Snider, J. B.,** An evaluation of mandibular fat as an indicator of condition in moose, in *Proc. 16th North Am. Moose Conf. Workshop*, 37, 1980.
43. **Ballard, W. B., Gardner, C. L., Westlund, J. H., and Miller, S. M.,** Use of mandible versus longbone to evaluate percent marrow fat in moose and caribou, *Alces*, 17, 147, 1981.
44. **Hanks, J., Cumming, D. H. M., Orpen, J. L., Parry, D. F., and Warren, H. B.,** Growth, condition, and reproduction in the impala ram (Aepyceros melampus), *J. Zool.*, 197, 421, 1976.
45. **Franzmann, A. W. and Arneson, P. D.,** Marrow fat in Alaskan moose femurs in relation to mortality factors, *J. Wildl. Manage.*, 40, 336, 1976.
46. **Bischoff, A. I.,** Limitations of the bone marrow technique in determining malnutrition in deer, in *Proc. West. Assoc. State Game Fish Comm.*, 205, 1954.

47. **Bear, G. D.,** Seasonal trends in fat levels of pronghorns *(Antilocapra americana)* in Colorado. *J. Mammal.*, 52, 583, 1971.
48. **Greer, K. R.,** A compression method indicates fat content of elk (wapiti) femur marrows, *J. Wildl. Manage.*, 32, 747, 1968.
49. **Neiland, K. A.,** Weight of dried marrow as indicator of fat in caribou femurs, *J. Wildl. Manage.*, 34, 904, 1970.
50. **Verme, L. J. and Holland, J. C.,** Reagent-dry assay of marrow fat in white-tailed deer, *J. Wildl. Manage.*, 37, 103, 1973.
51. **Kie, J. G.,** Femur marrow fat in white-tailed deer carcasses, *J. Wildl. Manage.*, 42, 66, 1978.
52. **Batcheler, C. L. and Clarke, C. M. H.,** Note on kidney weights and the kidney fat index, *N. Z. J. Sci.*, 13, 663, 1970.
53. **Dauphine, T. C.,** Kidney weight fluctuations affecting the kidney fat index in caribou, *J. Wildl. Manage.*, 39, 379, 1975.
54. **Robbins, C. T., Moen, A. N., and Reid, J. T.,** Body composition of white-tailed deer, *J. Anim. Sci.*, 38, 871, 1974.
55. **Ledger, H. P. and Smith, N. S.,** The carcass and body composition of the Uganda kob, *J. Wildl. Manage.*, 28, 825, 1964.
56. **Verme, L. J. and Ozoga, J. J.,** Effects of diet on growth and lipogenesis in deer fawns, *J. Wildl. Manage.*, 44, 315, 1980.
57. **Torbit, S. C.,** *In Vivo* Estimation of Mule Deer Body Composition, Ph.D. thesis, Colorado State University, Fort Collins, 1981.
58. **Monro, R. H. and Skinner, J. D.,** A note on condition indices for adult male impala *(Aepyceros melampus)*, *S. Afr. J. Anim. Sci.*, 9, 47, 1979.
59. **Ringberg, T., Reimers, E., and Sorumgaard, R.,** Growth and seasonal change in organ weights and carcass composition in Svalbard reindeer *(Rangifer tarandus platyrhynchus)*, in *Proc. 2nd Int. Reindeer/Caribou Symp.*, Reimers, E., Gaare, E., and Skjenneberg, S., Eds., Direktoratet for Vilt og Ferskvannsfisk, Trondheim, Norway, 1980, 333.
60. **Hofmann, R. R.,** Adaptive changes of gastric and intestinal morphology in response to different fibre content in ruminant diets, in *Proc. Symp. Dietary Fibre in Human and Anim. Nutr.*, Wallace, G. and Bell, L., Eds., The Royal Society of New Zealand, Bull. 20, 1982, 51.
61. **Child, G., Robbel, H., and Hepburn, C. P.,** Observations on the biology of tsessebe *(Damaliscus lunatus lunatus)* in northern Botswana, *Mammalia*, 36, 342, 1972.
62. **Albl, P.,** Studies on assessment of physical condition in African elephants, *Biol. Conserv.*, 3, 134, 1971.
63. **Robinson, W. L.,** Test of shelter requirements of penned white-tailed deer, *J. Wildl. Manage.*, 24, 364, 1960.
64. **Franzmann, A. W.,** Environmental sources of variation of bighorn sheep physiological values, *J. Wildl. Manage.*, 36, 924, 1972.
65. **Franzmann, A. W., LeResche, R. E., and Arneson, P. D.,** *Moose Productivity and Physiology*, Alaska Department Fish and Game, Juneau, 1976, 86.
66. **Huntley, B. J.,** Seasonal variation in the physical condition of mature male blesbok and kudu, *J. S. Afr. Wildl. Manage. Assoc.*, 1, 17, 1971.
67. **Franzmann, A. W. and Schwartz, C. C.,** *Moose Productivity and Physiology*, Alaska Department Fish and Game, Juneau, 1983.
68. **Kistner, T. P., Trainer, C. E., and Hartmann, N. A.,** A field technique for evaluating physical condition of deer, *Wildl. Soc. Bull.*, 8, 11, 1980.
69. **Franzmann, A. W. and LeResche, R. E.,** Alaskan moose blood studies with emphasis on condition evaluation, *J. Wildl. Manage.*, 42, 334, 1978.
70. **Faro, J. B. and Franzmann, A. W.,** *Alaska Peninsula Moose Productivity and Physiology Study*, Alaska Department Fish and Game, Juneau, 1978.
71. **Smith, C. A. and Franzmann, A. W.,** *Productivity and Physiology of Yakutat Forelands Moose*, Alaska Department Fish and Game, Juneau, 1979, 18pp.
72. **Doerr, J. G., Young, E. L. Jr., and Franzmann, A. W.,** Ecological and physiological aspects of a moose population in Thomas Bay, southeast Alaska, in *Proc. 16th North Am. Moose Conf. Workshop*, Prince Albert, Saskatchewan, 1980, 203.
73. **Seal, U. S., Verme, L. J., and Ozoga, J. J.,** Physiologic values, in *Disease and Parasites of White-tailed Deer*, Davidson, W. R., Hayes, F. A., Nettles, V. F., and Kellogg, F. E., Eds., Heritage Printers, Charlotte, N.C., 1981, chap. 3.
74. **Marler, R. J.,** Some hematologic and blood chemistry values in two herds of American bison in Kansas, *J. Wildl. Dis.*, 11, 97, 1975.
75. **Mehrer, C. F.,** Some hematologic values of bison from five areas of the United States, *J. Wildl. Dis.*, 12, 7, 1976.

76. **Barrett, M. W. and Chalmers, G. A.,** Hematological and clinochemical values for free-ranging pronghorn fawns, *Can. J. Zool.*, 57, 1757, 1979.
77. **Barrett, M. W. and Chalmers, G. A.,** Hematological values for adult free-ranging pronghorns, *Can. J. Zool.*, 55, 448, 1977.
78. **Seal, U. S. and Hoskinson, R. L.,** Metabolic indicators of habitat condition and capture stress in pronghorns, *J. Wildl. Manage.*, 42, 755, 1978.
79. **Seal, U. S. and Erickson, A. W.,** Hematology, blood chemistry and protein polymorphisms in the white-tailed deer *(Odocoileus virginianus)*, *Comp. Biochem. Physiol.*, 30, 695, 1969.
80. **Seal, U. S., Nelson, M. E., Mech, L. D., and Hoskinson, R. L.,** Metabolic indicators of habitat differences in four Minnesota deer populations, *J. Wildl. Manage.*, 42, 746, 1978b.
81. **Nieminen, M.,** The composition of reindeer blood in respect to age, season, calving, and nutrition, *Acta Univ. Ouluensis*, Pharmacol. Physiol. No. 11, University of Oulu, Finland, 1980, 67.
82. **Franzmann, A. W., Flynn, A., and Arneson, P. D.,** Serum corticoid levels relative to handling stress in Alaskan moose, *Can. J. Zool.*, 53, 1424, 1975.
83. **Seal, U. S., Ozoga, J. J., Erickson, A. W., and Verme, L. J.,** Effects of immobilization on blood analyses of white-tailed deer, *J. Wildl. Manage.*, 36, 1034, 1972.
84. **Presidente, P. J. A., Lumsden, J. H., Presnell, K. R., Rapley, W. A., and McCraw, B. M.,** Combination of etorphine and xylazine and captive white-tailed deer. II. Effects on hematologic, serum biochemical and blood gas values, *J. Wildl. Dis.*, 9, 342, 1973.
85. **Wesson, J. A., III, Scanlon, P. F., Kirkpatrick, R. L., and Mosby, H. S.,** Influence of chemical immobilization and physical restraint on packed cell volume, total protein, glucose, and blood urea nitrogen in blood of white-tailed deer, *Can. J. Zool.*, 57, 756, 1979.
86. **Wesson, J. A., III, Scanlon, P. F., Kirkpatrick, R. L., and Mosby, H. S.,** Influence of chemical immobilization and physical restraint on steroid hormone levels in blood of white-tailed deer, *Can. J. Zool.*, 57, 768, 1979.
87. **Mautz, W. W., Seal, U. S., and Boardman, C. B.,** Blood serum analyses of chemically and physically restrained white-tailed deer, *J. Wildl. Manage.*, 44, 343, 1980.
88. **Franzmann, A. W. and Thorne, E. T.,** Physiologic values in wild bighorn sheep *(Ovis canadensis canadensis)* at capture, after handling, and after captivity, *J. Am. Vet. Med. Assoc.*, 157, 647, 1970.
89. **Warren, R. J., Kirkpatrick, R. L., Oelschlaeger, A., Scanlon, P. F., Webb, K. E., Jr., and Whelan, J. B.,** Energy, protein, and seasonal influences on white-tailed deer fawn nutritional indices, *J. Wildl. Manage.*, 46, 302, 1982.
90. **Franzmann, A. W.,** Application of physiologic values to bighorn sheep management, in *1st Trans. North American Wild Sheep Conf.*, Decker, E., Ed., Colorado State University, Fort Collins, 1971, 87.
91. **Mould, E. D. and Robbins, C. T.,** Nitrogen metabolism in elk, *J. Wildl. Manage.*, 45, 323, 1981.
92. **Seal, U. S., Verme, L. J., and Ozoga, J. J.,** Dietary protein and energy effects on deer fawn metabolic patterns, *J. Wildl. Manage.*, 42, 776, 1978a.
93. **Robbins, C. T., Prior, R. L., Moen, A. N., and Visek, W. J.,** Nitrogen metabolism of white-tailed deer, *J. Anim. Sci.*, 38, 186, 1974.
94. **Kirkpatrick, R. L., Buchland, D. E., Abler, W. A., Scanlon, P. F., Whelan, J. B., and Burkhart, H. E.,** Energy and protein influences on blood urea nitrogen of white-tailed deer fawns, *J. Wildl. Manage.*, 39 692, 1975.
95. **Bahnak, B. R., Holland, J. C., Verme, L. J., and Ozoga, J. J.,** Seasonal and nutritional effects on serum nitrogen constituents in white-tailed deer, *J. Wildl. Manage.*, 43, 454, 1979.
96. **Hawley, A. W. L. and Peden, D. G.,** Effects of ration, season and animal handling on composition of bison and cattle blood, *J. Wildl. Dis.*, 18, 321, 1982.
97. **Abler, W. A., Buckland, D. E., Kirkpatrick, R. L., and Scanlon, P. F.,** Plasma progestins and puberty in fawns as influenced by energy and protein, *J. Wildl. Manage.*, 40, 442, 1976.
98. **Smith, S. H., Holter, J. B., Hayes, H. H., and Silver, H.,** Protein requirements of white-tailed deer fawns, *J. Wildl. Manage.*, 39, 582, 1975.
99. **Bahnak, B. R., Holland, J. C., Verme, L. J., and Ozoga, J. J.,** Seasonal and nutritional influences on growth hormone and thyroid activity in white-tailed deer, *J. Wildl. Manage.*, 45, 140, 1981.
100. **Seal, U. S.,** Assessment of habitat condition by measurement of biochemical and endocrine indicators of the nutritional, reproductive and disease status of free-ranging animal populations, in Classification, Inventory and Analysis of Fish and Wildlife Habitat, Marmelstein, A., Ed., U.S. Department of the Interior, Washington, D.C., 1977, 305.
101. **Franzmann, A. W., Arneson, P. D., and Ullrey, D. E.,** Composition of milk from Alaskan moose in relation to other North American wild ruminants, *J. Zoo Anim. Med.*, 6, 12, 1975.
102. **Luick, J. R., White, R. G., Gau, A. M., and Jenness, R.,** Composition changes in milk secreted by grazing reindeer. I. Gross composition and ash, *J. Dairy Sci.*, 57, 1325, 1974.

103. **Kirchgessner, M., Friesecke, H., and Koch, G., Eds.,** *Nutrition and the Composition of Milk*, Lippincott, Philadelphia, 1967.
104. **Gasaway, W. A. and Coady, J. W.,** Review of energy requirements and rumen fermentation in moose and other ruminants, *Nat. Can.*, 101, 227, 1974.
105. **Prins, R. A. and Geelen, M. J. H.,** Rumen characteristics of red deer, fallow deer, and roe deer, *J. Wildl. Manage.*, 36, 673, 1971.
106. **Short, H. L.,** Rumen fermentations and energy relationships in white-tailed deer, *J. Wildl. Manage.*, 17, 184, 1963.
107. **Short, H. L., Medin, D. E., and Anderson, A. E.,** Seasonal variations in volatile fatty acids in the rumen of mule deer, *J. Wildl. Manage.*, 30, 466, 1966.
108. **Bruggemann, J. D., Giesecke, J. D., and Walser-Karst, K.,** Methods for studying microbial digestion in ruminants *post mortem* with special reference to wild species, *J. Wildl. Manage.*, 32, 198, 1968.
109. **Klein, D. R. and Schonheyder, F.,** Variation in ruminal nitrogen levels among some cervidae, *Can. J. Zool.*, 28, 1436, 1970.
110. **Robinson, T. J.,** Influence of a nutritional parameter on the size differences of the three springbok subspecies, *S. Afr. J. Zool.*, 14, 13, 1979.
111. **Arman, P., Hopcroft, D., and McDonald, I.,** Nutritional studies of East Africa herbivores. II. Losses of nitrogen in faeces, *Br. J. Nutr.*, 33, 265, 1975.
112. **Hungate, R. E.,** *The Rumen and its Microbes*, Academic Press, New York, 1966, chap. 6.
113. **Robbins, C. T.,** *Wildlife Feeding and Nutrition*, Academic Press, New York, 1983, chap. 8.
114. **Eve, J. H. and Kellogg, F. E.,** Management inplications of abomasal parasites in southeastern white-tailed deer, *J. Wildl. Manage.*, 41, 169, 1977.
115. **Davidson, W. R., McGhee, M. B., Nettles, V. F., and Chappell, L. C.,** Haemonchosis in white-tailed deer in the southeastern United States, *J. Wildl. Dis.*, 16, 499, 1980.
116. **Franzmann, A. W.,** Condition assessment of Alaskan moose, in *Proc. 13th North Am. Moose Conf. Workshop*, 119, 1977.
117. **Bubenik, G. A.,** Chemical immobilization of captive white-tailed deer and the use of automatic blood samplers, in *Chemical Immobilization of North American Wildlife*, Nielsen, L., Haigh, J. C., and Fowler, M. E., Eds., The Wisconsin Humane Society, Milwaukee, 1982, 335.
118. **Kleinbaum, D. G. and Kupper, L. L.,** *Applied Regression Analysis and Other Multivariate Methods*, Duxbury Press, North Scituate, Massachusetts, 1978.
119. **Franzmann, A. W., Ballard, W. B., Schwartz, C. C., and Spraker, T. H.,** Physiologic and morphometric measurements in neonatal moose and their cows in Alaska, in *Proc. 16th N. Am. Moose Conf. Workshop*, Vol. 16, Lakehead University, Thunder Bay, Ontario, 1980, 106.
120. **Barrett, M. W. and Chalmers, G. A.,** Clinochemical values for adult free-ranging pronghorns, *Can. J. Zool.*, 55, 1252, 1977.
121. **Follis, T. B.,** *Reproduction and Hematology of the Cache Elk Herd*, Publ. 72-8, Utah State Division Wildlife Resources, Salt Lake City, 1972, 133.
122. **Weber, Y. B. and Bliss, M. L.,** Blood chemistry of Roosevelt elk *(Cervus canadensis roosevelti)*, *Comp. Biochem. Physiol.*, 43A, 649, 1972.
123. **Pedersen, R. J. and Pedersen, A. A.,** Blood chemistry and hematology of elk, *J. Wildl. Manage.*, 39, 617, 1975.
124. **Vaughn, H. W., Knight, R. R., and Frank, F. W.,** A study of reproduction, disease and physiological blood and serum values in Idaho elk, *J. Wildl. Dis.*, 9, 296, 1973.
125. **Knight, R. R.,** Some chemical characteristics of elk blood, *Bull. Wildl. Dis. Assoc.*, 5, 8, 1969.
126. **Anderson, A. E., Medin, D. E., and Bowden, D. C.,** Erythrocytes and leukocytes in a Colorado mule deer population, *J. Wildl. Manage.*, 34, 389, 1970.
127. **Rosen, M. N. and Bischoff, A. I.,** The relation of hematology to condition in California deer, in *Trans. N. Am. Wildl. Conf.*, Vol. 17, Wildlife Management Inst., Washington, D.C., 1952, 482.
128. **Browman, L. G. and Sears, H. S.,** Erythrocyte values, and alimentary canal pH values in the mule deer, *J. Mammal.*, 36, 474, 1955.
129. **Johnson, H. E., Youatt, W. G., Fay, L. D., Harte, H. D., and Ullrey, D. E.,** Hematologic values of Michigan white-tailed deer, *J. Mammal.*, 49, 749, 1968.
130. **Dommert, A. R., Tumbleson, M. E., Wescott, B., Murphy, D. A., and Korschgen, L. J.,** Hematologic values of dieldrin-treated white-tailed deer *(Odocoileus virginianus)* in Missouri, *Am. J. Vet. Clin. Pathol.*, 2, 181, 1968.
131. **Tumbleson, M. E., Wood, M. G., Dommert, A. R., Murphy, D. A., and Korschgen, L. J.,** Biochemical studies on serum from white-tailed deer in Missouri, *Am. J. Vet. Clin. Pathol.*, 2, 121, 1968.
132. **Tumbleson, M. E., Ticer, J. W., Dommert, A. R., Murphy, D. A., and Korschgen, L. J.,** Serum proteins in white-tailed deer in Missouri, *Am. J. Vet. Clin. Pathol.*, 2, 127, 1968.

133. **Sikes, D., Hayes, F. A., and Prestwood, A. K.,** Electrophoretic distribution of serum proteins of normal and an arthritic white-tailed deer *(Odocoileus virginianus), Am. J. Vet. Res.*, 30, 143, 1969.
134. **Coblentz, B. E.,** Serum cholesterol level changes in George Reserve deer, *J. Wildl. Manage.*, 39, 342, 1975.
135. **Franzmann, A. W.,** Comparative physiologic values in captive and wild bighorn sheep, *J. Wildl. Dis.*, 7, 105, 1971.
136. **Woolf, A. and Kradel, D. C.,** Mortality in captive bighorn sheep — clinical, hematological, and pathological observations, *J. Wildl. Dis.*, 9, 12, 1973.
137. **Woolf, A., Nadler, C. F., and Kradel, D. C.,** Serum protein electrophoresis in bighorn sheep with chronic pneumonia, *J. Wildl. Dis.*, 9, 7, 1973.
138. **Karns, P. D. and Crichton, V. F. J.,** Effects of handling and physical restraint on blood parameters in woodland caribou, *J. Wildl. Manage.*, 42, 904, 1978.
139. **McEwan, E. H.,** Hematological studies of barren-ground caribou, *Can. J. Zool.*, 46, 1031, 1968.
140. **Nieminen, M.,** Nutritional and seasonal effects on the hematology and blood chemistry in reindeer *(Rangifer tarandus L.) Comp. Biochem. Physiol.*, 66A, 399, 1980.
141. **Afanaser, V. P.,** Seasonal changes in protein and blood picture of reindeer, *Tr. Mosk. Vet. Akad.*, 47, 329, 1963.
142. **Vayrynen, P., Nieminen, M., and Hyvarinen, H.,** Seasonal changes in fatty acid composition of serum lipids in the reindeer, in *Proc. 2nd Int. Reindeer/Caribou Symp.*, Reimers, E., Gaare, E., and Skjenneberg, S., Eds., Direktoratet for Vilt og Ferskvannsfisk, Trondheim, Norway, 1979, 407.
143. **Latinen, M., Nieminen, M., and Hietanen, E.,** Glucuronidation in the reindeer during winter and summer feeding, in *Proc. 2nd Int. Reindeer/Caribou Symp.*, Reimers, E., Gaare, E., and Skjenneberg, S., Eds., Direktoratet for Vilt og Ferskvannsfisk, Trondheim, Norway, 1979, 373.
144. **Hyvarinen, H., Helle, T., Nieminen, M., Vayrynen, P., and Vayrynen, N.,** The influence of nutrition and seasonal conditions on mineral status in the reindeer, *Can. J. Zool.*, 55, 648, 1977.
145. **Nilssen, K. J. and Ringberg, T.,** Seasonal changes in body weight, food intake and thyroxine in free-ranging and captive Svalbard reindeer *(Rangifer tarandus platyrhynchus)*, in *Proc. 2nd Int. Reindeer/Caribou Symp.*, Reimers, E., Gaare, E., and Skjenneberg, S., Eds., Direktoratet for Vilt og Ferskvannsfisk, Roros, Norway, 1980.
146. **Yousef, M. K. and Luick, J. R.,** Estimation of thyroxine secretion rate in reindeer *(Rangifer tarandus);* effects of sex, age and season, *Comp. Biochem. Physiol.*, 40A, 789, 1971.
147. **Kitts, W. D., Bandy, P. J., Wood, A. J., and Cowan, I. McT.,** Effect of age and plane of nutrition on the blood chemistry of the Columbian black-tailed deer *(Odocoileus hemionus columbianus)*. A. Packed cell volume, sedimentation rate and hemoglobin, *Can. J. Zool.*, 34, 477, 1956.
148. **Bandy, P. J., Kitts, W. D., Wood, A. J., and Cowan, I. McT.,** The effect of age and the plane of nutrition on the blood chemistry of the Columbian black-tailed deer *(Odocoileus hemionus columbianus)*. B. Blood glucose, non-protein nitrogen, total plasma protein, plasma albumin, globulin, and fibrinogen, *Can. J. Zool.*, 35, 283, 1957.
149. **Tumbleson, M. E., Caneio, J. D., and Murphy, D. A.,** Serum biochemical and hematological parameters of captive white-tailed fawns, *Can. J. Comp. Med.*, 34, 66, 1970.
150. **McEwan, E. H. and Whitehead, P. E.,** Changes in the blood constituents of reindeer and caribou occurring with age, *Can. J. Zool.*, 47, 556, 1969.
151. **Hartsook, E. W., Whelan, J. B., and Ondik, M. A.,** Changes in blood proteins of deer during gestation and suckling, *J. Wildl. Manage.*, 39, 346, 1975.
152. **Youatt, W. G., Verme, L. J., and Ullrey, D. E.,** Composition of milk and blood in nursing white-tailed does and blood composition of their fawns, *J. Wildl. Manage.*, 29, 79, 1965.
153. **White, R. G. and Luick, J. R.,** Glucose metabolism in lactating reindeer, *Can. J. Zool.*, 54, 55, 1976.
154. **Wesson, J. A., III, Scanlon, P. F., Kirkpatrick, R. L., and Mosby, H., S.,** Influence of the time of sampling after death on blood measurements of the white-tailed deer, *Can. J. Zool.*, 57, 777, 1979.
155. **Keith, E. O., Ellis, J. E., Phillips, R. W., and Benjamin, M. M.,** Serologic and hematologic values of bison in Colorado, *J. Wildl. Dis.*, 14, 493, 1978.
156. **Seal, U. S., Verme, L. J., Ozoga, J. J., and Erickson, A. W.,** Nutritional effects on thyroid activity and blood of white-tailed deer, *J. Wildl. Manage.*, 36, 1041, 1972.
157. **Franzmann, A. W. and Bailey, T. N.,** Serial blood chemistry and hematology values from Alaskan moose *(Alces alces gigas), J. Zool. Anim. Med.*, 8, 27, 1977.
158. **Bubenik, G. A., Bubenik, A. B., Brown, G. M., Trenkle, A., and Wilson, D. I.,** Growth hormone and cortisol levels in the annual cycle of white-tailed deer *(Odocoileus virginianus), Can. J. Physiol. Pharmacol.*, 53, 787, 1975.
159. **Luick, J. R., Person, S. J., Cameron, R. D., and White, R. G.,** Seasonal variations in glucose metabolism of reindeer *(Rangifer tarandus)* estimated with [U-^{14}C] glucose and [3-^{3}H] glucose, *Br. J. Nutr.*, 29, 245, 1973.

160. **Yousef, M. K., Cameron, R. D., and Luick, J. R.,** Seasonal changes in hydrocortisone secretion of reindeer, *Rangifer tarandus*, *Comp. Biochem. Physiol.*, 40, 495, 1971.
161. **Whitehead, P. E. and McEwan, E. H.,** Seasonal variation in plasma testosterone concentration of reindeer and caribou, *Can. J. Zool.*, 51, 651, 1973.

Chapter 12

COMPUTER SIMULATION OF ENERGY BUDGETS

R. J. Hudson and R. G. White

TABLE OF CONTENTS

Carrying capacity is one of the most important but intractable concepts in wildlife management. Biologists know what it should represent but not how to assess it objectively. Since populations of wild herbivores appear to depend strongly on their forage supply, a step in the right direction has been to define carrying capacity on a bioenergetic basis (Chapter 2). However, dividing total energy or protein supplies by animal requirements is not entirely satisfactory because the way nutrients are packaged and distributed in the environment determines their availability. It has been difficult to incorporate considerations of nutritional status of individual animals and hence the ability of populations to attain levels expected on the basis of total nutrient supplies. This requires a more profound understanding of the behavioral and productive responses of herbivores to complex nutritional environments.

Nutritional requirements of domestic animals are well known. To serve the needs of an industry, agricultural scientists have developed feeding standards, simple factorial models for estimating the level and balance of energy and nutrients to support predetermined target levels of production.[1] This is an easier proposition than predicting the response of free-ranging animals. Although feeding standards can be applied to predict performance from feed characteristics and are amenable to computer implementation,[2] iterative calculations allowing for sequential changes in the environment usually are required. This demands a more mechanistic treatment of several processes and a clearer conception of feedback mechanisms.[3]

Computer simulation models have been developed in response to this need.[4,5] These fall into three major categories based upon scope and resolution; namely, physiological processes, whole animal bioenergetics, and grazing systems (Table 1). Although grazing systems models may prove most useful to managers, they usually are not very general and must be developed for specific environmental settings. They also contain many elements which are beyond the scope of this volume. Models of physiological processes provide insight into adaptive metabolic traits and make important linkages among the disciplines of physics, biochemistry, and bioenergetics. However, the problem at hand is one of assessing the productive response of individual animals and their behavioral and metabolic adaptations to the seasonal environment. Existing models of grazing ruminants which are perhaps most appropriate to this task are those by Rice et al.[26] and Swift.[27] But these must be updated with species-specific information and new developments in our understanding of basic processes.

This chapter formalizes the energy budget concept with a simple computer simulation of a representative large herbivore, the wapiti *(Cervus elaphus nelsoni)*. The specific objectives are to summarize current knowledge of the ecology, behavior, and bioenergetics of wapiti, and to evaluate the response of individuals of any age or reproductive status to seasonal environments, specifically those of the boreal aspen forest.

I. STRUCTURE AND IMPLEMENTATION

The model (BION) simulates bioenergetic transactions of individual female wapiti of any age from weaning to maturity which may or may not be pregnant and/or accompanied by a calf. Because of complications of antler growth and the rut, the current version does not provide a good representation of the bioenergetics of adult males.

The basic time step is one day although digestive kinetics and body pools of nonprotein nitrogen are calculated on an hourly time step. The program is designed to simulate changes over one year but can be cycled to reconstruct lifetime performance. Major tracking variables are summarized in graphic screens monitoring general parameters (weight, intake, conceptus weight, and milk production), activity budgets, feeding constraints, digestive kinetics, energy balance, and body pools of nonprotein and amino acid nitrogen.

Biological processes are summarized in nine computational blocks (subroutines) which interface environment, behavior, and bioenergetics (Figure 1). Calculations, repeated for

Table 1
SCOPE OF PUBLISHED MODELS RELEVANT TO ECOLOGICAL BIOENERGETICS OF LARGE HERBIVORES

Scope	Ref.
Process models	
Fiber digestion	6
Biochemical transformations	7,8
Ruminant digestion	9—15
Nitrogen metabolism	16,17
Energy metabolism	18
Joint metabolism of energy and nitrogen	19
Forage intake and selection	20—23
Thermal balance	24
Body composition	25
Whole animal models	
Grazing ruminants	26
Wild ruminants	27
Domestic sheep	28—32
Cattle	33—35
Roe deer	36
Grazing system models	
Domestic sheep	37—39
Beef Cattle	40

each simulated day, begin with the procedure CALENDAR which defines daily environmental conditions and increments days pregnant or lactating. ACTIVITIES estimates activity budgets and composes the daily diet in nutritional terms. POTENTIAL INTAKE selects the most limiting of metabolic, physical, and logistic controls on forage consumption. NUTRIENT SUPPLY determines rates of fermentation and passage and yields of energy and nitrogen from the digestive tract. NE/ME calculates the efficiency of utilization of metabolizable energy for maintenance, pregnancy, lactation, and growth. MAINTENANCE REQUIREMENTS determines the energy costs of free-existence by summing requirements for resting metabolism, activity, and thermoregulation. PREGNANCY calculates daily requirements for energy and nitrogen for conceptus growth while LACTATION executes similar calculations for milk production. Finally, GROWTH mobilizes tissues to meet energy or nitrogen demands if there is a shortfall or deposits surpluses as weight gain. Although both energy and nitrogen networks are included, energy is emphasized in accordance with the theme of this volume.

The program is coded in FORTH, a language which provides much faster computation speeds and more efficient use of memory than most other high-level languages. The model takes 40 sec (vs. 10 min for interpretive BASIC) to simulate 1 annual cycle and requires less than 64 Kbytes of memory. Therefore, it will run on microcomputers readily available to most researchers and managers.

II. ENVIRONMENTAL CONTEXT

The boreal aspen forest, the area for which this model was developed and applied, extends from southern Manitoba to northeastern British Columbia. This environment features five main habitat types arranged along a catenary sequence determined largely by topography and drainage. The numerous ponds are surrounded by concentric rings of sedge meadows, reedgrass meadows, willow fens, aspen forests, and open grasslands.

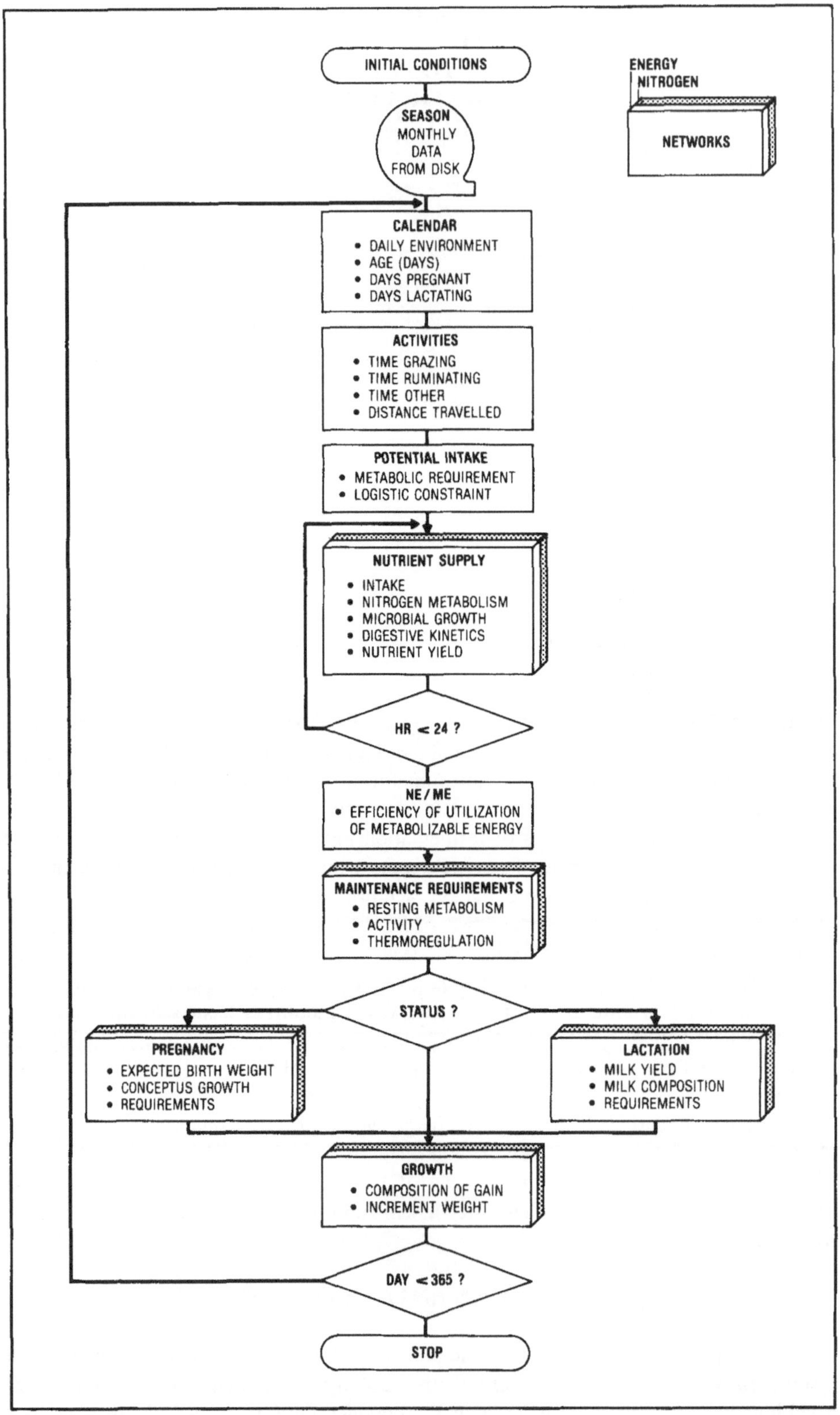

FIGURE 1. Computational blocks and sequences used to simulate the bioenergetics of female wapiti.

A. Seasonal Environment

By early April, most habitats are clear of snow and the spring flush commences. The first habitats to supply green forage are sedge wetlands followed closely by *Poa-Bromus* grasslands. Foliage appears in willow fens and poplar forests at the time of calving in mid-May. Forage quality is at its peak in May with protein contents averaging 20% and digestibility 65 to 70%. By early August, grassland herbage begins to mature. Leaf fall occurs in late September at the time of the rut. The permanent snow cover established by early November appears to protect herbage from weathering. In late winter, forage protein concentrations seldom drop below 7% except in sedge meadows and digestibilities generally remain above 45%.

B. Resource Use

In this heterogeneous fine-grained environment, movements of free-ranging wapiti are quite local and spatial behavior appears largely in response to foraging opportunities.[41] Open grasslands and meadows are used extensively from leaf drop until winter snow progressively forces greater use of shrub and forest communities. Open habitats are again used as the snow melts and sedge meadows receive heavy use during the spring flush. As grassland forage matures, use of shrub and forest habitats increases.

Seasonal diets of wapiti in the boreal forest have been described on a habitat-specific basis by Nietfeld.[42] In May, at the time of calving, the most important forages are upland grasses and forbs. As the grasslands begin to mature in August, an increasing proportion of foliage makes up the diet. With the rut and leaf fall, wapiti forage mainly on upland grasses and continue this pattern of selection well into winter with a brief period of use of lowland habitats which often provide a surge of autumn regrowth.

C. Input Parameters

An important application of bioenergetic models is evaluation of the consequences of alternative patterns of behavior. Prediction of patterns of resource use behavior is at the fringe of current understanding (Chapter 3) and it usually is easier to collect information on resource use directly than to describe the environment in sufficient detail to predict habitat and forage selection. Therefore, the model is driven by easily determined input parameters describing the nutritional and thermal environment to which animals are exposed.

Eight environmental parameters are entered as monthly means from which daily values are determined by linear interpolation (Table 2). Temperature, snow depth, and forage biomass estimated on a habitat-specific basis are weighted by the proportion of time animals spend foraging within each habitat.[43] Potential digestibility (asymptotic dry matter disappearance using the *in situ* nylon bag technique) and protein contents refer to diets actually selected. Since rate parameters for digestion and passage vary by forage class, dietary proportions of graminoids and twigs are entered from which the proportion of forbs/foliage can be calculated by difference. An option to provide supplemental feed of specified quality is provided to accommodate applications to commercial husbandry or other forms of intensive management.

III. FUNCTIONAL RELATIONSHIPS

The model consists of approximately 200 biological parameters. Descriptions are provided only of key relationships. A glossary of variables used in the text is provided in Table 3.

A. Activity Budgets

Like other large ruminants, wapiti spend most of their day either foraging or resting/ruminating.[43,44] Undisturbed sedentary populations in the boreal forest spend surprisingly

Table 2
ENVIRONMENTAL PARAMETERS FOR A TYPICAL YEAR IN THE ASPEN BOREAL FOREST

Month	Temperature (°C)	Snow (cm)	Forage biomass (kg/ha)	Potential digestibility (frac)	Forage protein (frac)	Natural diet: Graminoids	Natural diet: Twigs	Supplement (g/d)
Sep	10	0	1600	0.60	0.14	0.80	0.10	0
Oct	0	5	1400	0.55	0.11	0.80	0.10	0
Nov	− 10	10	1200	0.50	0.07	0.75	0.20	0
Dec	− 15	23	600	0.45	0.07	0.74	0.25	0
Jan	− 10	20	500	0.42	0.07	0.70	0.28	0
Feb	− 5	16	450	0.42	0.07	0.60	0.33	0
Mar	5	13	400	0.45	0.08	0.70	0.28	0
Apr	15	0	350	0.55	0.15	0.80	0.10	0
May	20	0	1000	0.70	0.23	0.90	0.05	0
Jun	20	0	1500	0.70	0.18	0.70	0.05	0
Jul	22	0	2500	0.65	0.16	0.70	0.05	0
Aug	20	0	1800	0.62	0.15	0.80	0.07	0

little time engaged in other activities. The current version of the model simulates the underlying rumen repletion-depletion cycle. Pre-emptive activity states in response to harassment by insects, predators, or humans must be entered explicitly.

Time available for grazing is limited ultimately by the rumination requirement plus several hours for miscellaneous activities.[44] This sums to about 800 min/day which seems to hold for a variety of domestic and wild herbivores.[49] Below this limit, grazing times to achieve satiation can be calculated as daily dry matter intake divided by the foraging rate *(vide infra)*.

Trudell-Moore and White[45] estimated a rumination requirement of 0.005 hr/g DM for caribou. Van Soest[46] cites data suggesting a value of 0.008 h/g cell walls for animals the size of wapiti. The best fit to data on rumination times in wapiti[47] is 0.005 h/g indigestible DM which gives values ranging from 3 hr in summer to 9 hr in winter. Since rumination time also depends on such factors as fracturing characteristics and particle size distributions,[48] this estimate may be quite specific to boreal habitats.

Social interaction seldom exceeds 1 hr/day except during the rut but varies widely according to nutritional condition.[50] The proportion of 1 hr spent in social interaction is considered proportional to the ratio of energy supply (ME) to metabolic rate. During the rut, an additional 2 hr are allowed. Whatever time is not spent foraging, ruminating, or interacting is assigned to idle time.

Travel among suitable forage patches is included in foraging time for estimating forage intake but is calculated separately in terms of distance for determining energy costs. Distance traveled (km/hr foraging) declines exponentially with increasing snow depths (SN, cm):[47]

$$\mathrm{DKM} = 0.24\, e^{-0.085\ \mathrm{SN}} \tag{1}$$

For environments in which resources are widely dispersed (e.g., mountains), a much more mechanistic treatment would be required.

B. Intake

Voluntary feed intake varies widely with forage quality, body condition, physiological status, and season. On medium quality forage, nonpregnant and nonlactating wapiti consume over 4000 g DM/day, the approximate equivalent of 2% of their body weight. Young animals and those recovering from periods of nutritional stress have higher relative levels of consumption.[51]

Table 3
GLOSSARY OF VARIABLES USED IN THE TEXT

	Units	Description
BIOMASS	kg/ha	Forage biomass
BUN	mg/100 mℓ	Blood urea nitrogen concentration
BW	kg	Birth weight
CF	KJ/hr	Energy cost of foraging
CG	KJ/kg	Energy content of gain
CP	KJ/hr	Energy cost of social interaction
CR	KJ/hr	Energy cost of ruminating
CW	kg	Conceptus weight
CYCLM	Scalar (0—1)	Seasonal metabolic rate
CYCLT	Scalar (0—1)	Seasonal appetite cycle
DKM	km	Distance walked while foraging
DL	days	Days lactating
DM	g	Forage dry matter
DP	days	Days pregnant
EA	KJ	Energy requirement for activity
EW	kg	Expected weight for well-fed animals
F	mm	Coat depth
FMR	KJ/d	Fasting metabolic rate
FMX	KJ/d	Maximum energy from tissues to support lactation
FRMX	g/min	Maximum feeding rate
FRSN	—	Slope of asymptotic feeding rate on snow depth
G	kg/d	Liveweight gain
H	W	Heat production
He	W/m^2	Evaporative heat loss
HT	KJ	Energy requirement of thermoregulation
Ie	$°C/m^2$	External insulation
It	$°C/m^2$	Tissue insulation
K	—	Slope of relationship between feeding rate and forage biomass
KB	—	Slope of feeding rate on biomass of browse
KF	—	Slope of feeding rate on biomass of forbs/foliage
KF+	Fraction	Efficiency of use of ME for growth
KG	—	Slope of feeding rate on browse of graminoids
KL+	Fraction	Efficiency of use of ME for lactation
KL−	Fraction	Efficiency of use of tissues for lactation
KM+	Fraction	Efficiency of use of ME for maintenance
KP+	Fraction	Efficiency of use of ME for pregnancy
KP−	Fraction	Efficiency of use of tissue energy for pregnancy
LTK	g/min	Feeding rate constrained by logistic factors
LVWT	kg	Liveweight
M	gN/kg OM	Microbial maintenance requirement
ME	KJ	Metabolizable energy
%ME	Fraction	Metabolizability of De from postscripted forage class
MEG	kJ	ME requirement for growth
MEL	kJ	ME requirement for lactation
MEM	kJ	ME requirement for maintenance
MEP	kJ	ME requirement for pregnancy
MTK	g/d	Dry matter intake to meet requirements for maintenance and production
MW	kg	Metabolic weight
MYP	kg/d	Potential milk yield
NH3	mg/100 mℓ	Ruminal ammonia concentration
NP	gN/d	Nitrogen retention in conceptus
P.	/h	Ruminal passage rates of postscripted component
PBROWSE	Fraction	Proportion of browse in the diet
PDG	Label	Potentially digestible fiber
PFAT	Fraction	Proportion of fat in gain
PG	kg/d	Target gain
PI.	/h	Fractional passage rate of postscripted forage class

Table 3 (continued)
GLOSSARY OF VARIABLES USED IN THE TEXT

	Units	Description
PNI	Fraction	Dietary nitrogen concentration
PR	KJ/d	Energy retention in conceptus
R.	g	Ruminal pools of postscripted component
RCAP	g	Rumen capacity
RDG	/h	Ruminal digestion rate
RFILL	g	Rumen fill
RMR	KJ/d	Resting metabolic rate
SA	m^2	Surface area
SN	cm	Snow depth
SOL	label	Forage cell contents
TC	°C	Critical temperature
TG	h	Time grazing
UR.EX	g/hr	Urea excretion in urine
UR.PO	g	Urea pool
UR.RE	g/hr	Urea recycled
UR.SP	ℓ	Urea space
UR.TE	g/hr	Urea entry
V	m/s	Wind speed
WC	kg	Weight change during pregnancy
WI	kg	Weight at conception
XDG	label	Undigestible fiber
Y	gN/kg OM	Microbial yield
Ymax	gN/kg OM	Maximum microbial yield

Although feed intake is not greatly stimulated by pregnancy, intake during lactation increases by 2-fold (wapiti)[52] to 2.5-fold (red deer)[53] relative to that at the end of pregnancy. Although feed intake and milk yield presumably are correlated in the long-term, feed intake lags by about 2 weeks. Intake is low after calving but rises to a peak at 6 weeks post partum, although milk production already has started to decline.[52]

Since high energy requirements stimulate feed intake, it might be expected that declining forage quality would also lead to compensation. This relationship holds only on very digestible rations. With most natural forages, voluntary feed intake tends to be directly related to quality because they are digested slowly.[54] The ability of ruminants to compensate is limited by the adaptive physical capacity of the digestive tract.

Although energy requirements and gut fill ultimately limit forage intake,[20,21] time available for grazing and rate of feeding in relation to forage availability may prevent wild ruminants from achieving these levels.

To accommodate these interactions, forage intake is modeled in a hierarchial fashion (Figure 2). First, forage intake to meet metabolic demands is determined. Secondly, the logistic constraint of searching for and consuming forage is applied. Thirdly, this is constrained by rumen fill. Forage intake is taken as the most limiting of metabolic, logistic, and physical controls *(vide infra)*.

These controls are sensitive to integration intervals; short intervals increase daily intake. Data for wapiti show six regular feeding bouts during winter which degrade to fewer broader peaks in summer.[43] Six bouts (i.e., integration interval of 4 hr) was considered most biologically representative and adopted in this model. With more information on thresholds of gut fill and energy deficit, it would be a simple matter to incorporate a variable number of feeding bouts each day.

1. Metabolic Controls

Metabolic control of dry matter intake (MTK, grams per foraging bout) is calculated from

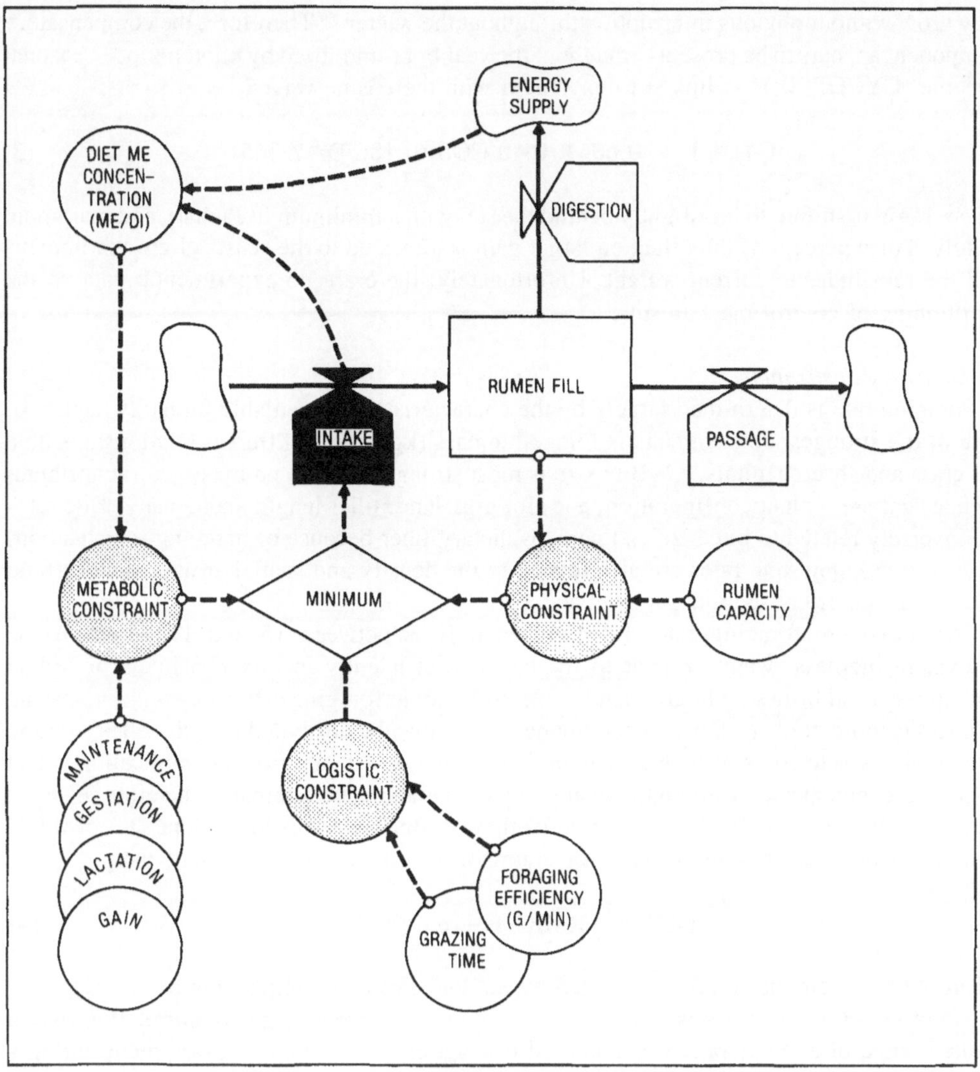

FIGURE 2. Calculation of forage intake during each feeding bout.

the sum of requirements (metabolizable energy, kJ/4 hr) for maintenance (MEM), activity (EA), thermoregulation (HT), potential milk production (MEL), gestation (MEP), and an allowance for growth (MEG). This sum is divided by the metabolizable energy concentration of the diet to obtain units of gDM.

The allowance for growth, a key feedback relationship, is calculated from estimates of growth targets and the energy content and efficiency of gain. The concept of target weights and hence gains probably is valid[55] although its symmetry and mechanism are not clear. Potential gains vary with both current body weight and the seasonal growth impetus. Just how these two determinants interact is not known. In general, during summer when the photoperiodic stimulus is high, rates of gain (PG, kg/d) vary linearly with spring weight (LVWT, kg) irrespective of whether weight is determined by age or nutritional history:[44]

$$PG = 1.4 + 0.0045 \text{ LVWT} \quad (2)$$

During winter, animals given unlimited access to good diets gain weight slowly except those that are larger than about 250 kg. Hand-reared calves weighing less than 80 kg by October may grow without obvious interruption throughout the winter.[56] Therefore, the compensatory component appears to be present throughout the year but is modified by photoperiod. Seasonal appetite (CYCLT, 0-1) is linked to daylength with the cosine wave:

$$\text{CYCLT} = 0.60 + 0.40 \text{ COS } (6.186 \text{ DAY}/365) \qquad (3)$$

where DAY is shifted to lag slightly behind the sun with a minimum in January and maximum in July. Forty percent of the effect on target gain is allocated to the seasonal growth impetus and the remainder to current weight. Unfortunately, there are no experimental data on the partitioning of control mechanisms.

2. Logistic Constraints

Foraging rate is determined largely by the characteristics of available forage (Chapter 4). One of the strongest determinants is forage biomass (kg/ha) but different relationships hold for grass and shrub habitats.[57-59] Bite size is most strongly influenced by structural attributes such as leaf area, plant configuration, and, for grassland, tiller length and density. Bite rates are inversely related to bite size and perhaps dietary fiber because of handling (mastication) time. But cropping bite rates are also related to the density and spatial arrangement of food items and to barriers such as snow cover.

The maximum foraging rate (FRMX) generally is between 15 and 17 g/min. At an equivalent biomass, wapiti appear to forage more efficiently in mixed habitats providing both grasses and browse. On grasslands, improvement in foraging efficiency with increasing biomass is more gradual. Differences among studies may be attributed to techniques, satiety, structural characteristics of habitat, and/or forage quality. An additional complication is that poor forage interspersed with higher quality material may require animals to be more selective and thus eat more slowly. Until the role of these factors is resolved, a slope (K) of 0.001 is used for browse/forbs and 0.0005 for graminoids in the asymptotic relationship:

$$\text{LTK} = \text{FRMX} \ (1 - e^{-\text{K}\cdot\text{BIOMASS}}) \qquad (4)$$

where LTK is potential intake controlled by the logistics of foraging (g/min).

Effective forage biomass is reduced by snow cover. A weak negative correlation over a limited range of snow depths was recorded by Nietfeld.[42] An indirect assessment during a year of deeper snow can be extracted from the slope of the relationship between the proportion of browse in the diet (PBROWSE) and snow depth.[47] During the time period in which these measurements were taken, forage biomass and quality of herbage and browse varied little. The shift can be attributed to the effect of snow cover (SN, cm) which is expected to markedly influence herbage availability but minimally influence browse availability. On this basis, maximum foraging rate on herbaceous forage is estimated:

$$\text{FRMX} = 15 \ (1 - 0.014 \text{ SN}) \qquad (5)$$

This approach allows some interaction of snow cover and range structure but the true picture is undoubtedly much more complex.

3. Rumen Distention

On lower quality forages, intake may be constrained by physical distention of the digestive tract, particularly the rumen. Potential intake during each foraging bout is determined as the difference between rumen capacity and rumen fill. Rumen fill is regulated by digestive kinetics as described in the next section. Rumen capacity is estimated as follows.

The initial estimate (g wet weight) is based on an allometric relationship with liveweight (EW, kg) derived for a variety of wild ruminants:[60]

$$RCAP = 105\ EW^{1.046} \quad (6)$$

Since loss of condition should not lead to a restriction of gut capacity, average weight expected on the basis of age is used rather than current weight.[61]

$$EW = 12.18\ AGEDAYS^{0.42} \quad (7)$$

Although dry matter is expected to vary seasonally, an average of 15% is assumed. There is evidence of alimentary hypertrophy in lactating animals.[62] This may be why increased forage intake lags several weeks behind the metabolic demands of milk production. The correction factor for lactation is calculated from potential milk yield *(vide infra)* with a 14-day lag. This fixed lag accounts for average conditions. It is likely that the lag will be longer for fat animals and shorter for lean animals.

Gut capacity may change adaptively in response to other factors such as photoperiod, seasonal levels of rumen fill, and diet quality.[63] Milne et al.[64] provided information that hints of increased capacity during the spring months. Unfortunately, information on wapiti and red deer is too limited to include such adjustments.

C. Digestive Kinetics

This computational block is structured to accommodate the competitive rates of digestion and passage and operates on a hourly time-step. There are two interrelated networks. The first deals with dry matter and energy (Figure 3). The second monitors nitrogen transactions (Figure 4).

1. Energy Network

The kinetics of dry matter digestion are modeled by partitioning the digestive tract into the rumino-reticulum and lower gut (abomasum, intestines, and caecum). In each segment, three physicochemical digesta pools are represented, namely, indigestible fiber (XDG), potentially digestible fiber (PDG), and cell contents (SOL). Indigestible fiber is primarily lignin and protected complex carbohydrates, corresponding roughly to 2.4 times the acid detergent lignin fraction[15] and more precisely to the residue remaining after prolonged (>48 hr) ruminal suspension of forage samples in dacron bags. The soluble fraction is defined as that which is immediately and completely fermented in the rumino-reticulum. Under most circumstances, it corresponds to cell contents according to the detergent fiber system. The potentially digestible cell-wall fraction is the difference between soluble and indigestible dry matter.

Digestion rates of the potentially digestible fraction vary with the nature of the forage and are assumed to obey first-order kinetics. Estimates of appropriate coefficients are provided by nylon bag experiments with ruminally fistulated wapiti.[65] Values of 0.12/hr are accepted for browse, forbs, and foliage and 0.06/hr for graminoid plants. Although coefficients are assigned to forage classes according to the structure of plant tissues, values are expected to vary with the degree of lignification and hence plant phenology. Ruminal digestion rates (RDG/hr) are modified by the size of the microbial population (R.MCR, gN) in relation to the amount of potentially digestible substrate (R.PDG, g):[26]

$$RDG = RDG' + RDG'\ (R.MCR/R.PDG) \quad (8)$$

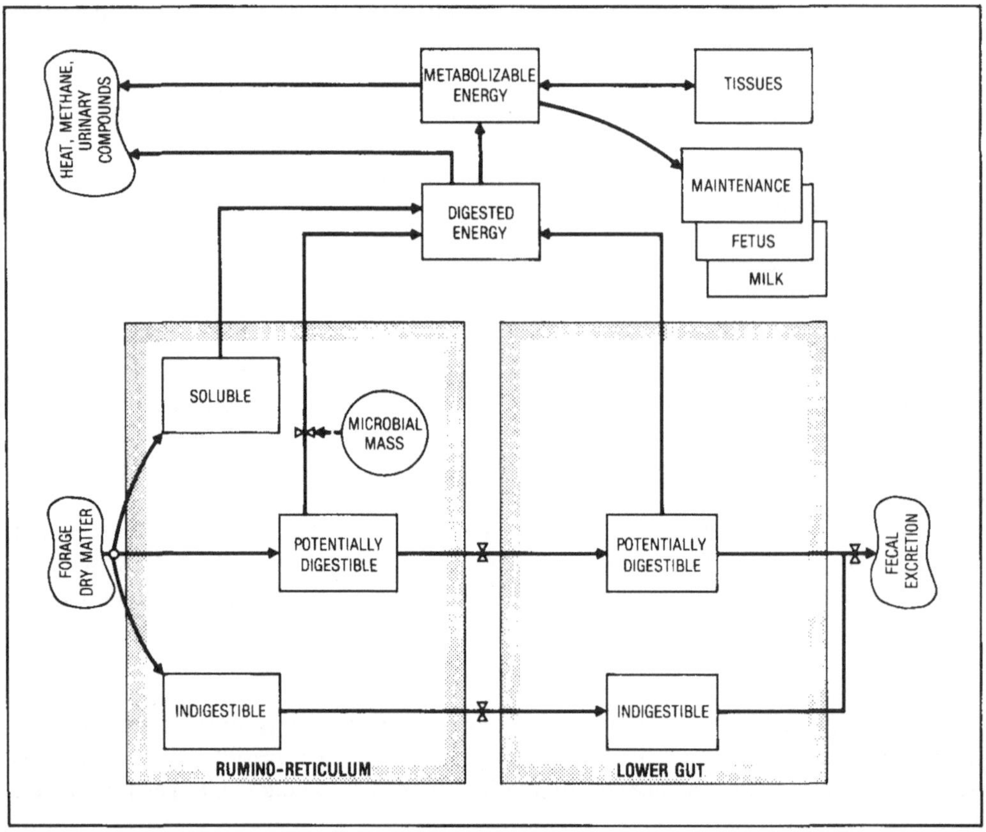

FIGURE 3. Components and interrelationships used to model dry matter digestion and energy assimilation.

where RDG′ is a rate constant based upon the proportions of forage classes comprising the diet. An upper limit of 0.15/hr is applied.

Passage rates are estimated from a similarly limited number of studies. The most comprehensive work has been conducted on the closely related European red deer.[64,66,67] On most diets, passage rates in red deer are more rapid than in sheep leading to less complete digestion of fibrous diets. Because of their larger size, rates of passage in wapiti probably are slightly lower.[68,69] On this basis, nominal passage rates of the potentially digestible phase (P.PDG′) from the reticulo-rumen are estimated to be 0.04/hr for graminoids and 0.05/hr for foliage, forbs, and browse.

As well as the physical nature of the feed, level of intake and hence rumen fill influence rate of passage. Data from Westra[56] do not reveal obvious effects in wapiti but this is not surprising for pelleted diets. From studies with sheep on forage diets, fractional flow rates for the rumino-reticulum are related to rumen dry matter by a power equation.[70] Mudgal et al.[71] provide more complete information on fractional rates for liquid, particulate, lignin, and microbial pools at two levels of feeding.

These findings can be generalized to animals with different rumen sizes by expressing the rumen dry matter pool as a ratio to rumen capacity (RFILL/RCAP). Fitting a power equation[70] to the data of Mudgal et al.,[71] fractional flow rates for the potentially digestible pool can be expressed:

$$\text{P.PDG} = \text{P.PDG}'\ e^{(1.46 \cdot \text{RFILL/RCAP})} \qquad (9)$$

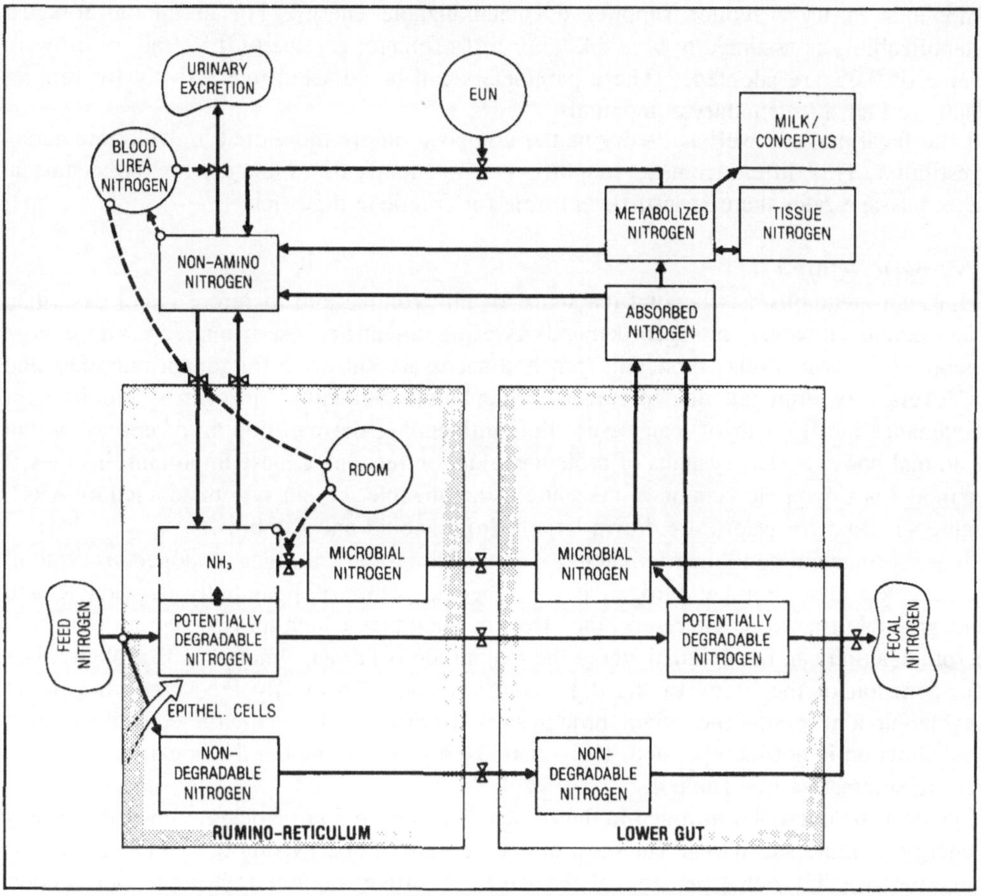

FIGURE 4. Components and interrelationships used to model nitrogen digestion, assimilation, and metabolism.

The indigestible pool is assumed to pass through the tract at a rate which is 80% of that for the potentially digestible pool while the liquid phase is assigned a value of 150%.[72,73] On the assumption that approximately half of the microbial population is associated with the particulate phase and half with the fluid phase, the microbial passage rate is taken as the average. Since sorting of particles in the lower tract is expected to be unimportant, the same fractional passage rate (0.07/hr) is used for all pools.

Another factor known to influence passage rates is rumen motility as influenced by thyroid hormones. This has been invoked as an explanation for the higher rates of passage of cold-stressed animals.[74] However, Westra and Hudson[75] were unable to detect an effect of cold exposure on thyroid hormones and digestive function of wapiti and rate constants for a pelleted ration were only weakly related to levels of circulating T3 and T4 hormones. Therefore, this aspect was ignored.

Total energy supplies are calculated as the daily sum of the soluble diet fraction and digested portion of potentially fermentable carbohydrate (allowing for competitive rates of digestion and passage). The proportion of digested dry matter which arises from the rumen is calculated separately in order to determine energy available for microbial synthesis. No attempt is made to trace the incorporation of energy-yielding materials into microbial cells and their subsequent digestion. Digested dry matter is assumed to be the same for all plant classes and to have an energetic density of 18.4 kJ/g.

Losses of heat and combustible gases during fermentation and incompletely oxidized compounds in urine reduce supplies of metabolizable energy. For graminoid forages, metabolizability is assumed to be 0.86.[76] For forbs/foliage, a value of 0.82 and for browse, a value of 0.72 are adopted.[77] These parameters can be adjusted interactively for forages which are high in secondary compounds.

Total fecal mass as well as its dry matter composition are monitored to determine actual digestibility. This differs from the proportion of potentially digestible dry matter because at higher passage rates there is insufficient time for complete digestion.

2. Nitrogen Network

Nitrogen metabolism is beyond the scope of this volume and therefore is not described in any detail. However, nitrogen demands brief treatment because it interacts with energy metabolism at two levels. First, nitrogen is a necessary substrate for rumen microbes and low levels may limit the rate of rumen fermentation. Secondly, nitrogen is required for maintenance and growth of lean tissue, thus influencing the partitioning of energy within the animal body. Newer systems of protein evaluation recognize these important linkages.[78] The model is a dynamic version of the static metabolizable protein system adopted by ARC[1] (Figure 4). Data for wapiti are drawn largely from Mould and Robbins.[79]

Sources contributing to the ruminal nitrogen pool are forage nitrogen, endogenous protein, and recycled urea. Forage nitrogen is described in terms of potentially degradable and nondegradable fractions. Therefore, the effect of phenolics which may protect protein from microbial action can be included where the magnitude is known. Endogenous protein arises from abrasion of the lining of the digestive tract and inefficiently recovered enzymes.[80] Recycled urea may enter the rumen through saliva or directly across the rumen wall although this distinction is not incorporated. Key relationships used to model the nonamino nitrogen pool are summarized in Table 4.

Synthesis of microbial nitrogen in the rumen depends, to a large degree, upon the supply of energy-yielding substrates. The most direct measure of energy supply is ATP. However, estimation of ATP yields per unit of fermented substrate requires rather complete understanding of biochemical pathways in the rumen cnrivonment. Usual values are 4 to 5 moles per mole of fermented hexose but there are major uncertainties about requirements for reactions such as transport against concentration gradients. In this light, Ørskov[78] recommends the use of ruminally fermented carbohydrate as the most useful base for calculating microbial yields. Mean values for forage diets consumed by cattle and sheep are slightly over 30 g N/kg ruminally digested organic matter (RDOM).[46] However, passage (dilution) rates have a significant effect on yields since microbial populations have a maintenance as well as growth requirement and are partly redigested if retained too long. Microbial yield (Y) can be related to dilution rate (Kp) by:[46]

$$1/Y = (M/Kp) + (1/Ymax) \tag{10}$$

where M is the maintenance requirement (0.15 g N/kg OM) and Ymax is the maximum yield (71.1 g N/kg OM).

With low supplies of degradable nitrogen, synthesis may be impaired. The most important nitrogenous substrate for bacteria is ammonia. Although the ammonia pool depends upon relative rates of release by fermentation and net exchange through the rumen epithelium, there is insufficient information to model these dynamics. A simple empirical approach is adopted which correlates ammonia concentrations (NH_3, mg/100 mℓ) to dietary nitrogen concentration (P.NI, frac.):[81]

$$NH_3 = 10.57 - 1562.5\ P.NI + 0.159\ (625\ P.NI)^2 \tag{11}$$

Table 4
RELATIONSHIPS DESCRIBING THE DYNAMICS OF THE UREA POOL

Urea space (ℓ)	URSP = 0.46 LVWT
Urinary excretion (g/hr)	UREX = (1/24) MW $(0.0005\ \text{BUN})^{1.7}$
Urea entry (g/hr)	URTE = 0.1 PDN + 0.032 WTLOSS/24
Urea recycled (g/hr)	URRE = 0.8 URTE (1-0.0083 BUN)
Urea pool (g)	URPO = URPO + DT (URTE - UREX - URRE)
Blood urea (mg%)	BUN= URPO/URSP

Note: LVWT = liveweight, MW = metabolic weight, WTLOSS = weight loss (kg).

Calculated from Mould and Robbins.[79]

Table 5
PREDICTION OF EFFICIENCIES OF UTILIZATION OF METABOLIZABLE ENERGY

Productive function	Energy from feed	Energy from tissues
Maintenance	km = 0.287 DG + 0.503	—
Pregnancy	Kp+ = 0.372 DG − 0.005	Kp− = 0.25
Lactation	Kl+ = 0.287 DG + 0.420	Kl− = 0.85
Gain	Kf+ = 0.64 DG + 0.006	—

Microbes respond to specific amino acids.[82] However, NH3 should be a satisfactory correlate on natural diets. This quantitative relationship will be strongly influenced by supplies of readily fermented carbohydrate but we have insufficient information to model the rumen ammonia pool more explicitly.

Nitrogen of microbial and feed origin entering the lower tract is subject to the competitive processes of digestion and passage. Following ARC,[1] rates are adjusted to provide approximately 80% digestion on a daily basis. The proportion of nitrogen in the form of amino acids (metabolizable nitrogen) is assumed to be 80% and absorbed amino acids are assumed to be utilized with an average efficiency of 75%.[1]

D. Efficiency of Metabolizable Energy Utilization

Metabolizable energy is used for maintenance and production with varying efficiencies. In general, these efficiencies (but not ME requirements) appear similar in most ruminants. Sheep and cattle are considered to be the same[1] and red deer are similar to sheep.[83] Efficiencies are influenced by the balance of metabolic substrates and tend to be highest for digestible forages. Based on work with domestic animals summarized by ARC,[1] the relationships in Table 5 were chosen.

E. Maintenance Requirements

Maintenance requirements for energy can be factored into resting metabolism, thermoregulation, and activity.

1. Metabolic Rate

Fasting metabolic rates are assumed to be similar and constant in most adult eutherian mammals (Chapter 7). Although correspondence to a simple allometric rule holds when a wide range of species are compared, relationships within species are less clear. Age and physiological state influence metabolic rates. In many species, seasonal cycles are quite striking, but it is unclear to what degree these are in direct response to photoperiod rather than a reflection of seasonal appetite.

Several workers have attempted to quantify factors influencing maintenance requirements. Graham et al.[84] considered the effects of feeding level (prior to fasting) and rates of gain in domestic sheep to derive factors for the "work of production". Chappel and Hudson[85] used a similar approach in a study of metabolic rates of bighorn sheep. Comparable studies on wapiti and red deer are not available so our approach had to be rather arbitrary.

The first step was to select a pivotal value for fasting metabolism to serve as an appropriate intercept for an equation relating metabolic rate to body weight, energy intake, and gain. Metabolic rates of red deer and wapiti have been measured in a number of studies (Chapter 7) but usually not under standardized conditions. From measurements of resting and fasting metabolic rates provided by Pauls et al.[86] and more recent unpublished data, we derived the following approximate relationship based on metabolic weight (MW, $kg^{0.75}$) with adjustments for photoperiod (CYCLM), gain (G), and previous ME intake:

$$\text{FMR} = \text{CYCLM}\ (400\ \text{MW} + 5600\ \text{G} + 46\ \text{ME}/0.86) \qquad (12)$$

where CYCLM is a sine wave with an amplitude of 0.22. ME required for maintenance was calculated by dividing FMR by the efficiency of utilization of metabolizable energy for maintenance (FMR/KM).

Heat production (RMR) for animals at rest in a thermoneutral environment is the sum of energy required for maintenance, dietary-induced thermogenesis (0.07 ME), and heat of nutrient metabolism specific to pregnancy, lactation, and growth.

2. *Thermoregulation*

Calculation of energy requirements for thermoregulation generally followed Christopherson and Young.[87] Heat production in response to cold is usually expressed as watts (J/sec) per unit surface area. Surface area (SA, m^2) is calculated:

$$\text{SA} = 0.09\ \text{LVWT}^{0.66} \qquad (13)$$

Although more precise relationships for cervids are provided by Moen,[88] this equation was used to determine insulation in the studies selected to develop this part of the model.

The air temperature below which heat production must be increased to maintain homeothermy is termed the lower critical temperature (Tc):

$$\text{Tc} = \text{Tr} + \text{He} \cdot \text{Ie} - \text{H}(\text{It} + \text{Ie}) \qquad (14)$$

where Tr is rectal temperature (39°C). He is minimum evaporative heat loss, H is total heat production, and Ie and It are external and tissue insulations (°C·m^2/W), respectively. Heat of thermoregulation is calculated:

$$\text{Ht} = (\text{Tc} - \text{Ta})/(\text{Ie} + \text{H} \cdot \text{It}/(\text{H} - \text{He})) \qquad (15)$$

Minimum evaporative heat loss is similar in red deer, sheep, and cattle with measured values for red deer of 19.2 W/m^2.[76] Tissue insulation varies from 0.09 at 8°C to 0.12 at 4°C for red deer weighing about 45 kg.[76] Obviously, tissue insulation is greater in large animals. Equations relating It of cattle to body weight and ambient temperature overestimate values for red deer because of their different conformation. Therefore, the following adjusted equation was used:

$$\text{It} = 10^{-3}(100 + 0.378\ \text{LVWT} + 2.07\ (\text{Ta} - 8)) \qquad (16)$$

External insulation is a complex function of coat depth and wind speed. Values (per unit coat depth) for red deer and cattle appear similar but are considerably lower than for sheep. The following relationship derived for cattle was adopted from Webster:[89]

$$Ie = 0.23 + 0.05F - 0.09V^{0.5} \tag{17}$$

where F is coat depth (CM) and V is wind speed (m/s). This is certain to be a very conservative estimate for winter coat of wapiti (Chapter 8).

3. Activity Costs

Activity costs are determined directly as expenditures rather than work accomplished divided by an efficiency term. Because it can be easily studied under controlled conditions, most work has been conducted on the energy costs of locomotion (Chapter 7). Energy costs of horizontal locomotion (CL) are calculated from:[90]

$$CL = 2.05\ LVWT \cdot TG \cdot DKM \tag{18}$$

where TG is time spent foraging and DKM is distance traveled per hour of foraging. This is arbitrarily doubled to account for variation in topography and substrate. There is considerable scope for refinement, particularly in relation to the effects of snow cover.[91]

Costs of other activities are derived from calibrated heart rates because of difficulties determining oxygen consumption in natural settings.[86] Contrary to the usual view, costs for each activity (per unit time) vary with resting metabolic rate.[44] When metabolic rates are low because of season or nutritional restriction, corresponding activity increments are low. The following costs are assigned to foraging (CF), resting/ruminating (CR), and social interaction (CP):

$$CF = -7.1615 + 1.7363\ MR \tag{19}$$

$$CR = -\ 6.44 + 1.3616\ MR \tag{20}$$

$$CP = 2\ MR \tag{21}$$

Total energy costs are determined as the sum of incremental costs of locomotion, foraging, rumination, and social interaction.

4. Nitrogen Losses

Maintenance requirements for protein are considered to be the amount required to replace endogenous urinary N excretion, endogenous fecal excretion, and dermal losses. The constant endogenous component is assumed to arise from the degradation and replacement of protein and simple nitrogenous components of tissues. It is considered equal to the minimum rate of urinary N excretion of animals maintained for a period on a diet with little or no protein. More commonly, it is estimated from the value of the intercept of urinary N on apparently digested N.

Mould and Robbins[79] estimated EUN to be 0.6 $gN/LVWT^{0.75}$ while metabolic fecal nitrogen was 5.58 gN/kg dry matter intake. In ruminants, these two quantities are not strictly independent and perhaps cannot simply be added to give total requirements.[78] However, in the absence of relevant data for wapiti, the factoral approach[79] is adopted.

F. Gestation

The gestation length of wapiti is approximately 260 days; i.e., several weeks longer than that of red deer. Calves are dropped in late May following conception in September.

Average birth weights (BW) of calves are about 16 kg but vary with the size and nutritional status of the dam. Data from Robbins and Robbins[92] indicate that wapiti conform to interspecies allometric relationships and that this may even hold for wapiti of different sizes:

$$BW' = 0.2143\ LVWT^{0.79} \tag{22}$$

But Thorne et al.[93] suggest that the pattern of weight change during pregnancy may modify this basic relationship. The following approximation is used:

$$BW = BW' + 12\ (WC/WI) \tag{23}$$

where WC is maternal weight change (gain or loss) and WI is body weight at conception.

Conceptus weight (fetus plus associated uterine tissues) at any stage of gestation is estimated:[94]

$$CW = 1.54\ BW\ e^{(0.0195\ DP - 5.122)} \tag{24}$$

Daily energy retention (PR) is[94]

$$PR = 4.184\ BW\ e^{(0.0193\ DP - 1.7938)} \tag{25}$$

and nitrogen required for development of the conceptus is[94]

$$NP = BW\ e^{(0.01969\ DP - 1.7274)} \tag{26}$$

G. Lactation

Several studies have been conducted on lactation in *Cervus elaphus*.[52,53,95,96] Arman et al.[53] studied the composition and yield of milk from red deer and Robbins et al.[52] provided comparable data for wapiti. Using the lactation curve form of Wood[97] and fitting data for well-fed red deer,[53] potential milk yield (MYP, kg/d) is calculated from days lactating (DL):

$$MYP = MW \cdot B3\ (DL^{0.4556})\ (e^{-0.02424\ DL}) \tag{27}$$

where B3 is a coefficient related to body condition (expected weight minus liveweight):

$$B3 = 0.027 + 0.01\ (EW - LVWT) \tag{28}$$

When applied to the larger-bodied wapiti, these equations give results consistent with the empirical measurements of Robbins et al.[52]

The energy content of milk (NC, kJ/kg) varies with the stage of lactation. Using data from Arman et al.:[53]

$$NC = 4327\ DL^{0.096} \tag{29}$$

There is a limit to which animals can mobilize tissues during negative energy balance which probably depends on amino nitrogen absorption from the intestines. Following Christian et al.,[38] the maximum daily draw on body reserves (FMX) is set at:

$$FMX = 300\ MW \tag{30}$$

This rule is applied to calculate actual milk yield. Nitrogen required for lactation is calculated using a value of 0.97% (6.2% protein) for the nitrogen content of wapiti milk.[52]

H. Growth

Lean tissues (wet weight) have energy contents of approximately 5,000 kJ/kg while fat has an energy content of 39,300 kJ/kg. As animals mature, priorities for tissue deposition change.

Most of the variation in the energy content of gain in immature animals is explained by liveweight. Irrespective of age or nutritional history, there is a tendency for animals of the same weight to have similar body compositions. The most relevant illustration of this principle is work on white-tailed deer by Robbins et al.[98] After scaling for differences in mature weights of deer (80 kg) and wapiti (310 kg), the relationship between the energy content of gain (CG, kJ/kg) and liveweight (kg) can be estimated:

$$CG = 2000\ LVWT^{0.37} \tag{31}$$

which corresponds to a fat proportion (PFAT) of:

$$PFAT = 0.00885\ LVWT^{0.635} \tag{32}$$

These relationships predict energy contents of gain of 11,000 kJ/kg (18% fat) at weaning and 16,500 kJ/kg (33% fat) at maturity. Where nitrogen is limiting, surplus energy is deposited solely as fat.

Integrated over the growth trajectory, these rules give realistic body compositions at all stages of development.[51] However, both environmental and physiological factors may cause short-term deviations. Animals on exceptionally high planes of nutrition tend to be fatter at similar liveweights because lean tissue growth is rate limited. Homeorhetic controls also may influence the seasonal partitioning of energy between lean and fat tissue growth. A further complication is the proportion of water associated with lean tissue growth which changes with maturity and with the seasons.[99] At the present time, these effects are too poorly documented to include in this version of the model.

Empty body weight is determined by integration of daily gains. Liveweight is calculated from empty body weight by adding the weight of digesta pools and, for pregnant females, the developing conceptus.

IV. EVALUATION AND APPLICATION

Computer simulation models usually are tested in two ways. First, they are exposed to sensitivity analysis to determine how precisely parameters must be known. Secondly, predictions are tested against empirical data or at least the subjective knowledge of experience. Often, independent data sets are not available to develop and test models, so this exercise may reduce to one of uncovering programming errors. Only once models faithfully accommodate a wide range of environmental conditions can probing questions be asked about bioenergetic and behavioral adaptations.

A. Sensitivity Analysis

The paucity of good estimates of many bioenergetic parameters for wapiti (and red deer) required use of crude approximations. Sensitivity of the model to these estimates obviously depends on environmental conditions since, in better environments, animals can invoke numerous compensatory mechanisms. For example, higher energy expenditures could be

Table 6
SENSITIVITY ANALYSIS OF KEY PARAMETERS: EFFECT OF 10% INCREASE OF EACH PARAMETER ON SIMULATED WEIGHT IN APRIL AND SEPTEMBER

			Simulated weight (kg)		Simulated DM intake (g)	
Parameter	**Description**	**Original value**	**April 10**	**September 10**	**April 10**	**September 10**
Default values		—	202	241	5817	8321
Intake						
FRSN	Effect of snow on feeding rate	0.014	199	239	5746	8301
KG	Effect of graminoid biomass on feeding rates	0.0006	209	249	6114	8370
KB	Effect of browse on feeding rate	0.0012	207	244	5942	8347
KF	Effect of forb/foliage biomass on feeding rate	0.0012	202	241	5817	8321
CYCLT	Amplitude of seasonal appetite cycle	0.40	201	240	5817	8293
Digestion						
P1.G	Fractional passage rate of grasses	0.04	202	241	5817	8362
P1.B	Fractional passage rate of browse	0.05	202	241	5817	8324
P1.F	Fractional passage rate of forbs/foliage	0.05	202	241	5817	8324
RDG.G	Fractional digestion rate of grasses in rumen	0.09	206	243	5817	8282
RDG.B	Fractional digestion rate of browse in rumen	0.12	203	241	5817	8311
RDG.F	Fractional digestion rate of forbs/foliage in rumen	0.12	202	241	5817	8314
IDG	Fractional digestion rate in lower gut	0.06	203	241	5817	8319
%ME.G	Metabolizability of DE from grasses	0.86	214	250	5817	7831
%ME.B	Metabolizability of DE from browse	0.78	205	243	5817	8282
%ME.F	Metabolizability of De from forbs/foliage	0.82	203	242	5817	8257
Expend						
BMR[1]	Mean annual fasted metabolic rate	400	180	230	5817	8624
CYCLA	Amplitude of seasonal metabolic cycle	0.20	210	245	5817	8319

offset by higher feed intake or fasting metabolic rates could be reduced in response to nutritional restriction. Therefore, the following analysis was conducted by exposing the model animal to relatively harsh winter conditions (Table 2).

We evaluated the performance of a pregnant and subsequently lactating 2-year-old female with an initial rut weight of 240 kg. Each selected parameter was subjected to a 10% increase while all others assumed default values. Criteria of performance were simulated weights and dry matter intakes on April 10 and September 10 to reflect the spring period of poorest nutritional condition and the pre-rut period when fat reserves are high.

Because of the complexity of feedback mechanisms, most parameters are relatively insensitive (Table 6). Parameters describing the logistics of foraging are most responsive to the conditions tested and become even more so in larger animals where forage availability constrains intake. Differences for parameters relating to forage classes are due simply to their proportional representation in the diet.

One critical area is digestive kinetics. The model is not that sensitive to a 10% change in rates of passage or digestion but estimates from the literature vary as much as 50%. Parameters describing metabolizability are sensitive but are known more precisely except for certain plants which contain quantities of secondary compounds.

As expected, the mean and amplitude of seasonal cycles of appetite and energy expenditure are important determinants of weight dynamics. A number of studies have been conducted on *Cervus elaphus*, but adaptive changes in response to photoperiod and the nutritional environment have not been studied in sufficient detail.

Although the model may appear to be robust, it is partly an expression of its complexity. As the number of feedback relationships increases, so do opportunities for compensation. Whether these are considered artifacts of complexity or adaptive mechanisms is a matter of perception. In nature, there are probably a number of designs that work almost as well except under extreme conditions.

B. Corroboration

Models are never really validated since critical proof cannot be obtained. They are not right or wrong, simply more or less useful in ordering current information and guiding future inquiries. One can do little more than gain confidence as models provide realistic predictions under a growing range of circumstances. This section displays predictions of the bioenergetics and productivity of wapiti under conditions for which we have at least some empirical data. We begin with a consideration of the growth trajectory throughout life and then take a closer look at bioenergetic transactions through the annual cycle.

1. Lifetime Performance

Simulated lifetime growth curves of reproductive and nonreproductive females on pasture are compared with available weight records in Figure 5. Predictions of the performance of animals given high quality supplemental feed *ad libitum* are also shown. Although corresponding weight data are not available for periods extending from weaning to maturity, recorded rates of gain during winter for animals of different fall weights are consistent with predicted values.[100]

In general, the model provides a reasonable representation of lifetime performance. However, empirical data are not hard to fit given the wide individual variation which seems to characterize wapiti and probably most herbivores. Annual weight cycles are well developed given the severity of the boreal winter and larger animals seem to lose more weight on an absolute and relative basis.[44] However, compensatory summer gains are striking and inversely proportional to the weight of animals at the lowest point at the end of winter.[100] This ensures a gradual approach to an asymptotic weight at 3 to 4 years of age.

Under the range of conditions studied, costs of reproduction are not clearly evident in

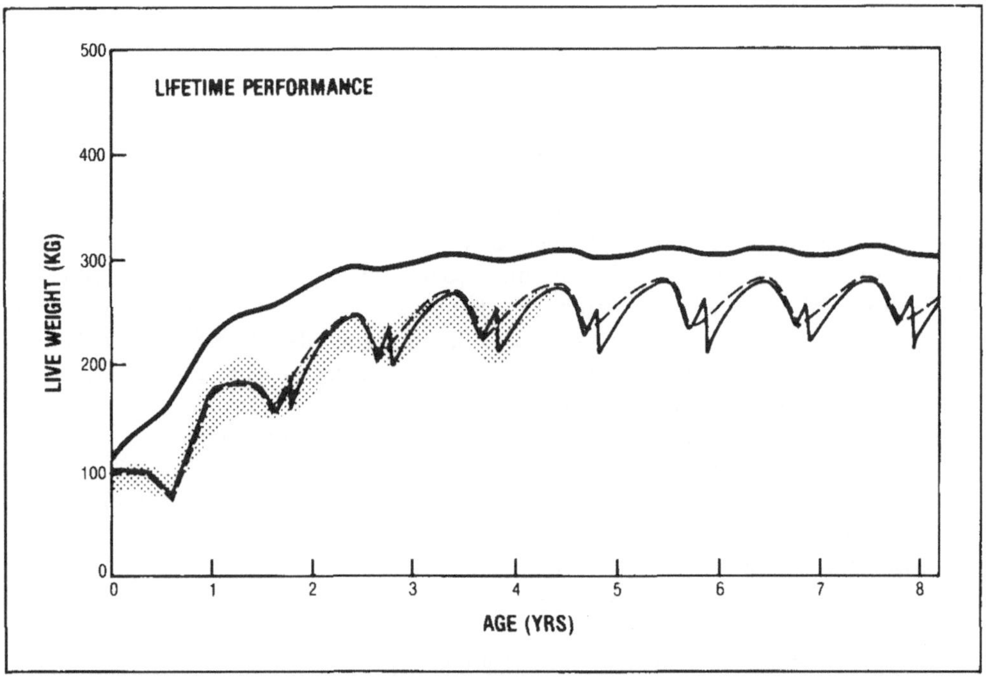

FIGURE 5. Projection of seasonal weight dynamics from weaning to maturity for wapiti (not bred) on *ad libitum* supplemental feeding (upper trajectory) or on pasture (lower trajectories). Consequences of reproduction are implied by the difference between lower trajectories for bred (solid line) and not bred (dashed line) females. The shaded area denotes weight ranges for a research herd receiving limited winter supplemental feeding.

weight dynamics. This version of the model does not include special adaptive mechanisms which would result in higher energetic efficiency of pregnant females as suggested by Clutton-Brock et al.[101] Hence, minimum winter weights are about the same for pregnant and non-pregnant females with the weight of the developing fetus balancing the greater loss of fat. Demands of lactation are insufficient to prevent compensation and pre-rut weights are only slightly lower than in nonlactating individuals. Although the demands of reproduction might be expected to be greater in younger females, lower birth weights and milk production minimize bioenergetic stresses. Although development of the calf has not been modeled, one would expect inferior performance of calves born to young or underweight females. Such patterns are consistent with our experience with wapiti. But studies on red deer[96,101] suggest that the picture may be considerably more complex.

The model suggests that the genetic potential for mature body size is not achieved by free-ranging wapiti in this environment. The period of plant growth is too short to allow full compensation for winter weight loss even when animals are not supporting a calf. Maximum weights have not been determined empirically but patterns of compensation suggest a target of about 310 kg and several older females in our research herd have attained fall weights of over 300 kg following supplemental winter feeding.

2. Annual Cycle

During winter, free-ranging animals lose weight in the face of declining quality and availability of native forage. The impact in both absolute and relative terms is greatest for larger individuals (Figure 6). Weight loss occurs mainly in late winter as snow deepens and forage quality declines to critical levels. Spring comes abruptly and weights, particularly of pregnant females, increase rapidly. Even at peak lactation, weight gains are positive and small females are able to narrow the contrast in weight with large females over the summer.

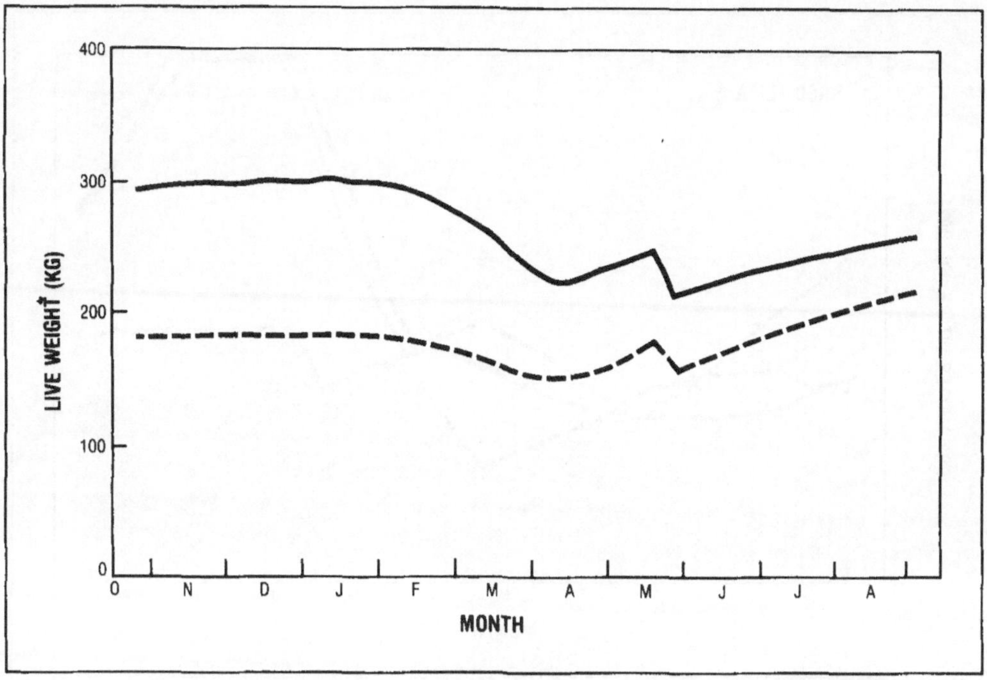

FIGURE 6. *Simulated annual weight dynamics for large and small bred female wapiti.* Weight loss in late winter is greater on an absolute and relative basis for large individuals when forage availability is low.

Rumen distention, metabolic requirements, and the logistics of foraging interact in a complex way to give rather different voluntary forage intakes of small and large animals through the seasonal cycle (Figure 7). In reproductive adult females, rumen-fill threatens to constrain intake only in autumn. Particularly in large individuals, weight loss appears to be due to the logistic constraint operating in late winter. With snowmelt, intake increases rapidly in compensation. However, as the spring flush of vegetation progresses, intake declines until lactation imposes additional demands after calving. The decline in requirements to support lactation is offset by declining forage quality so intake remains high for much of the summer.

A rather different pattern of constraints operates on weaned calves. Gut fill is predicted to be a major constraint on forage intake until the following spring and summer. Since logistic constraints appear less important, forage quality would be expected to be more important than availability. However, differences in foraging efficiency of large and small animals under harsh winter conditions have received only cursory attention.[42]

Much of the energy supplied by forage is dissipated as heat. The model predicts energy expenditures associated with basal (fasting) metabolism, thermoregulation, activity, and digestion/nutrient metabolism. Thermoregulation costs are minimal for wapiti in an environment with ample opportunity for sheltering.[86] Activity costs are low during winter because animals are relatively inactive but increase to about $^1/_3$ of resting metabolism in early summer.[86] Resting metabolic rates which include the basal component and heat increment of feeding are predicted to rise abruptly with the spring flush and decline gradually until the following March. This prediction for an average year is compared with measurements taken in each of two years in Figure 8. Although the prediction is broadly consistent with results obtained one year, it does not reproduce the sharp increase observed in the other. Differences between these data sets may be related to the severity of winter conditions. The abrupt rise (underestimated by the model) followed a harsh winter with little supplemental

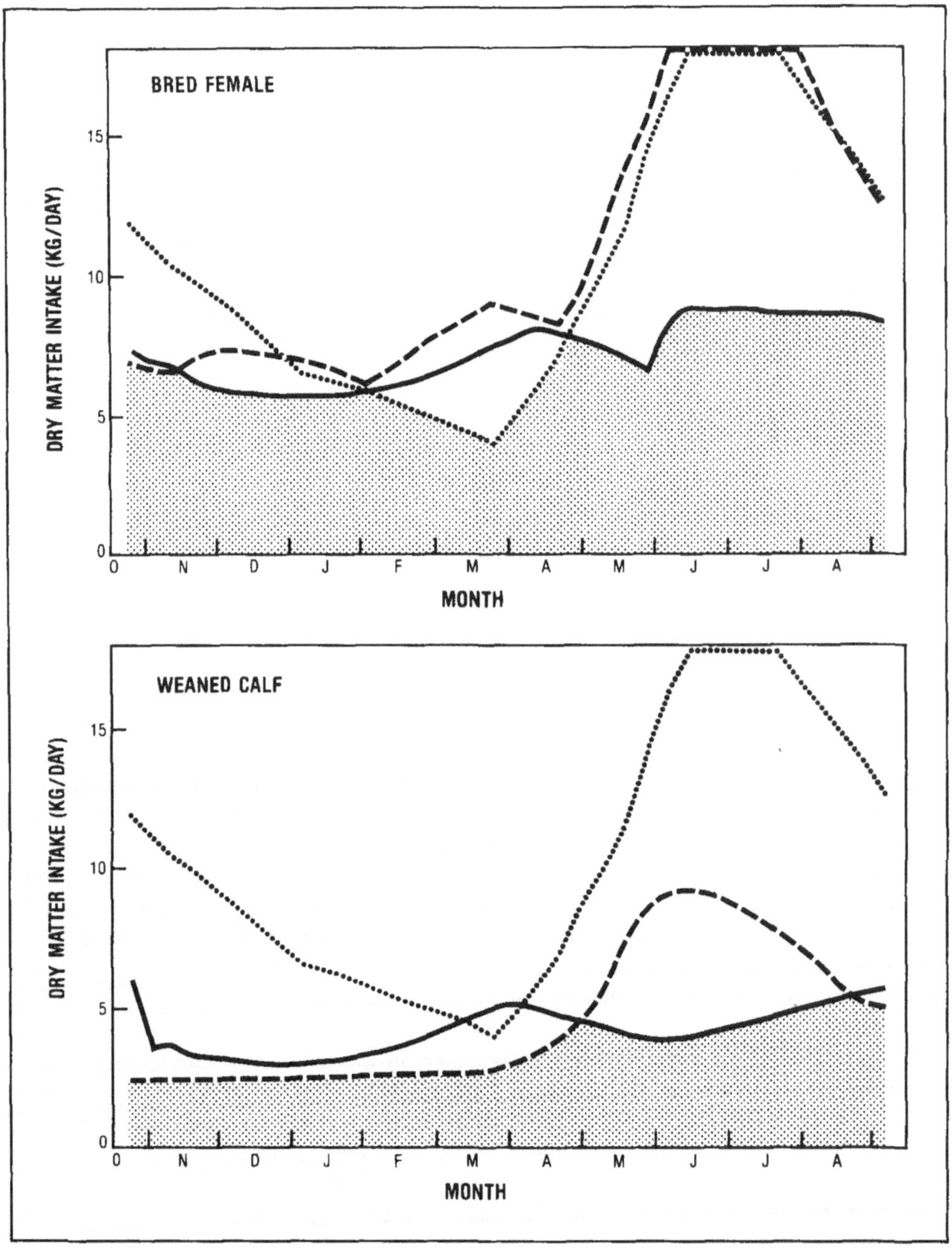

FIGURE 7. Simulated annual cycle of physical (dashed line), metabolic (solid line), and logistic (dotted line) constraints on daily forage intake. Realized dry matter intake is denoted by the shaded area for a bred adult (above) and nonreproductive calf from weaning to 1 year of age.

feeding and may reflect greater compensatory forage intake once pasture growth resumed. However, attempts to reproduce such conditions give predictions which are neither as abrupt nor as high. This points either to a deficiency of the model or problems with the metabolic measurements.

C. Bioenergetic Implications of Body Size

The theme of adaptive limits to body size was raised in Chapter 1. Large body size may

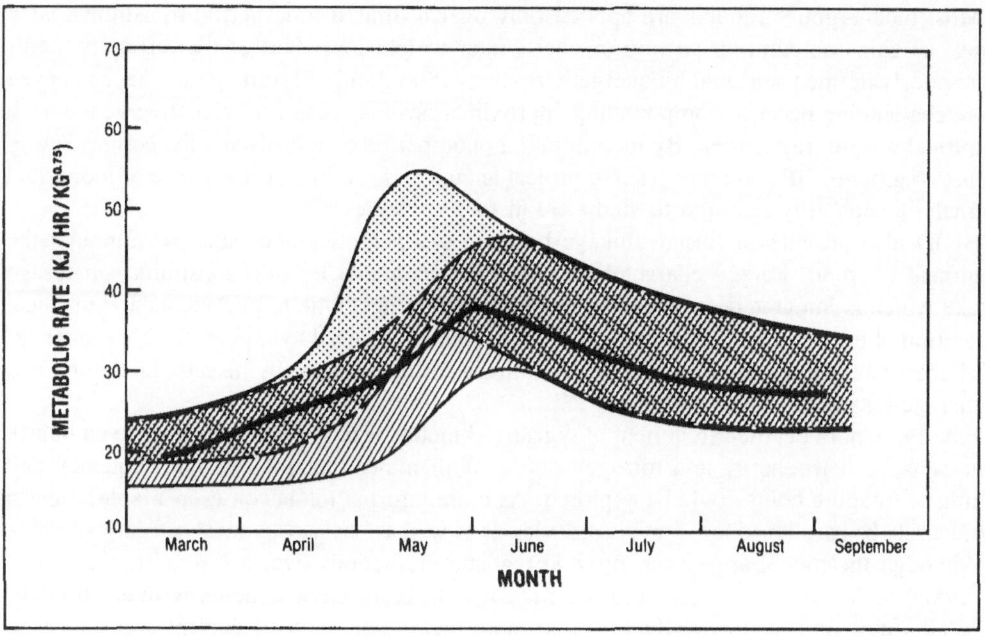

FIGURE 8. Simulated resting metabolic rates of adult female wapiti during late gestation and lactation compared with observations made following a harsh winter (upper) and moderate winter supplementation (lower).

carry both advantages and disadvantages; each environment may have an optimum, or possibly, a stable polymorphism of types adapted to specific recurring patterns. Clutton-Brock and Harvey[102] feel that harsh environments select for smaller individuals pointing to the competitive displacement of wapiti by red deer in New Zealand and to the greater bioenergetic embarrassment of large red deer stags on Rhum.[103] From results presented above, large bodied wapiti appear to experience similar hardship during the boreal winter.

However, our conception of "harsh" requires clearer definition. By exposing the model wapiti to different types of winter hardship, we conclude that large animals are adapted to conditions of extreme cold, low forage quality, and situations where forage is completely unavailable (e.g., under crusted snow) for *short* periods. In these circumstances, large body size provides thermal insulation, a more favorable balance between digestive capacity and metabolic requirements and a store of body reserves which confer nutritional "inertia". Small animals are adapted to long seiges of limited forage availability. This derives from the more modest logistic constraint on obtaining their daily forage requirement. However, limitations of gut fill make them more sensitive to deterioration of forage quality. Since grazing reduces forage biomass but often increases forage quality, we should expect smaller phenotypes to be more competitive in dense populations.

V. FUTURE DIRECTIONS

Compared with published bioenergetic models for wild ruminants,[27] this model (BION) sacrifices generality by drawing heavily upon existing information for a single species. Although sharpening focus, it provides a number of more general enhancements.

First, it considers reproductive processes and therefore can be used for females of any age or reproductive status. Males can be approximated as nonreproductive females of larger mature body size but this ignores complex physiological and behavioral events associated with the rut and pronounced changes in body fluids.[99]

Most bioenergetics models are applied only over a limited time period to estimate such things as gains on summer pasture or over-winter weight loss. Part of the difficulty stems from poorly defined seasonal bioenergetic rhythms. Simulating lifetime performance is even more challenging because compounding approximations may result in wide divergence from empirical weight trajectories. By incorporating photoperiodic controls and the compensatory nature of growth, BION can be used to project annual and lifetime performance of individual animals, a capability required to study life history strategies.[104]

BION also provides a formal linkage between bioenergetics and behavior although the treatment is crude. Forage characteristics are used to model logistic constraints on forage intake which is integrated with feedback controls from metabolic requirements and physical distention of the digestive tract. This allows the model to be used to explore the consequences of alternative resource-use strategies or the impacts of disturbance by insects, predators, and human activity.

Finally, a more detailed treatment of digestive kinetics is provided by moving to an hourly time step for both energy and nitrogen pools. With minor elaboration, the frequency and timing of foraging bouts could be explored. As more information becomes available, digesta could easily be partitioned into particle-size classes as well as physicochemical categories.[48,105]

Although the model appears to provide realistic predictions over a limited range of circumstances, its ability to accommodate radically different environments is likely limited. Like all simulation models, BION is little more than a summary of current understanding which must be continually modified in the light of new information. This synthesis identified a number of key feedback relationships which are critical to understanding the behavior and performance of wild herbivores.

The first relates to adaptive changes in digestive anatomy and function. Physical capacity, rates of passage, and perhaps sorting of digesta particles respond to seasons, productive processes, and diet.[106,107] The present version recognizes the influence of lactation but empirical relationships have not been derived.

A second area is adaptive modification of metabolic rate (Chapter 6). Seasonal cycles of appetite, growth, metabolic rate, and blood levels of various hormones and metabolites are marked in northern cervids. Photoperiod and environmental effects interact in a manner which is difficult to unravel experimentally[108] and available data do not allow us to conclude which factors drive the cycle. Time-lags further complicate calculation of maintenance requirements when energy supplies change abruptly.[109]

A third question relates to the complex homeorhetic controls operating on the composition of gain.[110] The assumption that this is dictated by current body weight is an over simplification. The rate of gain and supplies of nitrogen undoubtedly have an effect. The existence of lipotrophic and lipolytic phases in the seasonal growth cycle need to be explored.[111,112]

Contributions to this volume have summarized a great deal of information that could be used to update this model, adapt it to other species, and perhaps apply generalized versions to the problems of adaptive radiation. Historically, simulation models have served largely as research tools. However, with the growing availability of microcomputers, we anticipate greater application in management.

ACKNOWLEDGMENTS

This study was made possible by grants from the Natural Sciences and Engineering Research Council, Alberta Environment Research Trust, Alberta Fish and Wildlife Division, and the Alberta Agricultural Research Council. Drs. J. Kennelly and R.N.B. Kay suggested many improvements of the preliminary model and commented on the manuscript. W. G. Watkins provided unpublished data on weight dynamics and metabolic rates of wapiti.

REFERENCES

1. Agricultural Research Council, *The Nutrient Requirements of Ruminant Livestock*, Commonwealth Agricultural Bureau, Farnham Royal, England, 1980.
2. **Zulberti, C. A. and Reid, J. T.,** An equation, suitable for computer use, based on the ARC feeding system to determine the energy requirements of growing and fattening cattle, *Anim. Prod.*, 14, 17, 1972.
3. **Black, J. L., Faichney, G. J., Beever, D. E., and Howarth, B. R.,** Alternative systems for assessing the nitrogen value of feeds for ruminants, in *Forage Protein in Ruminant Animal Production*, Thomson, D. J., Beever, D. E., and Gunn, R. G., Eds., Occ. Publ. 6, British Society of Animal Production, Thames Ditton, Surrey, 1982, 107.
4. **Van Dyne, G. M. and Abramsky, Z.,** Agricultural systems models and modelling: an overview, in *Study of Agricultural Systems*, Dalton, G. E., Ed., Applied Science, London, 1975, 23.
5. **Van Dyne, G. M., Brockington, N. R., Szocs, Z., Duek, J., and Ribic, C. A.,** Large herbivore subsystem, in *Grasslands, Systems Analysis and Man*, Breymeyer, A. I. and Van Dyne, G. M., Eds., Cambridge University Press, 1980, chap. 4.
6. **Mertens, D. R. and Ely, L. O.,** A dynamic model of fiber digestion and passage in the ruminant for evaluating forage quality, *J. Anim. Sci.*, 49, 1085, 1979.
7. **Baldwin, R. L., Koong, L. J., and Ulyatt, M. J.,** The formation and utilization of fermentation end-products: mathematical models, in *Microbial Ecology of the Normal Gut*, Clark, R. T. J. and Bauchop, T., Eds., Academic Press, New York, 1977, 347.
8. **Baldwin, R. L., Christ, K., Waghorn, G., and Smith, N. E.,** The synthesis of models to describe metabolism and its integration, *Proc. Nutr. Soc.*, 40, 139, 1981.
9. **Baldwin, R. L. and Denham, S. C.,** Quantitative and dynamic aspects of nitrogen metabolism in the rumen: a modelling analysis, *J. Anim. Sci.*, 49, 1631, 1979.
10. **Beever, D. E.,** Simulation model of rumen fermentation, in *Computers in Animal Production*, Occ. Publ. 5, British Society of Animal Production, Thames Ditton, Surrey, 1981, 89.
11. **Black, J. L., Beever, D. E., Faichney, G. J., Howaith, B. R., and Graham, M. McC.,** Simulation of the effects of rumen fermentation on the flow of nutrients from the stomach of sheep. I. Description of a computer program, *Agric. Syst.*, 6, 195, 1981.
12. **France, J., Thornley, J. H. M., and Beever, D. E.,** A mathematical model of the rumen, *J. Agric. Sci.*, (Cambridge), 99, 343, 1982.
13. **Koong, L. J., Baldwin, R. L., Ulyatt, M. J., and Charlesworth, T. J.,** Iterative computation of metabolic flow and stoichiometric parameters for alternate pathways in rumen fermentation, *Comput. Prog. Biomed.*, 4, 209, 1975.
14. **Reichl, J. R. and Baldwin, R. L.,** Rumen modelling: rumen input-output balance models, *J. Dairy Sci.*, 58, 879, 1975.
15. **Van Soest, P. J., Sniffen, C. J., Mertens, D. R., Fox, D. G., Robinson, D. M., and Krishnamoorthy, U.,** A net protein system for cattle: the rumen submodel for nitrogen, in *Protein Requirements for Cattle: Symposium*, Owen, F. N., Ed., Oklahoma State University, MP-109, 1982, 265.
16. **Morris, J. G., Baldwin, R. L., Maeng, W. J., and Maeda, B. T.,** Basic characteristics of a computer simulation model of nitrogen utilization in the grazing ruminant, in *Tracer Studies on Non-Protein-Nitrogen for Ruminants*, Int. Atomic Energy Agency, Vienna, 1974.
17. **Mazanov, A. and Nolan, J. V.,** Simulation of the dynamics of nitrogen metabolism in sheep, *Br. J. Nutr.*, 35, 149, 1976.
18. **Smith, N. E., Ely, L., Baldwin, R. L., Dolkas, C., Koong, L. J., Moberg, G., and Plucinski, T.,** Models of energy metabolism of intact animals: data tabulation, representation of basal metabolic activities and endocrine effects upon metabolism, *Comput. Biol. Med.*, 5, 315, 1975.
19. **Koong, L. J., Falter, K. H., and Lucas, H. L.,** A mathematical model for the joint metabolism of nitrogen and energy in cattle, *Agric. Syst.*, 9, 301, 1983.
20. **Forbes, J. M.,** Inter-relationships between physical and metabolic control of voluntary food intake in fattening, pregnant and lactating mature sheep: a model, *Anim. Prod.*, 24, 91, 1977.
21. **Forbes, J. M.,** Development of a model of voluntary food intake and energy balance in lactating cows, *Anim. Prod.*, 24, 203, 1977.
22. **Forbes, J. M.,** A model of the short-term control of feeding in the ruminant: effects of changing animal or feed characteristics, *Appetite*, 1, 21, 1980.
23. **Sibbald, A. R., Maxwell, T. J., and Eadie, J.,** A conceptual approach to the modelling of herbage intake by hill sheep, *Agric. Syst.*, 4, 119, 1979.
24. **Vera, R. R., Koong, L. J., and Morris, J. G.,** A model of heat flow in the sheep exposed to high levels of solar radiation, *Comput. Prog. Biomed.*, 4, 214, 1975.
25. **Loewer, O. J., Smith, E. M., Taul, K. L., Turner, L. W., and Gay, N.,** A body composition model for predicting beef animal growth, *Agric. Syst.*, 10, 245, 1983a.

26. **Rice, R. W., Morris, J. G., Maeda, B. T., and Baldwin, R. L.,** Simulation of animal functions in models of production systems: ruminants on the range, *Fed. Proc. Fed. Am. Soc. Exp. Biol.*, 33, 188, 1974.
27. **Swift, D. M.,** A simulation model of energy and nitrogen balance for free-ranging ruminants, *J. Wildl. Manage.*, 47, 620, 1983.
28. **Dryden, G. McL.,** The simulation of growth rate in grazing sheep — a comparison of the NRC and ARC feeding standards, *Agric. Syst.*, 7, 189, 1981.
29. **Geisler, P. A. and Jones, C. M.,** A model for calculation of the energy requirements of the pregnant ewe, *Anim. Prod.*, 29, 339, 1979.
30. **Geisler, P. A. and Neal, H.,** A model for the effects of energy nutrition on the pregnant ewe, *Anim. Prod.*, 29, 357, 1979.
31. **Graham, N. McC., Black, J. L., and Faichney, G. J.,** Simulation of growth and production in sheep — Model I: a computer program to estimate energy and nitrogen utilisation, body composition and empty liveweight change, day by day for sheep of any age, *Agric. Syst.*, 1, 113, 1976.
32. **Newton, J. E. and Edelsten, P. R.,** A model of the effect of nutrition on litter size and weight in the pregnant ewe, *Agric. Syst.*, 1, 185, 1976.
33. **Bywater, A. C.,** Simulation of the intake and partition of nutrients by the dairy cow. II. Yield and composition of milk, *Agric. Syst.*, 1, 261, 1976.
34. **Keener, H. M.,** Simulation of energy utilization of bovine animals, *Agric. Syst.*, 4, 79, 1979.
35. **Loewer, O. J., Smith, E. M., Benock, G:, Gay, N., Bridges, T., and Wells, L.,** Dynamic simulation of animal growth and reproduction, *Trans. ASAE*, 23, 131, 1980.
36. **Weiner, J.,** Model of the energy budget of an adult roe deer, *Pol. Ecol. Stud.*, 1, 103, 1975.
37. **Vickery, P. J. and Hedges, D. H.,** Mathematical relationships and computer routines for a productivity model of improved pasture grazed by merino sheep, Tech. Pap. 4, Commonwealth Scien. Indust. Res. Org., Aust., 1972.
38. **Christian, K. R., Freer, M., Donnelly, J. R., Davidson, J. L., and Armstrong, J. S.,** *Simulation of Grazing Systems*, Centre Agric. Publ. Doc., Wageningen, 1978.
39. **White, D. H., Bowman, P. J., Morley, E. H. W., McManus, W. R., and Filan, S. J.,** A simulation model of a breeding ewe flock, *Agric. Syst.*, 10, 149, 1983.
40. **Loewer, O. J., Smith, E. M., Gay, N., and Fehr, R.,** Incorporation of environment and feed quality into a net energy model for beef cattle, *Agric. Syst.*, 11, 67, 1983.
41. **Gates, C. C. and Hudson, R. J.,** Habitat selection of wapiti in a boreal forest enclosure, *Nat. Can.*, 108, 153, 1981.
42. **Nietfeld, M. T.,** Foraging Behaviour of Wapiti in the Boreal Mixed-Wood Forest, Central Alberta, M.Sc. thesis, University of Alberta, Edmonton, 1983.
43. **Gates, C. C. and Hudson, R. J.,** Foraging behaviour of wapiti in a boreal forest enclosure, *Nat. Can.*, 110, 197, 1983.
44. **Hudson, R. J., Watkins, W., and Pauls, R.,** Bioenergetics and behaviour of wapiti in the boreal forest, in *Biology of Deer Production*, Drew, K. R. and Fennessy, P., Eds., MAFF, New Zealand, in press.
45. **Trudell-Moore, J. and White, R. G.,** Physical breakdown of food during eating and rumination in reindeer, *Acta Zool. Fenn.*, 175, 47, 1983.
46. **Van Soest, P. J.,** *Nutritional Ecology of the Ruminant*, O & B Books, Corvallis, Oregon, 1982.
47. **Gates, C. C.,** Patterns of Behaviour and Performance of American Elk in the Boreal Mixed-Wood Forest, Ph.D. thesis, University of Alberta, Edmonton, 1980.
48. **Welch, J. G.,** Rumination, particle size and passage from the rumen, *J. Anim. Sci.*, 54, 885, 1982.
49. **Trudell, J. and White, R. G.,** The effect of forage structure and availability on food intake, biting rate, bite size and daily eating time in reindeer, *J. Appl. Ecol.*, 18, 63, 1981.
50. **Muller-Schwarze, D., Stagge, B., and Muller-Schwarze, C.,** Play behavior: persistence, decrease and energetic compensation during food shortage in deer fawns, *Science*, 215, 85, 1982.
51. **Suttie, J. M., Goodall, E. D., Pennie, K., and Kay, R. N. B.,** Winter food restriction and summer compensation in red deer stage *(Cervus elaphus)*, *Br. J. Nutr.*, 50, 737, 1983.
52. **Robbins, C. T., Podbielancik-Norman, R. S., Wilson, D. L., and Mould, E. D.,** Growth and nutrient consumption of elk calves compared to other ungulate species, *J. Wildl. Manage.*, 45, 172, 1981.
53. **Arman, P., Kay, R. N. B., Goodall, E. D., and Sharman, G. A. M.,** The composition and yield of milk from captive red deer *(Cervus elaphus* L.), *J. Reprod. Fert.*, 37, 67, 1974.
54. **Ammann, A. P., Cowan, R. L., Mothershead, C. L., and Baumgardt, B. R.,** Dry matter and energy intake in relation to digestibilities in white-tailed deer, *J. Wildl. Manage.*, 37, 195, 1973.
55. **Ryg, M.,** Relationships between hormone-induced and compensatory weight changes in reindeer *(Rangifer tarandus tarandus)*, *Comp. Biochem. Physiol.*, 72A, 33, 1983.
56. **Westra, R.,** Nutrition of Rocky Mountain Elk, Ph.D. thesis, University of Alberta, Edmonton, 1978.
57. **Collins, W. D., Urness, P. J., and Austen, D. D.,** Elk diets and activities on different lodgepole pine habitat segments, *J. Wildl. Manage.*, 42, 799, 1978.

58. **Collins, W. B. and Urness, P. J.,** Feeding behavior and habitat selection of mule deer and elk on northern Utah summer range, *J. Wildl. Manage.*, 47, 646, 1983.
59. **Hudson, R. J. and Nietfeld, M. T.,** Effect of forage depletion on the feeding rate of wapiti, *J. Range Manage.*, 38, 80, 1985.
60. **Parra, R.,** Comparison of foregut and hindgut fermentation in herbivores, in *Ecology of Arboreal Folivores*, Montgomery, G. G., Ed., Smithsonian Inst. Press, Washington, D.C., 1978, 205.
61. **Moen, A. N.,** *Wildlife Ecology*, W. H. Freeman, San Francisco, 1973.
62. **Tulloh, N. M.,** Physical studies of the alimentary tract of growing cattle. IV. Dimensions of the tract in lactating and non-lactating cows, *N. Z. J. Agric. Res.*, 9, 999, 1966.
63. **Hofmann, R. R.,** Adaptive changes of gastric and intestinal morphology in response to different fiber content in ruminant diets, in *Proc. Symp. Dietary Fibre in Human and Anim. Nutr.*, Wallace, G. and Bell, L., Eds., Royal Society of New Zealand, Bull. 20, 1982, 51.
64. **Milne, J. A., Macrae, J. C., Spence, A. M., and Wilson, S.,** A comparison of voluntary intake and digestion of a range of forages at different times of the year by the sheep and the red deer *(Cervus elaphus)*, *Br. J. Nutr.*, 40, 347, 1978.
65. **Renecker, L., Hudson, R. J., and Berzins, R.,** Comparative digestive physiology of moose, wapiti and cattle, North Am. Moose Conf. Workshop, Whitehorse, Canada, 1982.
66. **Kay, R. N. B. and Goodall, E. D.,** The intake, digestibility and retention time of roughage diets by red deer *(Cervus elaphus)* and sheep, *Proc. Nutr. Soc.*, 35, 98A, 1976.
67. **Milne, J. A., MacRae, J. C., Spence, A. M., and Wilson, S.,** Intake and digestion of hill-land vegetation by the red deer and the sheep, *Nature (London)*, 263, 763.
68. **Dean, R. E., Thorne, E. T., and Moore, T. D.,** Passage rate of alfalfa through the digestive tract of elk, *J. Wildl. Manage.*, 44, 272, 1980.
69. **Westra, R. and Hudson, R. J.,** Digestive function of growing wapiti, *J. Wildl. Manage.*, 45, 148, 1981.
70. **Grovum, W. L. and Williams, V. J.,** Rate of passage of digesta in sheep. VI. The effect of level of food intake on mathematical predictions of the kinetics of digesta in the reticulo-rumen and intestines, *Br. J. Nutr.*, 38, 425, 1979.
71. **Mudgal, V. D., Dixon, R. M., Kennedy, P. M., and Milligan, L. P.,** Effect of two intake levels on retention time of liquid, particulate and microbial markers in the rumen of sheep, *J. Anim. Sci.*, 54, 1051, 1982.
72. **Evans, E.,** An evaluation of the relationships between dietary parameters and rumen liquid turnover rate, *Can. J. Anim. Sci.*, 61, 91, 1981.
73. **Evans, E.,** An evaluation of the relationships between dietary parameters and rumen solid turnover rate, *Can. J. Anim. Sci.*, 61, 97, 1981.
74. **Kennedy, P. M., Young, B. A., and Christopherson, R. J.,** Studies on the relationship between thyroid function, cold acclimation and retention time of digesta in sheep, *J. Anim. Sci.*, 2, 1084, 1977.
75. **Westra, R. and Hudson, R. J.,** Urea recycling in wapiti, in *North American Elk: Ecology, Behavior and Management*, Boyce, M. and Hayden-Wing, L., Eds., University of Wyoming, Laramie, 1979, 236.
76. **Simpson, A. M., Webster, A. J. F., Smith, J. S., and Simpson, C. A.,** Energy and nitrogen metabolism of red deer *(Cervus elaphus)* in cold environments; a comparison with cattle and sheep, *Comp. Biochem. Physiol.*, 60, 251, 1978.
77. **Robbins, C. T.,** *Wildlife Feeding and Nutrition*, Academic Press, New York, 1983.
78. **Ørskov, E. R.,** *The Protein Nutrition of Ruminant Animals*, Academic Press, London, 1982.
79. **Mould, E. D. and Robbins, C. T.,** Nitrogen metabolism in elk, *J. Wildl. Manage.*, 45, 323, 1981.
80. **Kennedy, P. M. and Milligan, L. P.,** The degradation and utilisation of endogenous urea in the gastrointestinal tract of ruminants: a review, *Can. J. Anim. Sci.*, 60, 205, 1980.
81. **Satter, L. D. and Roffler, R. E.,** Influence of nitrogen and carbohydrate inputs on rumen fermentation, in *Recent Developments in Ruminant Nutrition*, Haresign, W. and Cole, D. J. A., Eds., Butterworths, London, 1981, 115.
82. **Maeng, W. J. and Baldwin, R. L.,** Factors influencing rumen microbial growth rates and yields: effects of urea and amino acids over time, *J. Dairy Sci.*, 59, 643, 1975.
83. **Simpson, A. M., Webster, A. J. F., Smith, J. S., and Simpson, C. A.,** The efficiency of utilization of dietary energy for growth in sheep *(Ovis ovis)* and red deer *(Cervus elaphus)*, *Comp. Biochem. Physiol.*, 59A, 95, 1978.
84. **Graham, N., McC., Searle, T. W., and Griffiths, D. A.,** Basal metabolic rate in lambs and young sheep, *Aust. J. Agric. Res.*, 25, 957, 1974.
85. **Chappel, R. W. and Hudson, R. J.,** Prediction of energy expenditures by Rocky Mountain bighorn sheep, *Can. J. Zool.*, 58, 1908, 1980.
86. **Pauls, R. W., Hudson, R. J., and Sylven, S.,** Energy expenditures of free-ranging wapiti, *Univ. Alberta Feeders' Day Rep.*, 68, 87, 1981.
87. **Christopherson, R. J. and Young, B. A.,** Heat flow between large terrestrial animals and the cold environment, *Can. J. Chem. Eng.*, 59, 181, 1981.

88. **Moen, A. N.,** *Biology and Management of Wild Ruminants,* Part 1, Corner Brook Press, Lansing, N.Y., 1980, chap. 1.
89. **Webster, A. J. F.,** Prediction of heat losses from cattle exposed to cold outdoor environments, *J. Appl. Physiol.,* 30, 684, 1971.
90. **Gates, C. C. and Hudson, R. J.,** Energy costs of locomotion in wapiti, *Acta Theriol.,* 23, 265, 1978.
91. **Parker, K. L., Robbins, C. T., and Hanley, T. A.,** Energy expenditures for locomotion by mule deer and elk, *J. Wildl. Manage.,* 48, 474, 1984.
92. **Robbins, C. T. and Robbins, B. L.,** Fetal and neonatal growth patterns and maternal reproductive effort in ungulates and subungulates, *Am. Nat.,* 114, 101, 1979.
93. **Thorne, E. T., Dean, R. S., and Hepworth, W. G.,** Nutrition during gestation in relation to successful reproduction in elk, *J. Wildl. Manage.,* 40, 330, 1976.
94. **Robbins, C. T. and Moen, A. N.,** Uterine composition and growth in pregnant white-tailed deer, *J. Wildl. Manage.,* 39, 684, 1975.
95. **Loudon, A. S. I., Darroch, A. D., and Milne, J. A.,** The lactation performance of red deer on hill and improved species pastures, *J. Agric. Sci.* (Cambridge), 102, 149, 1984.
96. **Loudon, A. S. I., McNielly, A. S., and Milne, J. A.,** Nutrition, milk yield and suckling behaviour in red deer: implications for lactational control of fertility, *Nature (London),* 302, 145, 1983.
97. **Wood, P. D. P.,** The biometry of lactation, *J. Agric. Sci.,* (Cambridge), 88, 333, 1977.
98. **Robbins, C. T., Moen, A. N., and Reid, J. J.,** Body composition of white-tailed deer, *Anim. Sci.,* 38, 871, 1974.
99. **Drew, K. R.,** Meat production from farmed deer, in *Biology of Deer Production,* Drew, K. R. and Fennessy, P., Eds., MAFF, New Zealand, in press.
100. **Watkins, W. G. and Hudson, R. J.,** Weight dynamics wapiti in relation to winter supplemental feeding, *Univ. Alberta Feeder's Day Rep.,* 63, 60, 1984.
101. **Clutton-Brock, T. H., Guinness, F. S., and Albon, S. D.,** The costs of reproduction to red deer *(Cervus elaphus)* hinds, *J. Anim. Ecol.,* 52, 367, 1983.
102. **Clutton-Brock, T. H. and Harvey, P. H.,** The functional significance of variation in body size among mammals, in *Advances in the Study of Mammalian Behavior,* Eisenberg, J. F. and Kleiman, D. G., Eds., Spec. Publ. 7, American Society of Mammalogists, Shippensburg, Pa., 1983, 632.
103. **Clutton-Brock, T. H., Guiness, F. E., and Albon, S. D.,** *Red Deer: Behavior and Ecology of Two Sexes,* University of Chicago Press, Chicago, 1982.
104. **Clutton-Brock, T. H.,** Reproductive effort and terminal investment in iteroparous animals, *Am. Nat.,* 123, 212, 1984.
105. **Mertens, D. R. and Ely, L. O.,** Relationships of rate and extent of digestion to forage utilization — A dynamic model evaluation, *J. Anim. Sci.,* 54, 895, 1982.
106. **Hartnell, G. F. and Satter, L. D.,** Determination of rumen fill, retention time and ruminal turnover rates of ingesta at different stages of lactation in dairy cows, *J. Anim. Sci.,* 48, 381, 1979.
107. **Staalund, H., Jacobsen, E., and White, R. G.,** Comparison of the digestive tract in Svalbard and Norwegian reindeer, *Arc. Alp. Res.,* 11, 457, 1979.
108. **Forbes, J. M.,** Effects of lighting patterns on growth, lactation and food intake of sheep, cattle and deer, *Livestock Prod. Sci.,* 9, 361, 1982.
109. **Turner, H. G. and Taylor, C. S.,** Dynamic factors in models of energy utilization with particular reference to maintenance requirement of cattle, *World Rev. Nutr. Diet.,* 42, 135, 1983.
110. **Bauman, D. E., Eisemann, J. H., and Curie, W. B.,** Hormonal effects on partitioning of nutrients for tissue growth: role of growth hormone and prolactin, *Fed. Proc. Fed. Am. Soc. Exp. Biol.,* 41, 2538, 1982.
111. **McMillin, J. M., Seal, U. S., and Karns, D. D.,** Hormonal correlates of hypophagia in white-tailed deer, *Fed. Proc. Fed. Am. Soc. Exp. Biol.,* 39, 2964, 1980.
112. **Verme, L. J. and Ozoga, J. J.,** Effects of diet on growth and lipogenesis in deer fawns, *J. Wildl. Manage.,* 44, 315, 1980.

INDEX

A

B

C

D

E

F

G

H

I

K

L

N

O

P

Q

R

S

T

Y

Z